D1270497

CELLULAR TECHNOLOGIES FOR EMERGING MARKETS

CELLULAR TECHNOLOGIES FOR EMERGING MARKETS

2G, 3G AND BEYOND

Ajay R. Mishra
Nokia Siemens Networks

A John Wiley and Sons, Ltd., Publication

This edition first published 2010

Registered office
John Wiley & Sons Ltd, The Atrium, Southern Gate, Chichester, West Sussex, PO19 8SQ, United Kingdom

For details of our global editorial offices, for customer services and for information about how to apply for permission to reuse the copyright material in this book please see our website at www.wiley.com.

Library of Congress Cataloging-in-Publication Data

Mishra, Ajay R.
Cellular technologies for emerging markets : 2G, 3G, and beyond / Ajay R Mishra.
p. cm.
Includes bibliographical references and index.
ISBN 978-0-470-77947-7 (cloth)
1. Cellular telephone systems. I. Title.
TK5103.2.M567 2010
384.5′35–dc22

2010005780

A catalogue record for this book is available from the British Library.

ISBN 9780470779477 (HB)

Typeset in 10/12pt Times by Aptara Inc., New Delhi, India
Printed and Bound in Singapore by Markono Print Media Pte Ltd

Dedicated to
The Lotus Feet of my Guru

Dedication:

The Inner Voice of my Country

Contents

Foreword 1:

Role of Technology in Emerging Markets

Telecom wireless technology has been progressing rapidly over the last two decades. Initial introduction of the GSM platform created global standards in the 1980s and provided opportunities to innovate new business models to reduce costs and increase affordability, leading to substantial growth and expansion in the emerging countries. In the process, GSM technology was enhanced through several new features and functionalities to add data capabilities. In the 1990s, third generation wireless technology was introduced in advanced countries of the western world and Japan. At the same time China and India witnessed an unpredicted growth with over 700 million subscribers in China and over 500 million subscribers in India. Similar growth in many other emerging markets of Latin America, Africa and Asia pushed the number of global mobile phone users to over 4 billion worldwide.

The expansion of mobile phones in the emerging markets has been critical in the overall development of the rural areas and the people at the bottom of the pyramid. This has provided a unique access to basic telephone services and a variety of new SMS based applications related to entertainment, news, agriculture, payments, etc. It has been shown by the OECD and other studies that a 10 % increase in the mobile phone coverage increases the GDP of the country by 0.6 %. This offers hope for new features and functionalities with more data capabilities and applications related to education, health, governance, etc. to benefit the poor in the emerging markets.

All of this was possible because we were able to make a business case for affordable technology and bring down the total cost of ownership for the people. This is where Ajay Mishra's book steps in. It provides a comprehensive coverage of many technologies that will give the readers a quick understanding of the upcoming new opportunities. A basic understanding of the evolution of technologies will help make the right choices for future network capabilities.

Once we are able to bring down the total costs of ownership by placing the right technology, we can provide an opportunity for real economic development and growth to the community. The key is to continue to focus on lowering the cost of mobile services where basic voice services will become a commodity and the future revenue for the operators will come from novel and useful applications and transaction services. Only then the real potential of the mobile revolution will be realized.

Sam Pitroda
Advisor to the Prime Minister of India
Former/First Chairman Telecom Commission of India

Foreword II

Role of Technology in Emerging Markets

Foreword 2:

Connecting the Unconnected

The world now has more than 4B telephone lines – thanks to wireless connectivity as more than 65 % are mobile connections. The increase has been tremendous in emerging markets such as India where mobile connections are now happening in double digit millions every month. It has been a phenomenal journey of perhaps one technology (i.e. wireless/mobile) that has not only outgrown the vision of the founding fathers but has been quite successful in touching the lives of people living in the remotest of locations. We have many studies that have very strongly pointed to the fact that an increase mobile penetration would impact the lives of people and this is absolutely amazing.

As we talk about 'connecting the un-connected' and reducing the digital divide, it is absolutely necessary that the benefits of technology reach to people living in the remotest places on this planet. Many of the emerging markets, although immensely successful for highest connectivity growths, have not achieved similar success in making its people reap the benefits of being connected to the world.

Technology will play an important role in bringing down the total costs of ownership. With a host of technologies at the disposal of emerging markets, it would be even easier for operators and industry in general to bring connectivity to the door steps of people in the farthest of locations. I think that by giving the right overview of the technologies that will play a role in emerging markets, under one cover, this book will prove to be extremely useful to decision-makers in the cellular industry. The book brings technology and design aspects that one would need for day-to-day decision making in a simple and lucid way. Only when both connectivity and its benefits will reach every one single person would we say that we are living in a truly connected world.

Adel Hattab
Vice-President
Nokia Oy

Preface

Emerging markets have seen an unprecedented growth in the last few years. The operator focus has been on giving complete coverage to all regions (urban to rural) and to subscription to all – people from the highest to the lowest income groups. When the idea is taking coverage for the remotest of the regions and getting the 'unconnected–connected', technology and business modelling are two important focus areas. This book covers one of them – technology. Many of the mobile technologies find importance in one network. No more do we see networks that are working on just one or two technologies but we are seeing networks that are an amalgamation of technologies. Engineers and executives working in the field sometimes find it challenging to get hold of a single manual that gives them an overview of technologies that are existing in the mobile field. This book tries to address that challenge – providing an overview of technology, designing and applications of the few important technologies under one cover.

There are many books that are available dealing with individual technologies and so this book is not for in-depth reading of one technology but rather a quick overview of some key technologies. Experts of one technology can quickly understand what they can expect in other technologies. So, this book will be beneficial to beginners, experts, managers and technocrats at the same time.

Chapter 1 discusses the scenario in emerging markets and technologies that are making their mark. Chapter 2 focuses on GSM and EGPRS and includes a technology overview, details on network architecture and network planning/ optimization.

Chapters 3 and 4 are concerned with UMTS and CDMA, covering technology, network architectures and designing issues.

In Chapter 5 we go beyond the third-generation technology. Technologies that are sometimes called 3.5G (HSPA) and 3.9G (LTE) are discussed. These are of immense interest in current scenarios – both in the developed and emerging markets.

Going further, we look into OFDM and All-IP technologies in Chapter 6. Both of these have started to make an impact and are being studied with much greater interest by the technocrats of emerging markets.

We look into the world of Wi-Fi, WLAN and WiMAX in Chapter 7. Although Wi-Fi and WLAN have established places in the technology world, they are finding more importance as we move towards fourth-generation networks.

WiMAX and LTE are still being debated but leaving that for cellular operators to decide, we focus on looking into the technical aspects of WiMAX in this chapter.

Convergence is again a fascinating world and is covered along with the underlying technology of IMS in Chapter 8.

Although UMA has been more common in North America, it is briefly covered in Chapter 9 to give the reader an overview of the concept that is implemented in one of the biggest cellular markets in the world.

Chapter 10 deals with DVB-H, the underlying technology for mobile TV. This technology is now making inroads into emerging markets and has an impact on the life of 'common man' – taking TV to his/her handheld devices.

There are two appendices as well – one which covers VAS applications while the other one concentrates on highly important areas for anyone and everyone in the telecom industry – 'energy'.

Finally, at the end of this text, there is a Bibliography with a carefully chosen list of books and papers for further reading which I hope the interested reader will find useful. In conclusion, I would appreciate it if readers can give me feedback with respect to comments concerning this text and suggestions for improvement, via fcnp@hotmail.com.

Ajay R. Mishra

Acknowledgements

Writing this book has been nothing short of an exciting journey – and no words are sufficient to thank those people who have helped in various ways during the course of this project.

My big thanks go to Mark Hammond and Sarah Tilley from John Wiley & Sons, Ltd, Chichester, UK, who believed that this project would finally be completed in spite of numerous delays.

Special thanks are due to my following colleagues and friends for taking out the time to read the manuscript and give their valuable comments: Johanna Kahkonen, Mika Sarkioja, Sushant Bhargava, Shweta Jain, Pauli Aikio, Munir Sayyad (Reliance Communications) and Cameron Gillis.

Many thanks go to Sam Pitroda, Advisor to The Prime Minister of India and First Chairman of the Telecom Commission of India, and Adel Hataab Vice President, Nokia Oy for donating their precious time in writing the Forewords and sharing their vision with us.

Many thanks are due to Rauno Granath and Amit Sehgal for their contributions to Chapter 1 and to Sameer Mathur and Anne Larilahti for their contributions in writing the Appendices.

Thanks also to KanakShree Vats, Kanchan Agarwal, Shankar Shivram, C. Ravindranath Bharathy, Das Bhumesh Kailash, Dandavate Pushpak Ravindra, Abhishek Kumar and Kriti Vats for helping me during the last phases of the writing of this book.

My all-time thanks must go to my Professors/Mentors, G. P. Srivastava, K. K. Sood and J. M. Benedict, and to my colleagues, Antti Rahikainen, Reema Malhotra and Prashant Sharma, for their moral support during the course of my career.

Finally I would like to thank my parents, Mrs Sarojini Devi Mishra and Mr Bhumitra Mishra, who gave me the inspiration to undertake this project and deliver it to the best of my capability.

1

Cellular Technology in Emerging Markets

Rauno Granath
Nokia Siemens Networks

Amit Sehgal
Nokia Siemens Networks

Ajay R. Mishra
Nokia Siemens Networks

1.1 Introduction

From the remotest areas of the developing world to the most advanced areas of the developed world, connectivity has become a key issue. How to connect the 'unconnected' is an issue that is facing the governments of most of the developing countries, while mobile operators in advanced countries are looking towards connecting their consumers to enhanced services. While the developing world is trying various advanced technologies, it is not necessarily following the path taken by the developed world. They are trying out various permutations and combinations of technologies to reach their goal to connectivity and profits. In this context, it becomes important to understand the various technologies that would help technologists in the developing world realize their ultimate goal – getting the 'unconnected' connected in the shortest duration of time.

1.2 ICT in Emerging Markets

During year 2009 the global cellular industry was able to celebrate its 4th billionth subscription to its services. By any means this is a staggering figure. It is even more staggering to realize how short a time it has taken to achieve this. It is hard to come up with any other example

Cellular Technologies for Emerging Markets: 2G, 3G and Beyond Ajay R. Mishra

where a new technology has proliferated and diffused throughout the world, to all continents, countries and markets and among all consumer groups, cultures and socio-economic strata. How did this happen? Was it planned and designed into the specifications and implementations of early cellular technologies? It is quite safe to say that the huge success of the most common and used cellular technologies has taken the industry itself by a little bit of surprise. However the global ecosystems around the cellular technologies have not been 'stunned' by the success, rather the growth momentum and positive response have been used as strong levers to develop the next steps in the evolution towards even richer and more penetrated services.

Looking back 20 years, the first cellular or mobile services were clearly created for and targeted to the business segment. The clear value addition was the mobility itself. People who carry out businesses which are not tied to a fixed office desk and location obtained a great productivity boost by being connected all the time. One can think of some other examples where 'freeing people from a fixed place' will bring obvious economic benefits – at the macro level as well as at the individual level. One of these could be by comparing people having watches instead of a 'grandfather's clock' inside a house. Having a 'time with you' greatly enhanced the way one can plan and synchronize interactions with other people.

'Mobility' was the first phase of cellular penetration and while the actual number of users in the first phase was relatively low, it was as important because it demonstrated business viability as well as showing some of the main requirements. As the users were mainly from the business segment their requirements became very apparent in 2nd generation technology specifications and functionalit of the systems. Some of the seeds for future global success can be traced here: international roaming, globally harmonized frequencies allowing use of the same device – or a simpler device, certified interoperability between network and user devices, etc. All of this started to push the industry towards a truly global scale, enabling the immense cost benefits later.

The next phase of rapid penetration took place when individual consumers started to see a similar value in being connected. For the first time the concept of 'affordability' really kicked in. When the overall cost of getting and being connected became low enough compared to the perceived value there was a true mass market adoption – in any given market, throughout the world. One can only conclude that the basic demand – everybody's basic human need to communicate – is very universal.

In many mature markets that phase was reached during the early-2000s. Perhaps it's a better topic for a book about social behaviour but it became increasingly difficult – even impossible to participate the society without being individually connected – all the time. At this phase an additional boost for the mass market came through 'fixed-to-mobile substitution' – people actually gave up, or never subscribed to fixed services any more. It also meant that most households practically had a mobile device for every family member and market penetrations reached close to or above the 100 % mark.

Around the mid-2000s a similar development was already clearly seen in many developing markets as well. Here, the concept of 'affordability' comes out in the clearest way. There are three basic pillars for this which can be illustrated as shown in Figure 1.1.

Liberalization of the whole telecommunications sector – and the resulted regulatory environment – is at least as important an element in overall affordability as any of the technology-derived innovations and business models. This was actually one key element in, for example Western European mobile success. In most countries the telecom infrastructure was regarded as a natural monopoly, among other utilities, due to the costs of building and operating the

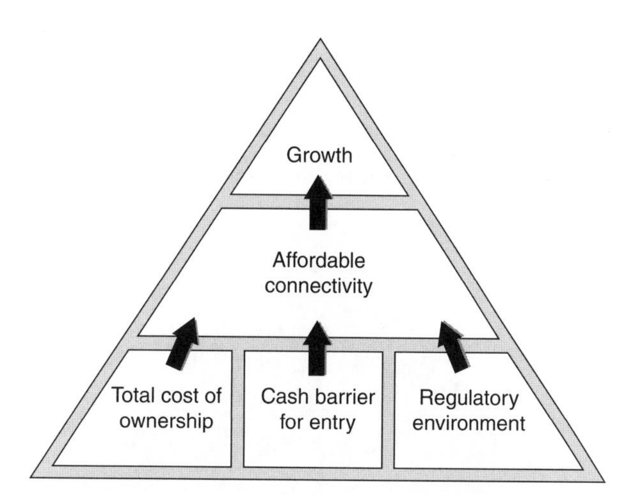

Figure 1.1 The three pillars of telecom development in emerging markets.

fixed telephony networks. In many cases it was a government-owned monopoly, and in some cases partly due to privately and partly government-owned set-ups. With the advent of the first cellular technologies and mobile telephony services the sector was ready for a drastic change. The cost dynamics and advantages of cellular technologies made it feasible to open the sector for competition, overseen by national regulatory bodies. Free competition in a transparent regulation environment is the best mechanism to really push all technological innovations and cost break-throughs to the end consumer.

Nothing highlights this better than an example from Nigeria. During the early part of the 2000s Nigeria licensed its first four mobile operators, three privately owned and one incumbent. In just 18 months the country's telephony penetration doubled (Trends in Telecommunications Reform, ITU, 2003). In other words, the mobile operators were able to provide, in 18 months, as many connections as the government-owned fixed telephony provider from the beginning of the country's independence!

Whereas regulatory environment is more of the industry topic in each country the other two elements of affordability are very much user- or consumer-centric. Cost, or rather the Total Cost of Ownership (TCO), is the obvious one. The TCO includes all the costs that it takes to get and stay connected: the cost of the handset, the cost of the subscription and the ongoing cost of the service itself. All of these typically also include government taxes. Technology innovations and a massive global scale have greatly reduced the TCO over the last few years.

Another important element is 'Cash', that is how do people finance the consumption of the service. One of the great business model innovations stemming from developing markets is the pre-paid model where services can be consumed in very small increments – matching the daily cash situation of particularly low-income segments.

Playing with the two aforementioned aspects – the universal human need to communicate and the concept of affordability being the main drivers for penetration – it is easy to model and understand the huge global success of mobile telephony services. Modelling with the well known 'income pyramid' one can readily see that each step downwards in 'affordability' brings in a larger potential customer segment (Figure 1.2).

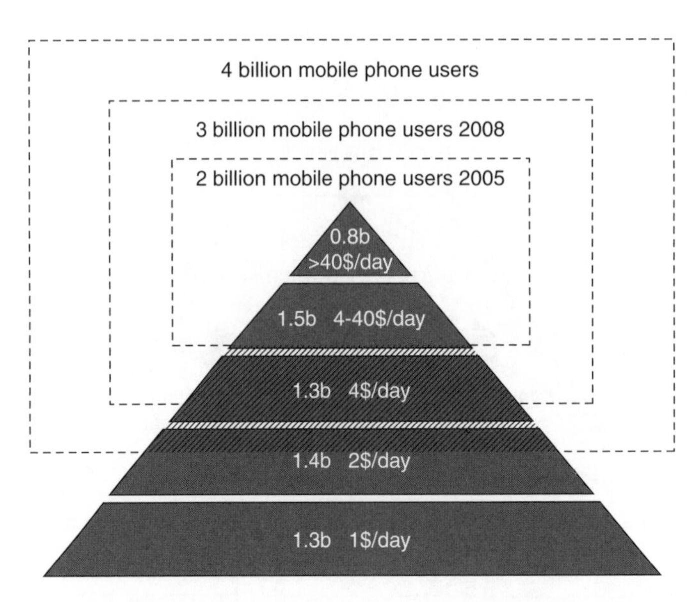

Figure 1.2 World population split according to income segment (USD/ capita/day).

The rapid development of connectivity through mobile technologies in developing countries throughout the 2000s was early on identified as one true opportunity to bridge the 'digital divide'. In fact, advancing the benefits of ICT technologies was adopted as one of the UN Millennium Development Goals.

Several international studies have come up with clear evidence between the mobile phone penetration and macroeconomic development. In a typical emerging market, an increase of 10 mobile phones per 100 people boosts the GDP growth by 0.6 percentage points (Vodafone policy paper, 2005). A 2006 study by McKinsey and Company (in cooperation with the GSMA) found that the indirect impact of mobile phone penetration is at least three times as great. In addition, the latest study by the World Bank (Quian, 2009) comes up with the figure of a 0.81 percentage GDP boost for low- and middle-income economies.

Lately, the focus of research has been in broadband, instead of pure voice services. The same World Bank study shows clearly that the 0.81 %-unit boost will increase to 1.12 with usage of the Internet and all the way up to 1.38 %-units in the case of broadband connectivity for the services and the Internet.

While the basic mobile connectivity continue to increase beyond the 4B mark it is now important to have a similar advance in broadband connections. Interestingly, very similar mechanisms and market behaviour seem to have now taken place in mature markets that led to the massive increase of mobile voice services 10 years ago. Mobile broadband services have become affordable – in terms of cost, cash and regulatory environment – so that there is a 'fixed-to-mobile' substitution going on in many markets. The industry has come up with the necessary technology (speed, latency and end-user devices) and business models (flat rate pricing) enabling rapid consumer acceptance. Several new services – like social networking – are once again extending the social dimension to the picture. People want to get into their services independent of the place and time.

While the technology can't provide all the answers to unlock the potential of broadband in developing markets, it surely has a key role as well. The industry knows what it takes to give broadband connectivity a similar success in all parts of the world – and for all people. Affordability and access, relevant services for people to enhance their business, social or personal interests will truly make the whole ICT as 'the biggest democratizer of opportunities ever seen'.

1.3 Cellular Technologies

Mobile operators use the radio spectrum to provide their services. Spectrum is a scarce resource and has been allocated as such. It has traditionally been shared by a number of industries, including broadcasting, mobile communications and the military. Before the advent of cellular technology, the capacity was enhanced through a division of frequencies and the resulting addition of available channels. However, this reduced the total bandwidth available to each user, affecting the quality of service. Introduced in the 1970s, cellular technology allowed for the division of geographical areas (into cells), rather than frequencies, leading to a more efficient use of the radio spectrum. Figure 1.3 details the evolution of cellular technologies and the dominant ones at the present time and for the coming years.

Based on usability, cost and quality and quantity of services etc, the evolution of cellular technology has been divided into generations.

1.3.1 First Generation System

Also referred to as 1G, this period was characterized by analogue telecommunication standards and supported basic voice services. The development started in the late 1970s with Japan taking a lead in deployment of the first cellular network in Tokyo, followed by the deployment

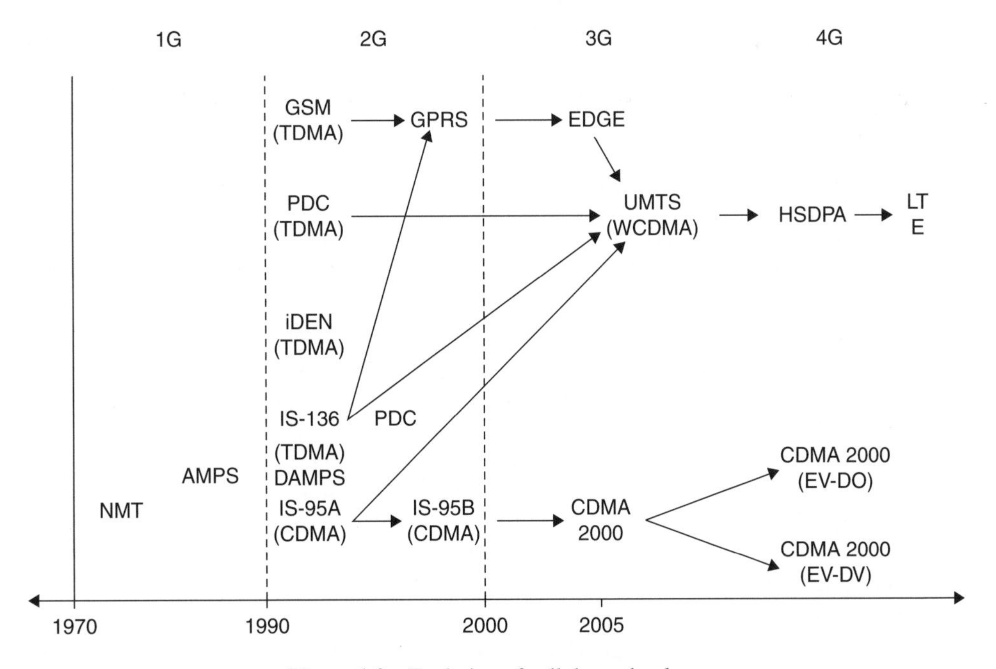

Figure 1.3 Evolution of cellular technology.

of NMTs (Nordic Mobile Telephones) in Europe, while the 'Americas' deployed AMPS (Advanced Mobile Phone Service) technology.

Each of these networks implemented their own standards – with features such as roaming between continents non-existent. This technology also had an inherent limitation in terms of channels, etc. The handsets in this technology were quite expensive (more than $1000).

1.3.2 Second Generation System

As we have seen above, the various systems were incompatible with each other. Due to this, work towards development of the next technology was implemented that would lead to a more harmonized environment. Such work was commissioned by the European Commission and resulted, in the early-1990s, in the next generation technology known as the 'Second Generation Mobile Systems', which were also digital systems as compared to the first generation's analogue technology. Key 2G systems in these generations included GSMs (Global Systems for Mobile Communications), TDMA IS-136, CDMA IS-95, PDC (Personal Digital Cellular) and PHSs (Personal Handy Phone Systems).

IS 54 and IS 136 (where IS stands for Interim Standard) are the second generation mobile systems that constitute D-AMPS. IS-136 added a number of features to the original IS-54 specification, including text messaging, circuit-switched data (CSD) and an improved compression protocol. CDMA has many variants in the cellular market. CDMAone (IS-95) is a second-generation system that offered advantages such as increase in coverage, capacity (almost 10 times that of AMPS), quality, an improved security system, etc.

GSM was first developed in the 1980s. It was decided to build a digital system based on a narrowband TDMA solution and having a modulation scheme known as GMSK. The technical fundamentals were ready by 1987 and the first specifications by 1990. By 1991, GSM was the first commercially operated digital cellular system with Radiolinja in Finland. With features such as pre-paid calling, international roaming, etc., GSM is by far the most popular and widely implemented cellular system with more than a billion people using the system (by 2005).

1.3.3 Third Generation System

This improvement in data speed continued and as faster and higher quality networks started supporting better services like video calling, video streaming, mobile gaming and fast Internet browsing, it resulted in the introduction of the 3rd generation mobile telecommunication standard (UMTS). These third generation cellular networks were developed to offer high speed data and multimedia connectivity to subscribers. Under the initiative IMT-2000, ITU has defined 3G systems as being capable of supporting high-speed data ranges of 144 kbps to greater than 2Mbps.

The Universal Mobile Telecommunications System (UMTS) is one of the third-generation (3G) mobile phone technologies. It uses W-CDMA as the underlying standard. This was developed by NTT DoCoMo as the air interface for their 3G network FOMA. Later, ITU accepted W-CDMA as the air-interface technology for UMTS and made it a part of the IMT-2000 family of 3G standards.

CDMA2000 has variants such as 1X, 1XEV-DO, 1XEV-DV and 3X. The 1XEV specification was developed by the Third Generation Partnership Project 2 (3GPP2), a partnership consisting of five telecommunications standards bodies: CWTS in China, ARIB and TTC in Japan, TTA in Korea and TIA in North America.

1.3.4 Fourth Generation System

In the 18th TG-8/1 in 1999, a new working group WP8F was established for looking into the efforts to develop the systems beyond the IMT-2000. As IMT-2000 was not able to solve the problems related to higher data rates and capacity, next generation systems (also called as 4G) development was give that mandate. A 4G system will be a complete replacement for current networks and be able to provide a comprehensive and secure IP solution where voice, data, and streamed multimedia can be given to users on an 'Anytime, Anywhere' basis, and at much higher data rates than previous generations. Some features of 4G include the following:

- The intention of providing high-quality video services leading to data-transfer speeds of about 100 Mbps.
- The 4G technology offers transmission speeds of more than 20 Mbps.
- It will be possible to roam between different networks and different technologies.
- 4G basically resemble a conglomeration of existing technologies and is a convergence of more than one technology.

1.4 Overview of Some Key Technologies

Let us now have a look at the some of the key technologies.

1.4.1 GSM

GSMs (Global Systems for Mobile Communications) was the first commercially operated digital cellular system. Developed in the 1980s through a pan-European initiative, The European Telecommunications Standards Institute (ETSI) was responsible for GSM standardization.

Today it is the most popular cellular technology. By mid-2009, GSMs have a user base of over 3.9 billion in more than 219 countries and territories worldwide; with a market share of more than 89 % (the global wireless market is more than 4.3 billion). In addition, GSM has the widest spectral flexibility for any wireless technology – 450, 850, 900, 1800 and 1900 MHz bands; tri- and quad-band GSM phones are common. Thus it is rare that users will ever travel to an area without at least one GSM network to which they can connect.

GSM uses TDMA (Time Division Multiple Access) technology and is the legacy network leading to the third-generation (3G) technologies, the Universal Mobile Telecommunication System (UMTS) (also known as WCDMA) and High Speed Packet Access (HSPA). GSM differs from its predecessors in that both signalling and speech channels are digital and thus is considered a second generation (2G) mobile phone system.

GSM is a very secure network. All communications (voice and data) are encrypted to prevent eavesdropping. GSM subscribers are identified by their Subscriber Identity Module (SIM) card. This holds their identity number and authentication key and algorithm. Thus it's the card rather than the terminal that enables network access, feature access and billing.

1.4.2 EGPRS

Enhanced GPRS (EGPRS) is another 3G technology that allows improved data transmission rates. Here EDGE ('Enhanced Data Rates for GSM Evolution' – a new radio interface technology with enhanced modulation) is introduced on top of the GPRS and is used to transfer data in a packet-switched mode on several time slots, as an extension on top of the standard GSM. This leads to almost an increase in data rates of almost three-fold.

The major advantage of EDGE is that it does not require any hardware or software changes in the GSM core networks. No new spectrum is required and thus EDGE can effectively be launched under the existing GSM license. WCDMA (including HSPA) and EDGE systems are complimentary. There is a wide range of EDGE capable user devices in the market, including USB modems, modules for PCs, phones, routers, etc.

EDGE was first deployed by Cingular (now AT&T) in the United States in 2003. By mid-2009 there were more than 440 GSM/EDGE networks in 181 countries, from a total of 478 mobile network operator commitments in 184 countries.

1.4.3 UMTS

The Universal Mobile Telecommunications System (UMTS) is a voice and high-speed data technology that is 'part' third-generation (3G) wireless standards. Wideband CDMA (WCDMA) is the radio technology used in UMTS. Furthermore, UMTS borrows and builds upon concepts from GSM and most UMTS handsets also support GSM, allowing seamless dual-mode operation. Therefore, UMTS is also marketed as 3GSM. UMTS is based on Internet Protocol (IP) technology with user-achievable peak data rates of 350 kbps.

UMTS builds on GSM and its main benefits include high spectral efficiency for voice and data, simultaneous voice and data for users, high user densities supportable with low infrastructure costs, high-bandwidth data applications support and migration path to VoIP in future. Operators can also use their entire available spectrum for both voice and high-speed data services.

UMTS has been in commercial usage since 2001, in Japan. As of April 2009, it was available with 282 operators in more than 123 countries and enjoyed a subscriber base of 330 million.

As for GSM, UMTS can also work over a wide range of spectrum bands – 850, 900, 1700, 1800, 1900, 2100 and 2600 MHz bands (450 MHz and 700 MHz are expected to be added soon). Thus transparent global roaming is an important aspect of UMTS. Also, UMTS operators can use a common core network that supports multiple radio-access networks, including GSM, EDGE, WCDMA, HSPA, etc. This is called the UMTS multi-radio network (shown in Figure 1.4) and provides great flexibility to operators.

UMTS networks can be upgraded with High-Speed Downlink Packet Access (HSDPA), sometimes known as 3.5G. Currently, HSDPA enables downlink transfer speeds of up to 21 Mbs.

1.4.4 CDMA

Code Division Multiple Access (CDMA) was originally known as IS-95. It is the major competing technology to GSM. There are now different variations, but the original CDMA is now known as cdmaOne.

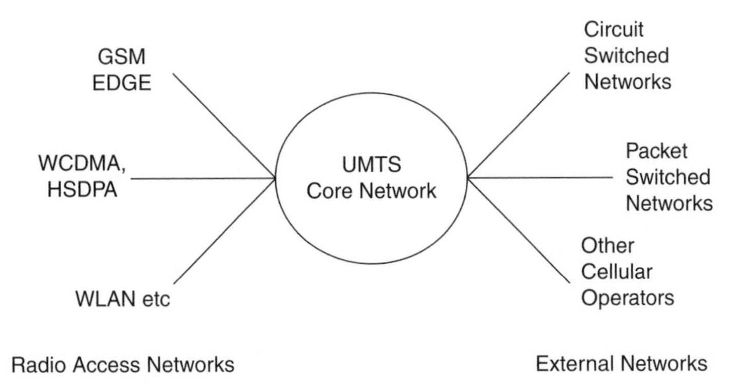

Figure 1.4 UTMS multiradio network.

Currently there is cdma2000 and its variants like 1X EV, 1XEV-DO and MC 3X. The technology is used in ultra-high-frequency (UHF) cellular telephone systems in the 800-MHz and 1.9-GHz bands. CDMA employs spread-spectrum technology along with a special coding scheme and is characterized by high capacity and a small cell radius.

CDMA was originally developed by Qualcomm and enhanced by Ericsson. However, QUALCOMM still owns a substantial portfolio of CDMA patents, including many patents that are necessary for the deployment of any proposed 3G CDMA system. It has now been granted royalty-bearing licenses to more than 75 manufacturers for CDMA and, as part of these licenses, has transferred technology and 'know-how' in assisting these companies to develop and deploy CDMA products.

CDMA was adopted by the Telecommunications Industry Association (TIA) in 1993. In September 1998, only three years after the first commercial deployment, there were 16 million subscribers on cdmaOne systems worldwide. By mid-2009, there were around 500 million subscribers on CDMA (including variants).

Another variant of CDMA is TDS-CDMA. Time Division Synchronous Code Division Multiple Access (TD-SCDMA) or UTRA/UMTS-TDD, also known as UMTS-TDD or IMT 2000 Time-Division, is an alternative to W-CDMA. Although the name gives an impression of simply a channel access method based on CDMA, its applicability is to the whole-air interface specification.

The technology is promoted by the China Wireless Telecommunication Standards group (CWTS) and was approved by the ITU in 1999. It is being developed by the Chinese Academy of Telecommunications Technology, Datang, and Siemens AG, and is China's country's standard of 3G mobile telecommunication. However, it is expected to remain as a niche market technology as it lacks a large ecosystem and would muster limited research and development. In addition, necessary competition and economies of scale to reduce investments and generate demand might be missing, besides the fact that its delayed arrival has given rival 3G technologies a good head start. TD-SCDMA came under spotlight as one of the technologies used in the 2008 Olympics at Beijing, China.

1.4.5 HSPA

High Speed Packet Access (HSPA) is a collection of two mobile telephony protocols, namely High Speed Downlink Packet Access (HSDPA) and High Speed Uplink Packet Access

(HSUPA). It is basically an extension/improvement of the performance of existing WCDMA protocols. HSPA improves the end-user experience by increasing peak data rates up to 14 Mbps in the downlink and 5.8 Mbps in the uplink, according to network and user device capabilities.

Mobile broadband is a key part of the commercial offering of most mobile network operators today and the strong market uptake which has been seen in every market is boosting revenues and profits. The path to mobile broadband began with WCDMA and has grown globally with HSPA to boost capacity and user data speeds. Several operators have positioned HSPA as an alternative to fixed broadband, with the added value of mobility.

HSPA has been commercially deployed by over 270 operators in more than 110 countries, as of 2009. Data traffic and revenues are growing strongly with HSPA. According to GSA surveys of the mobile broadband market, WCDMA has a 72 % market share of commercial 3G networks. More than 90 % of the 275 commercial WCDMA network operators have launched HSPA.

1.4.6 LTE

LTE (Long Term Evolution) is marked as the 4th generation of mobile technology designed to provide uplink peak rates of at least 50 Mbps and downlink peak rates of at least 100 Mbps. The specifications support both Frequency Division Duplexing and Time Division Duplexing. Designed as a flat IP-based network architecture it can replace the GPRS Core Network and ensure support for, and mobility between, some legacy or non-3GPP systems such as GPRS and WiMax.

LTE has been designed to offer ‘rich’ broadband user experience and will further enhance mobile value-added services and applications supporting banking, gaming, health categories, etc. Even experience with more demanding applications such as interactive TV, mobile video blogging, advanced games, etc will also be significantly improved.

The main advantages with LTE are high throughput, low latency, ‘plug-and-play’, besides improved end-user experience and simple architecture resulting in low-operating expenditures. LTE will also support seamless integration with older network technologies, such as GSM, CDMA, UMTS and CDMA2000. LTE is the natural migration choice for GSM/HSPA operators. LTE is also the next generation mobile broadband system of choice of leading CDMA operators, who are expected to be in the forefront of service introduction.

With over 39 LTE commitments in 19 countries, at least 14 networks are expected to be commercially deployed by 2010. It is expected that there will be nearly 34 million users worldwide by 2010 that are expected to reach 400–450 million users by 2015. Some of the first operators intending to deploy LTE include Verizon Wireless, MetroPCS Wireless and US Cellular in the United States, NTT-DOCOMO and KDDI in Japan, TeliaSonera, Tele2 and Telenor in Europe, China Mobile in China, and KT and SK Telecom in Korea, in 2010.

1.4.7 OFDM

OFDM (Orthogonal Frequencies Division Multiplexing) is a broadband technique like CDMA. In this, instead of modulating a single carrier as is the case with FM or AM, a number of carriers are spread regularly over a frequency band. Orthogonal FDMs (OFDM) spread-spectrum techniques distribute the data over a large number of carriers that are spaced apart at precise frequencies. This spacing provides the 'orthogonality' in this technique which prevents the demodulators from seeing frequencies other than their own.

OFDM has been successfully used in DAB and DVB systems. For DAB, OFDM forms the basis for the Digital Audio Broadcasting (DAB) standard in the European market. For ADSL, OFDM forms the basis for the global ADSL (Asymmetric Digital Subscriber Line) standard. For Wireless Local Area Networks, development is ongoing for wireless point-to-point and point-to-multipoint configurations using OFDM technology. In a supplement to the IEEE 802.11 standard, the IEEE 802.11 Working Group published IEEE 802.11a (details of these standards are described later in Chapter 7 of this book) which outlines the use of OFDM in the 5.8 GHz band. This technology is starting to play an important role in development of fourth-generation networks.

1.4.8 All IP Networks

NGNs (Next-Generation Networks) are 'packet-based' networks, based upon Internet Protocol. Complementary to LTE, another project under development at 3GPP is SAE (System Architecture Evolution). While LTE aims at an evolved radio access network, SAE deals with core network, with a focus on packet domain. Thus the developments of the 3GPP system are compliant with Internet protocols. It is an evolution of the 3GPP system to meet the growing demands of the mobile telecommunications market and is designed to make use of multiple broadband technologies and other 'Quality of Service'-enabled transport technologies where service-related functions are independent from underlying transport-related technologies. This will ensure generalized mobility which will allow consistent and ubiquitous provision of services to users, besides the capability to deliver telephony, television, data and a host of other services at lower marginal cost then the current networks. In 2004, 3GPP proposed IP as the future for next-generation networks and began feasibility studies into All IP Networks (AIPNs).

1.4.9 Broadband Wireless Access

Broadband wireless technologies have opened up possibilities of high-speed, affordable Internet access anywhere and at any time. Although this technology has been available for quite some time, however, 'islands' of proprietary deployment has significantly increased the cost of service and hindered its global expansion. Around about 2003, broadband wireless began to emerge as the key to resolving connectivity bottlenecks. A typical BWA spectrum in shown in Figure 1.5.

Several governments started appreciating the importance of broadband connectivity for social, economic and educational development. They started initiatives to support sustainable broadband services in various regions. Meanwhile, standardization bodies such as IEEE and ITU also started work towards standardization of technologies and harmonization of regulatory frameworks worldwide.

Some key technologies that fall under BWA are described in the following sections.

1.4.9.1 WiMAX

WiMAX stands for 'Worldwide Interoperability for Microwave Access'. It was developed by the WiMAX Forum; formed in June 2001 to promote conformity and interoperability of

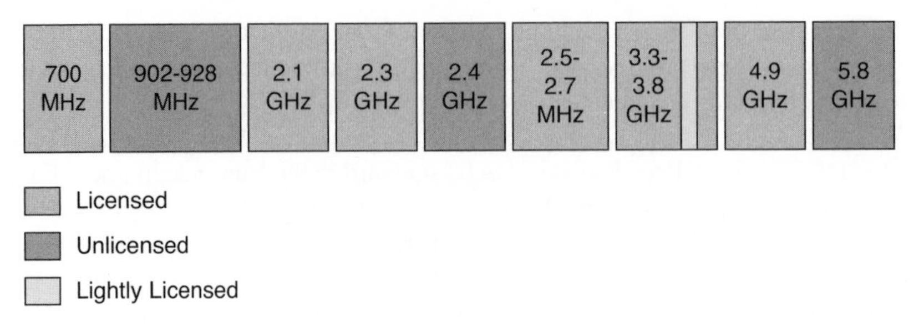

Figure 1.5 Broadband wireless spectrum.

the standard. WiMax was designed as an alternative to cable and DSL, to enable the 'last-mile delivery' of wireless broadband access and has been considered as a wireless 'backhaul technology' for 2G, 3G and 4G networks. However, it is also a possible replacement candidate for other telecommunication technologies such as GSM and CDMA, and can also be used as an overlay to increase capacity.

WiMAX's main competition comes from existing and widely deployed wireless systems such as UMTS and CDMA2000. Also, 3G/LTE technologies are being touted as 'WiMax killers'. By 2008, WiMax had a subscriber base of more than 2.5 million with 450 WiMAX networks deployed in over 130 countries.

1.4.9.2 Wi-Fi

Wi-Fi stands for 'Wireless Fidelity'. With Wi-Fi, it is possible to create high-speed wireless local area networks, provided that the computer to be connected is not too far from the access point. In practice, Wi-Fi can be used to provide high-speed connections (11 Mbps or greater) to laptop computers, desktop computers, personal digital assistants (PDAs) and any other devices located within a radius of several dozen metres indoors (in general, 20–50 m away) or within several hundred metres outdoors.

1.4.9.3 Wireless LAN

WLAN (Wireless Local Area Network) is commonly known as Wireless LAN. Generally it is understood as being the technology which links two or more computers or devices without using wires. WLAN uses spread-spectrum or OFDM modulation technology based on radio waves to enable communication between devices in a limited area. In WLAN, users get the ability to be connected to a network while still be able to move around within a broad coverage area.

1.4.10 IMS

The IP Multimedia Subsystem (IMS) is an IP-based architectural framework for delivering voice and multimedia services. The specifications have been defined by the 3rd Generation Partnership Project (3GPP). This is based on the IETF Internet protocols and is 'access-independent'. It supports IP to IP sessions over 802.11, 802.15, wireline, CDMA, GSM, EGPRS, UMTS and other packet data applications.

IMS intends to make Internet technologies, such as web browsing, instant messaging, e-mail, etc, in addition to services such as WAP and MMS, ubiquitous. IMS is expected to lead to new business models and opportunities.

IMS has given the operators and service providers with the power to control and charge for the services they have provided. Some of the key services involve multi-media messaging services (MMSs), 'Push-to-talk', etc. There are 'Capex' and 'Opex' savings when using the converged IP backbone and open IMS architectures. There are also some hidden advantages such as usage of standardized interfaces which would prevent operators from being 'bounded' by single supplier's proprietary interfaces and the existing infrastructure can be used to create new services.

1.4.11 UMA

Unlicensed Mobile Access (UMA) is the commercial name of the 3GPP Generic Access Network (GAN) standard. This technology provides access to GSM and GPRS mobile services over unlicensed wireless networks such as Bluetooth and 802.11.

This technology enables its users to roam and handover between cellular networks and wireless LANs/WANs using dual-mode (GSM/Wi-Fi) mobile handsets, ensuring a consistent user experience for their mobile voice and data services. This is akin to convergence between mobile, fixed line and Internet telephony.

The fundamental idea behind UMA was to provide a high bandwidth and low-cost wireless access network integrated into operator cellular network. Features such as seamless continuity and roaming were a part of this. This led to development of the UMAC (Unlicensed Mobile Access Consortium0 that promoted the UMA technology. UMAC worked with the 3GPP and the first set of specifications appeared in 2004. 3GPP was defined by the UMA as a part of the '3GPP Release 6' (3GPP TS 43.318) under the name of GAN (Generic Access Network)

1.4.12 DVB-H

Digital Video Broadcasting (DVB) is a suite of internationally accepted open standards for digital television. This standard is led by a consortium of over 270 broadcasters, manufacturers, network operators, software developers, regulatory bodies and others in over 35 countries. It is intended as an open technical standard for the global delivery of digital television and data services. Services on this standard are currently available on every continent with more than 220 million DVB receivers deployed.

The concept of providing television on handheld devices led to the development of DVB technology for handheld or DVB-H. The Digital Video Broadcast (DVB) Project started research work related to mobile reception of DVB-Terrestrial (DVB-T) signals as early as 1998, accompanying the introduction of commercial terrestrial digital TV services in Europe. The EU sponsored projects, such as 'Motivate' and the 'Media Car platform' came up with various conclusions, for example transmissions possible on DVB-T networks, but more robustness needed and the addition of spatial diversity increases the reception performance which helped in the development of mobile TVs. In the year 2002, work started in the DVB Project to define a set of commercial requirements for a system supporting handheld devices. The technical work then led to a system called Digital Video Broadcasting-Handheld (DVB-H), which was published as a European Telecommunications Standards Institute (ETSI) Standard EN 302 304 in November 2004.

1.5 Future Direction

Radio technology and standards are still very much in an active development phase. Researchers are continuously coming up with advancements for optimally using the spectrum and in ways which are cost-efficient as well. Complex signals processing mathematics are employed to reconstruct a data stream from an encoded radio wave. With processing powers becoming cheaper, more complex algorithms can be used to improve performance. While there are theoretical limits to such improvements, it is still a long way off.

Similarly, work is ongoing to provide better user experience, for example if the same handset can work with multiple standards then it can be used as an extension within small premises and used as an handset when the person leaves the building.

Thus it is certain that in the coming years, radio technology will become more digital and smarter. It is then up to regulators and technologists as to how this advancement will be encapsulated within the existing and future regulatory and deployment frameworks.

2

GSM and EGPRS

2.1 Introduction

The limitations of the first generation analogue mobile telephony system led to the development of the second generation mobile systems. Systems such as Nordic Mobile Telephones (NMT) in (Scandinavian) Europe, AMPS (Advanced Mobile Phone Service) in the USA and TACS (Total Access Communication System) in the UK operated under the so-called first generation systems. However, they were incompatible with each other and covered a small geographic area. Though developments did take place within these systems, however, digital revolution paved the way for the next wave of mobile systems which went on to cover most of the planet. These second generation digital systems were more harmonized and of course had a digital technology at their foundation leading to better voice quality and spectrum utilization. Many variants of second generation systems came in different markets that included GSM (Global Systems for Mobile Communications), TDMA IS-136, CDMA IS-95, PDC (Personal Digital Cellular) and PHS (Personal Handy Phone System). However, in this chapter, we will only discuss in detail the GSM and EGPRS systems.

The GSM system is the most popular second generation technology with over a billion people connected through this system (in 2007, the world saw more than 3 billion people connected to voice telephony). The first GSM networks appeared up commercially in the early 1990s but, however, work on these systems started as early as during the 1980s. Based on the GSMK modulations scheme, GSM was a TDMA solution. With the specification ready by 1991, the stage was set for the first commercial networks. The popularity of this system, along with smaller handsets invading the market, boosted the mobile subscriber figures. The short messaging system (SMS) became a 'killer application', the popularity of which can be gauged by the fact that an estimated number of more than a trillion short messages were sent across the world in 2008 (almost 15 billion SMSs were sent in the year 2000 alone).

With data 'knocking on the doors' of the mobile world an extension to GSM networks in the form of GPRS (General Packet Radio Service) came into being. In simpler terms, to the already existing network the form of packet core was added to handle the data traffic. The theoretical speed of these networks was up to 171.2 kilo bits per second (kbps). An enhancement to the data speeds led to enhancements in GPRS networks and were known as EGPRS networks. We

Cellular Technologies for Emerging Markets: 2G, 3G and Beyond Ajay R. Mishra

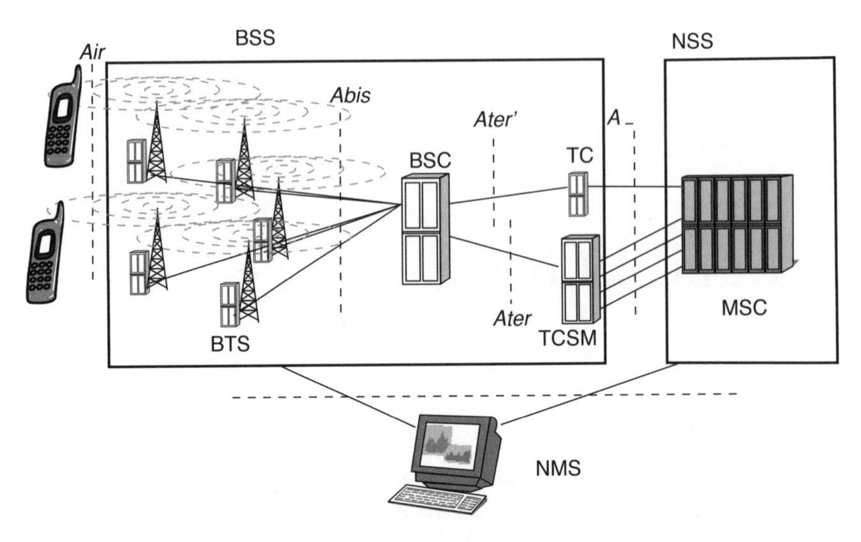

Figure 2.1 GSM architecture.

will discuss later the architecture of these networks and the reasons that led to greater data speeds.

2.2 GSM Technology

2.2.1 GSM Network

Let us understand the key components of the GSM network. The latter consists of three main domains: the Mobile Station (MS), the Base Station Sub-system (BSS), the Network Sub-system (NSS) and the Network Management System (NMS), as shown in Figure 2.1.

2.2.1.1 Mobile Station (MS)

The mobile station is perhaps the most important part of the whole system. Why is this? Simply because the subscriber uses this to talk. Plus the whole network quality could be well perceived by the subscriber based on the experience of using/talking on his/her mobile. There are various mobile devices available on the market, possibly with all possible permutation and combinations of voice, data, music, cameras, etc., ranging from a few dollars to thousands of dollars. However, let us try to understand what are the 'blocks' that make up a simple GSM mobile station. An MS is made up of two main parts, as shown in Figure 2.2: the handset itself and the subscriber identity module (SIM). The latter is personalized and is unique to the subscriber. The handset or the terminal equipment should have qualities similar to that of fixed phones in terms of quality, apart from being 'user-friendly' in usage. It also has functionalities such as GMSK modulation and demodulation up to channel coding/decoding. Plus it needs to possess dual tone multi-frequency generation and should have a long lasting battery.

The SIM or SIM card is basically a microchip operating in conjunction with the memory card. The SIM card's major function is to store the data for both the operator and subscriber. The SIM card fulfills the needs of the operator and the subscriber as the operator is able to

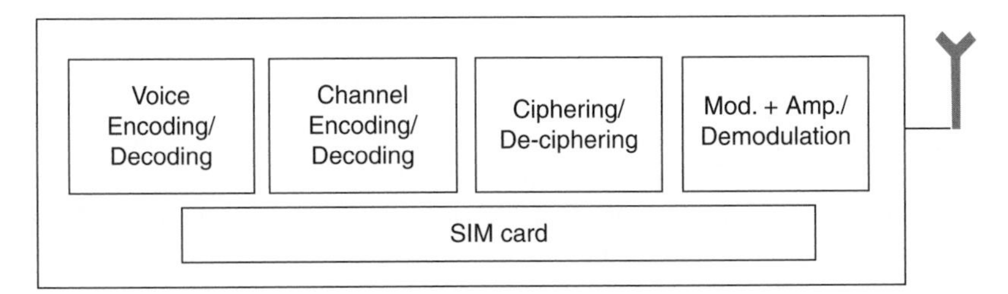

Figure 2.2 Block diagram of the GSM mobile station.

maintain control over the subscription and the subscriber can protect his personal information. Thus, the most important SIM functions include authentication, radio transmission security and storing the subscriber data.

2.2.1.2 Base Station Sub-system (BSS)

The BSS consists of the base transceiver station (BTS), base station controller (BSC) and trans-coder sub-multiplexer (TCSM). The TCSM is usually physically located at the MSC.

Base Transceiver Station (BTS)

The BTS is the interface between the BSC and the mobile station (or subscriber). Due to its position in the network, that is connecting the subscriber's mobile system to the network, this becomes an important element. As the name suggests, the BTS contains elements called 'transceivers' that have the capability to transmit and receive. These transceivers (or TRXs) are connected to the antennae through which information is transmitted/received to/from the mobile station, as shown in Figure 2.3. The antennae can be omni-directional or directional ones. The base stations usually have one to three sectors. Each sector is usually located at 120 degrees (the actual angle is dependent upon radio network plans), thus covering a 360 degree area around them. Each sector antenna is connected to the TRXs in the BTS, while the number

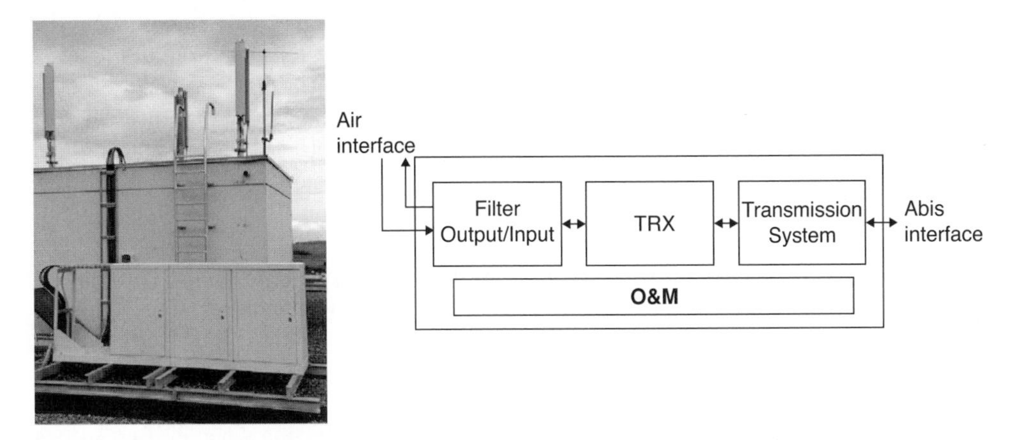

Figure 2.3 A BTS with sector antennas and the corresponding block diagram.

of TRXs' are dependent upon the subscribers that are needed to be catered in that particular direction.

The BTS maintains synchronization to the mobile station. It also consists of a transmission unit, known as the TRU. This is the unit that interacts with the various interfaces, such as A_{bis} and A_{ter}, and which is responsible for allocating the traffic and associated signalling to the correct TRX. Cross-connections are possible at the 2 Mbps level and have a 'drop-insert facility' at the 8 kbps level. Other functions include providing radio interface timing, detecting the access attempts of the mobile station, performing the frequency hopping function and RF signal processing functions such as combining, filtering, coupling etc., encryptions and de-encryption on the radio path, channel coding and decoding, interleaving on the radio path and forwarding the measurement data to the BSC based on which it (BSC) makes decisions related to the mobile station.

Base Station Controller (BSC)

As the name suggest, the BSC controls the base transceivers stations in a network. One BSC controls several BTSs in a network. In simpler terms, the BSC can also be called the 'brain' of the network. This can be well understood by the kind of functions that a BSC handles. As mentioned above, the BTS sends the measurements data to the BSC, based on which the BSC makes the decisions related to the mobile stations, that is the BSC is responsible for radio resource management and configuration. These include functions such as BCF, BTS, TRX management and channel allocation, channel release, radio link supervision (measurement handling) and power control (BTS and MS). To which cell a 'moving subscriber' needs to be connected, the BSC decides on this, based on the measurement reports from the BTS. This decision is based on inputs such as signal quality, signal level, interference, power budget calculations and distance. To improve the link quality between the BTS and MS, frequency hopping management is carried out by the BSC (including implementing no-frequency hopping', baseband-frequency hopping and synthesized frequency hopping). The BSC is also responsible for signalling between the BSC-MSC (CCS7) and the BSC-BTS (LAPD, TRXSIG, BCFSIG). Encryption management functions, such as storing the encryption parameters and forwarding them to the BTS, are conducted by the BSC. A block diagram of a typical BSC is shown in Figure 2.4. The switch matrix (SM) takes care of the relay functions and the inter-working of the A_{bis} and A interface signals coming from the BTS an BSC. Connections to the BTS are established through the Terminal Control Elements (TCEs) which also provides the control functions. On the A interface side as well, the TCE provide similar functions. The database maintains the status of the whole BSS, including the BTS operations and BSS information such as frequency, quality, etc. The central functions are responsible for tasks such as handover decisions, power control, etc.

Transcoder and Sub-Multiplexer (TCSM)

Although the TCSM is physically located at the MSC it is controlled by the BSC and hence it is a part of the BSS. Therefore, planning for this is a part of a transmission planning engineer's job description. The transcoder converts the 64 kbps signal to 16 kbps. It converts the 160 A-law PCM samples of an 8-bit speech channel (20 ms) into a 'vocoder block'. This creates a TRAU frame that is assigned on the PCM signal towards the A-interface direction. In the downlink direction, the TC also performs the speech activity function wherein if no speech

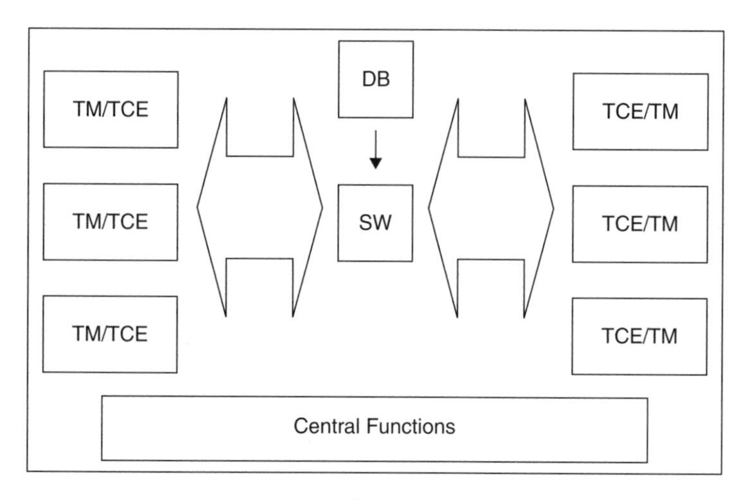

Figure 2.4 Block diagram of the BSC.

is detected then the comfort background noises are transmitted to the mobile station. The sub-multiplexer rearranges the 16 kbps signals more effectively, thus increasing the possibility of reducing the number of 2 Mbps links on the A_{ter} interface (typically by a factor of 3 to 4) – see Figure 2.5.

2.2.1.3 Network Sub-System (NSS)

The network sub-system acts as an interface between the GSM network and the public networks, PSTN/ISDN. The main components of the NSS are the MSC, HLR, VLR, AUC and EIR.

Mobile Switching Centre (MSC)

The MSC or 'Switch' as it is generally called, is the single most important element of the NSS as it is responsible for the switching functions that are necessary for the inter-connection between the mobile users and that of the mobile and the fixed network users. For this purpose,

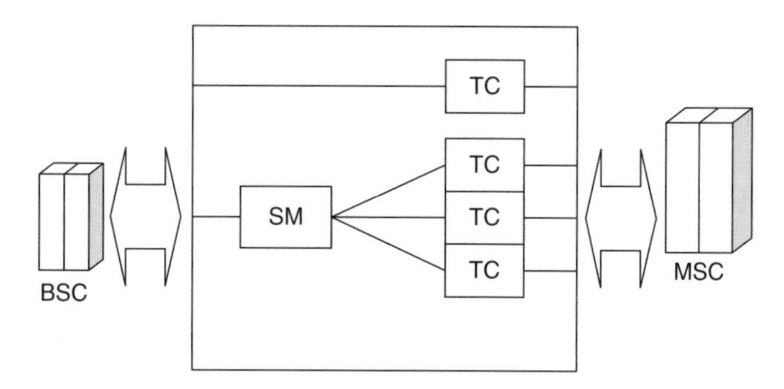

Figure 2.5 Transcoder and sub-multiplexer.

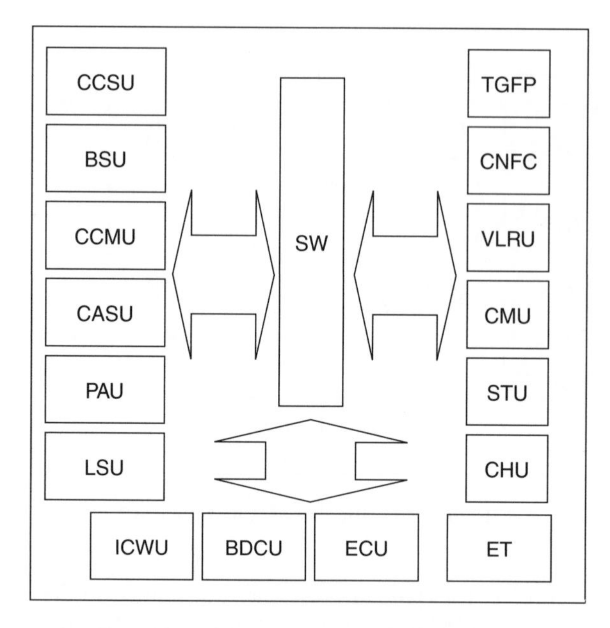

Figure 2.6 Block diagram of the MSC/VLR.

the MSC makes use of the three major components of the NSS, that is the HLR, VLR and AUC. A block diagram of the MSC is shown in Figure 2.6.

CCSU (Common Channel Signalling Unit)
This handles trunk signalling (SS7) towards the HLR, other MSCs and PSTN exchanges.

BSU (Base Station Signalling Unit)
This handles SS7 signalling towards the BSC and call control for mobile-originated calls.

CCMU (Common Channel Signalling Management Unit)
This handles the centralized functions of the SS7 signalling system (needed in large exchanges while in smaller exchanges the CCMU functions are carried out by the CM and STU).

CASU (Channel Associated Signalling Unit)
This performs R2 signalling.

PAU (Primary Access Unit)
This handles DPNSS signalling towards the PABXs.

LSU (Line Signalling Unit)
This controls the announcement machine (ANM).

IWCU (Inter-Working Control Unit)
This controls the Compact Data Services Unit (CDSU) and Echo Cancellers (ECs).

GSW (Group Switch)
This performs the basic function of the MSC which is switching of telephone calls.

TG (Tone Generator)
This is responsible for generating various types of tones such as dial tone, busy tone, information tone, etc.

DTMFG (Dual Tone Multi-Frequency Generator)
This is used for generating DTMF signals.

CNFC (Conference Circuit)
This is used for enabling multi-party conferences.

VLRU (Visitor Location Register Unit)
This performs the VLR functions.

CMU (Cellular Management Unit)
As the name suggests, this controls and supervises the cellular network and handovers.

CHU (Charging Unit)
This collects the charging data.

STS (Statistical Unit)
This collects the exchange-specific statistical data.

OMU (Operation and Maintenance Unit)
This performs the operation and maintenance tasks.

BDCU (Basic Data Communications Unit)
This contains all communication links to O and M networks (terminals for X.25 packet networks and/or for the time slots of PCM links) and to the 'Billing Centre'.

ECU (Echo Canceller Unit)
This is needed in interworking to the PSTN.

ET (Exchange Terminal)
This is the unit which handles the external 2 Mb PCM circuits.

Home Location Register (HLR)
The HLR contains the information related to each mobile subscriber. Each subscriber mobile has some information that contains data such as the kind of subscription, services that the user can use, the subscriber's current location and the mobile equipment status. The database in the HLR remains intact and unchanged until the validity of the subscription.

Visitor Location Register (VLR)

The VLR (Visitor Location Register) comes into action once the subscriber enters the coverage region. Unlike the HLR, the VLR is 'dynamic' in nature and interacts with the HLR when recoding the data of a particular mobile subscriber. When the subscriber moves to another region, the database of the subscriber is also shifted to the VLR of the new region.

Authentication Centre (AUC)

The Authentication Centre (AUC or AC) is the 'responsibility' for the policing actions in the network. This has all the data that is required to protect the network against false subscribers and protection of the calls of the regular subscribers. There are two major keys in the GSM standards, for example one which is the encryption of the mobile users and the other which is 'authentication' of the mobile users. The encryption keys are held both in the mobile equipment and the AUC and the information is protected against non-authorized access.

Equipment Identity Registers (EIRs)

Each item of mobile equipment has its own personal identification, which is denoted by a number known as the International Mobile Equipment Identity (IMEI). This number is installed during the manufacture of the equipment itself, stating the conformation to the GSM standards. Thus, whenever a call is made, the network would check the identity number and if this number is not found on the approved list of the authorized equipment, the access is denied. The EIR contains this list of the authorized numbers and allows the IMEI to be verified.

2.2.1.4 Network Management System (NMS)

The main task of the network management system is to ensure that the running of the network is smooth. For this purpose, it has four major tasks to perform: network monitoring, network development, network measurements and fault management. Once the network is up and running, the NMS takes over the responsibility to monitor the performance of the network. If it sees some faults, it would generate relevant alarms. Some of the faults may be corrected through the NMS itself (mostly software-oriented) while for some sites, visits would be required. The NMS is also responsible for the collection of data and analyses its performance, thereby leading to accurate decisions related to the optimization of the network. The capacity and the configuration of the NMS are dependent upon the size (both in terms of capacity and geographical area) and the technological needs of the network.

2.2.2 Signalling and Interfaces in the GSM Network

Let's start from the MSC to the MS. Signalling between the MSC and BSC is the CCS#7 and is transferred using 64 kpbs timeslots. It passes through the transcoder but the sub-multiplexer arranges the signalling channel from multiple A interface PCM frames to the end of the Ater interface PCM. LAPD signalling is used between the BSC and BTS. These are of two types: TRXSIG and OMUSIG. Some information coming from the BSC in TRXSIG is transferred to the MS using the LAPDm protocol in the air-interface. The TRXSIG is used for performing functions such as call control, both the uplink and downlink measurements, paging, handover, etc., while the OMUSIG is used for downloading of the BTS software from the BSC, monitoring and control purposes.

2.2.2.1 Air-Interface

The interface between the MS and BTS is the air-interface. The speech signal generated by the MS is encoded by the A or the μ law to 8-bit samples at a 8 kHz sampling rate. This speech is coded in the transcoder of the GSM system to bring the rate from 64 kbps to an effective bit rate of 13 kbps. Some additional coding is required for protecting the signal against noise and distortion and this requires an additional (approximately) 3 kpbs, thus making a total in all of 16 kbps (traffic plus signalling). These traffic signals are allocated on the TDMA frame, having 8 TS. With a TS of 0 being used for signalling such as PCH, RACH, etc., the remaining seven are used to carry the traffic. As mentioned before, LADm is used between the MS and BTS.

2.2.2.2 Abis-Interface

The Abis-interface exists between the BTS and BSC. Usually, 16 kbps signalling is used on this interface. The number of the TRXs supported by this interface depends upon the type of signalling. The higher the signalling, the lower the amount of TRX it would be able to support. The traffic channels are 16 kbps. Usually, 96 traffic channels per 2 Mbps are supported with one TRXSIG per TRX and one OMUSIG per BTS on the Abis-interface.

2.2.2.3 Ater-Interface

The interface between the BSC and TCSM is the Ater-interface. The traffic and signalling channels coming from the PCMs from the MSC are re-allocated in the TCSM. The 'Ater' contains 16 kbps traffic channels. Each 2 Mpbs contains up to 116 (or 120) traffic channels. It contains 64 kbps CCS#7 signalling channels and 64 kbps channel for X.25.

2.2.2.4 A-Interface

The A-interface exists between the MSC and TCSM. It contains 30 and 64 kbps traffic channels within each 2 Mbps frame. Also present is 64 kbps CCS#7 signalling and X.25. A TS16 in the A-interface is generally used for CCS#7 signalling purposes.

These interfaces are shown in Figure 2.7.

2.2.3 *Channel Structure in the GSM*

The frequency spectrum for GSM systems is limited. It combines both the TDMA and FDMA techniques for 'best use' of the frequency bands. The frequency bands of 890–915 MHz for downlink and 935–960 MHz for uplink are used for the GSM system. Both of these are divided into 124 carrier frequencies of 200 kHz each. The time division of each carrier frequency creates 'time slots', each numbered from 0 to 7. One burst in the TDMA frame

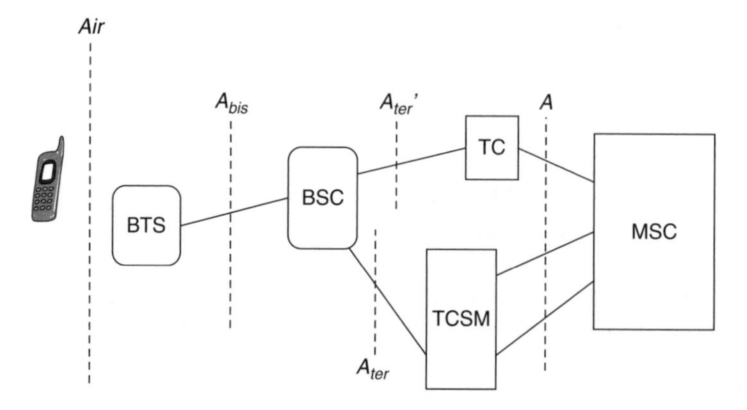

Figure 2.7 Air-, A_{bis}-, A_{ter}- and A-interfaces.

equals one time slot which equals one logical channel. There are two types of channels: physical and logical channels. The formers ones are actual time slots between the BSS and MS while the latter ones are needed for structuring and signalling purposes. The physical channels are of two types: Half-Rate (HR) and Full-Rate (FR). While the FR channel is a 13 kbps coded speech or data channel with a raw data rate of 9.6, 4.8 and 2.6 kbps, the HR supports 7, 4.8 and 2.4 kbps. The logical channels are also of two types: Traffic Channels (TCHs) and Control Channels (CCHs).

The bit rates of the traffic channels (TCHs) can be 13 kbps (FR), 5.6 kbps(HR) or an Enhanced Full Rate (EFR) of 12.2 kbps. The EFR is a dedicated channel for one user. AMR (Adapative Multirate Coding) is another codec type that allows the adaptation of source and channel coding according to network conditions, thus resulting in better quality and capacity savings.

Logical control channels are basically of two types: Common Control Channels (CCCHs) and Dedicated Control Channels (DCHs). Table 2.1 summarizes the logical control channels in a GSM network along with their function.

Table 2.1 Logical control channels in GSM

Channel	Abbreviation	Function/Application
Access Grant Channel (*DL*)	AGCH	Resource allocation, that is subscriber access authorization
Broadcast Common Control Channel (*DL*)	BCCH	Dissemination of general information
Cell Broadcast Channel (*DL*)	CBCH	Transmits the cell broadcast messages
Fast Associated Control Channel	FACCH	For user network signalling
Paging Channel (*DL*)	PCH	Paging for a mobile terminal
Random Access Channel (*DL*)	RACH	Resource request made by mobile terminal
Slow Associated Control Channel	SACCH	Used for transport of radio layer parameters
Standalone Dedicated Control Channel	SDCCH	For user network signalling
Synchronization Channel (*DL*)	SCH	Synchronization of mobile terminal

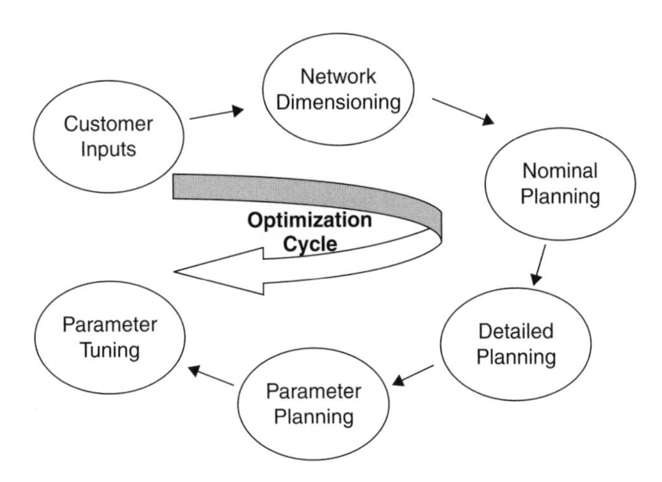

Figure 2.8 Network planning and optimization process.

2.3 Network Planning in the GSM Network

2.3.1 Network Planning Process

As shown in Figure 2.8, the network planning process starts with the customer/network operator. The 'operator' is the one who decides on the fundamental specifications/outlines of the network and gives this information to the vendor who is able to deliver a better quality network, in fewer amounts of time and cost to itself. This selection process is usually based on the 'tendering concept'. The vendor then carries on with the task of network dimensioning where the output is BOQ (Bill of Quantity) or in general terms, a list that contains the type and number of equipments, for example MSC, BSC, BTS, etc. that would be used to install the network. In the GSM network, there are fundamentally three types of planning: radio network planning, transmission network planning and core network planning. Radio network planning concerns planning of the air interface, that is up to the BTS (including coverage and capacity). Transmission planning concerns the access part of the network, that is between the BTS and MSC (including the 'Line of Sight'), while the core network planning considers planning the core network elements in terms of capacity, element numbers and parameters. Apart from traffic planning, all these individual planning aspects also cover the interface and signalling planning as well. Once the number, location and capacity of the BTS, BSC and MSC elements are decided in preliminary calculations, a detailed planning phase starts wherein the site selection process, detailing capacity-coverage planning and parameter planning is carried out. This is followed by an installation and commission phase at which the site 'goes live'. After this, parameter tuning is carried out in order to achieve the best coverage, capacity and quality from the network. After a few months/years, the network optimization cycle starts wherein all the steps are followed again to cater to changing number, behaviour of the subscriber and the changing landscape of the city/region.

2.3.2 Radio Network Planning and Optimization

Coverage, capacity and quality form the three most important aspects of network planning. Of course, cost is also an equally (and in some cases more important than the rest) important factor but as it is dependent on factors beyond technical ones as well, hence it is not discussed here. Coverage comes from the distance covered by the signal propagating from the radio antennae. Thus, not only the signal strength but also the atmospheric conditions and ecosystem are important for the coverage predictions and actual coverage. There are three main types of area (based on habitat and topography) that need to be covered: urban, sub-urban and rural. For the coverage predictions, software is used by the planning engineers. In the pre-planning phase, the imaginary sites are located on the digital maps and coverage prediction software is then 'run'. The cells/sites are of three types: macro, micro and pico. When the antenna is placed above the average rooftop level, it is termed a macro site (this is coverage to maximum area but is also prone to interference). When the antenna is placed at below the average roof level, it is termed a micro site. These sites are less prone to interference but cover a small area. Pico sites are used for indoor coverage. The signal travels a complex but small path from the radio antenna to the mobile station and travels through various terrain and structures – both natural and man-made. When the signal travels from one antenna to another, it looses strength. This is called the 'free space loss and is calculated as follows:

$$L_{\mathrm{dB}} = 92.5 + 20\ \log(f) + 20\ \log(d)$$

where f is the frequency in GHz and d is the distance in km.

The signal, when travelling from one antenna to another, takes many paths. This includes the reflected signal and the diffracted signal: in the former the direction of propagation does not change while in the latter the direction of propagation changes. In both of these cases, the surface does not absorb (partly or completely) the signal. In some cases, the building or surface (such as the body of a vehicle) from where the signal is becoming reflected, absorbs the signal. This is known as 'building loss' and is one of the parameters in link budget calculations which are explained later in this chapter. When the signal travels through vegetation, it experiences a loss in strength known as 'foliage' loss. Another phenomenon to be understood is known as 'fading'. As the name suggests, this is the loss in signal strength, that is the signal fades as it travels from one antenna to another. There are two types of fading: multi-path and frequency-selective. In the former, the signal propagates at different directions and reaches the receiving antenna. Due to the multiple paths taken by the signal, the resultant signal is a summation of all of the signals reaching a point (receiving antennae). This resultant signal may be of lower strength as all of the signals reaching the receiving point may be out of phase with each other. The resultant signal strength depends upon the amplitude and the phase of the constituent signals at the receiving point. Multi-path fading causes fast fluctuations in the signal level. Atmospheric conditions may impact a certain frequency and are capable of bringing down the signal level. However, the atmospheric conditions can impact only one frequency at a given time. Another aspect that can impact signal strength significantly is 'interference'. This happens due to poor frequency planning, that is a signal of the same frequency or adjacent frequency acts as an interfering signal to the main signal, bringing its strength down. These concepts explained above are used in link/power budget calculations. However, let us understand a little more about some further concepts before looking into an example of link budget.

BTS and MS Sensitivity. As per the GSM Specifications 05.05, the BTS sensitivity is given to be –106 dBm. The BTS product manual has this value and the same can be used for link budget calculations. MS Sensitivity is different for each class of mobile station, for example for MS class 4, which means GSM 900, the recommended value is –102 dBm and correspondingly for MS class 1, GSM 1800, the value is –100 dBm. The MS sensitivity can also be calculated via information for the receiver noise figure F and minimum E_b/N_0.

Antenna gains. The BTS antenna gain is dependent upon the antenna type (directional or omni-directional). For a mobile station antenna, a gain of 0 dBi is used.

Diversity gain. This is used to balance the imbalance between the uplink and downlink and typically this is done by putting the diversity 'at the BTS reception'. Diversity gain is around 5 dB.

Cable and connector losses. The cables and connectors used in the system link design cause small amounts of losses which play a significant role in the link budget calculations and hence they need to considered. The types of cables and connectors used are responsible for the amount of losses. A 0.5 inch cable would give a loss of around 7 dB per 100 metre length at 800 MHz and 10 dB per 100 metres at 1800 MHz. However, a 1.5 inch cable gives a loss of 3 dB per 100 metre length at 800 MHz and 4.5 dB per 100 metres at 1800 MHz. Connectors give a loss of around 0.1dB.

Losses of around 2–3 dB come from isolators, combiners and filters used in the network. Some gain is added from the mast head amplifiers and boosters used in the network. The former is used for amplifying the received signal while the latter is used to amplify the transmitted signal. Due to the frequency 're-use', interference takes place in the network, resulting in signal degradation. This is known as the 'interference degradation margin'. An average of 3 dB is used as recommended by ETSI 3.30.

Based on software calculation results, primary site surveys are undertaken. Site survey teams consist of radio, transmission, installation planning engineers, plus civil and electrical engineers. The results from site surveys produce a list of sites that could be used for installing the BTS equipment. Of course, this is only possible after an agreement is signed between the operator and the land owner.

2.3.2.1 Detail Planning

Link Budget

Link budget calculations (see Table 2.2) form the fundamental of detail planning. They form the basis of coverage predictions. The output of link budget calculations is the path loss and received power. It shows the signal propagation in free space. If the path loss is higher, that is the received signal power is less, the coverage will take place over a smaller area and vice versa. Thus, the link budget calculations directly impact the coverage predictions. To increase the received signal power, various techniques such as frequency hopping (FH) and equipment enhancements were used. In FH, by using many frequency channels, C/I is improved. The link budget improves due to effects of the FH in the form of frequency and interference diversity. Improvements in the link budget can take place through equipment enhancements as well. The reception of power can be improved at the base station by using receiver diversity while the transmitted power can be increased by using power boosters. Low-noise amplifiers (LNAs) are also used to improve the signal strength. The LNA is placed at the receiver and it improves the signal strength by amplifying the signal and keeping a low noise figure. An LNA should be used at both antennae (main and diversity) at the receiver end.

Table 2.2 Example of link budget calculations

POWER BUDGET	Unit	UL	DL
RECEIVING END		BTS	MS
Noise figure	dB	0	0
Es/No	dB	0	0
RX RF -input sensitivity	dBm	−111.7	−102
Fast fading margin	dB	0	0
Body loss	dB	0	3
Cable loss + connector	dB	4.5	0
TMA/MHA gain	dB	2	0
RX-antenna gain	dBi	18.5	0
Diversity gain	dB	4.3	0
Isotropic power	dBm	−132	−99
TRANSMITTING END		MS	BTS
TX RF-output power (GMSK)	dBm	28	46
Backoff for 8-PSK	dB	0	0
Isolator + combiner + filter	dB	0	1.8
RF-peak output power (combiner output)	dBm	28	44.2
Cable loss + connector	dB	0	4.5
TMA/MHA insertion loss	dB	0	0.7
TX-antenna gain	dBi	0	18.5
Body loss	dB	3	0
Peak EIRP	dBm	25	57.5
Maximum path loss	dB	157	156.5

Coverage Planning

Coverage planning is carried out based on three aspects: the geographical area to be covered, coverage threshold and coverage probability. The area to be covered and the threshold are interlinked, for example for urban and sub-urban areas it is –75 dBm and –85 dBm, respectively, while for rural areas and a car it is –95 dBm and –90 dBm, respectively. The coverage probability is 90–95 %. We have seen some important aspects of coverage planning before. However, for detail coverage planning, propagation models are used. The most popular ones include the Okumara–Hata and the Walfish–Ikegami models. The former one is used for macro-cell planning to predict the median radio signal attenuation. The Walfish–Ikegami model is used for micro-cell planning (and can be used for macro-cells as well). Both of these are empirical models. The propagation loss can be calculated:

$$L = A + B \log f - 13.82 \log h_{bts} - a(h_m) + (44.9 - 6.55 \log h_b) \log d + Lother$$

by using the Okumara–Hata model where:

f is the frequency (MHz);
h is the BTS antenna height (m);
$a(h)$ is a function of the MS antenna height;
d is the distance between the BS and MS (km).

Lother – this is the attenuation due to land usage classes.

$$a(h_m) = [1.1\ \log(fc) - 0.7]\ h_m - [1.56\ \log(fc) - 0.8]$$

The value of $a(h_m)$ for a small or medium-sized city is:

$$a(h_m) = 8.25\ [\log\ (1.54h_m)]^2 - 1.1,\ \text{ for } fc \leq 200\ \text{MHz}$$

and for a large-sized city:

$$a(h_m) = 3.2\ [\log\ (11.75h_m)]^2 - 4.97,\ \text{ for } fc \leq 400\ \text{MHz}$$

where the constants A and B are:

$$A = 69.55;\ B = 26.16\ (\text{for } 150 - 1000\ \text{MHz})$$
$$A = 46.3;\ B = 33.9\ (\text{for } 1000 - 2000\ \text{MHz})$$

Using the Walfish–Ikegami model, the propagation loss is calculated as follows:

$$P = 42.6 + 26 \log\ (d) + 20 \log(f)$$

and for the NLOS conditions, the path loss is given as:

$$P = 32.4 + 20 \log\ (f) + 20 \log\ (d) + \text{L}rds + \text{L}ms$$

where the various parameters are as follows:

d is the distance (km);
f is the frequency (MHz);
L*rds* represents the rooftop–street diffraction and scatter loss;
L*ms* represents the multi-screen diffraction loss;
w is the road width;
b is the distance between the centres of the two buildings;
H_{bu} is the height of the building;
H_{bts} is the height of the BTS antenna.

The Okumara–Hata (OH) model is generally used for frequency ranges of 15–1500 MHz and 1500–2000 MHz. The Walfish–Ikegami (WI) model is used for frequency ranges of 800–2000 MHz. In the OH model, the range for the base station antenna height is from 30 to 200 metres, the mobile antenna height is from 1 to 10 metres and the cell range, that is the distance between the BTS and MS, is from 1 to 20 km. The WI model is used for heights (height of the BTS antenna from the ground level) of up to 50 metres for a distance of up to 5 km.

These models are not applied as they 'stand'. There is another process known as 'model tuning' during which the propagation models are customized for applying in that particular area (for which planning is being carried out). Inputs from two sources are used: measurement analysis from 'tool and field measurements'. For the former, an accurate digital map is a primary requirement. The field measurements are then carried out through 'drive testing'. What is the

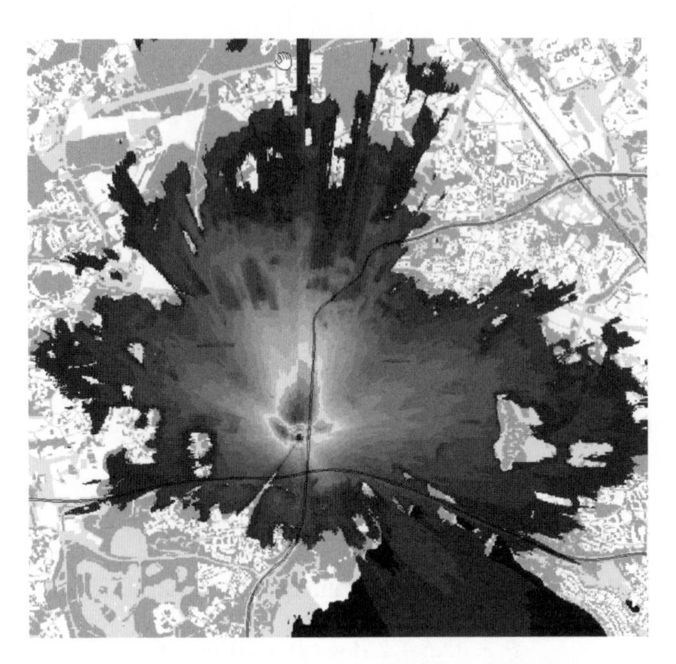

Figure 2.9 An example of a typical coverage plan.

output of this exercise? It is correction factors. This exercise gives correction factors for each 'clutter type'. A higher value of the correction factor means difficult propagation conditions while a lower value signifies smoother propagation conditions. Theoretically, the cell size as calculated by the planning tools is not possible. This is due to the fading phenomenon as discussed earlier. How is the coverage quality defined? It is through location probability which describes the probability for the receiver catching the signal. It can also be defined as the probability of field strength being above the sensitivity level. The practical assumption related to location area probability is of 50 %, and is equal to the sensitivity of the receiver in the target region. Planning engineers should know that the location area probability should be higher than 50 %. A typical coverage plan is shown in Figure 2.9.

Capacity Planning

As discussed in the sections above, in the pre-planning phase, coverage planning calculations lead to a number of base stations, while capacity planning calculations lead to a number of base stations. The allocation of capacity is based on coverage and the traffic estimates in a given region/area/cell. Traffic definition is in 'erlangs' which is defined mathematically as follows:

$$Erlang = \frac{(number\ of\ calls\ in\ hour) \times (averege\ call\ length)}{3600\ \text{seconds}}$$

There are two types of erlang formulae: erlang B and erlang C. The former one does not takes queuing into account while the latter ones does take queuing into account. Erlang B calculations involve three parameters: blocking probability, traffic and number of users.

Table 2.3 Example of an erlang B table

Blocking Probability # channels	1 %	2 %	3 %	5 %	10 %
1	0.01	0.02	0.03	0.05	0.11
2	0.15	0.23	0.28	0.38	0.60
3	0.46	0.60	0.71	0.90	1.27
4	0.87	1.09	1.26	1.53	2.05
5	1.36	1.66	1.87	2.22	2.88

Blocking probability states the amount of calls that are 'allowed' to be blocked, for example if the blocking probability is 2 %, then no more than 2 calls per 100 calls are allowed to be blocked, that is not being able to reach the dialled subscriber due to unavailable resources. A concise example of an erlang B table is shown in Table 2.3. With a given blocking probability and knowing the number of channels, the traffic generated can be estimated.

We have already discussed and understood about channel structures and time slots. It is important to understand this from the capacity planning perspective as it gives a radio planning engineer information of the amount of subscribers than can talk at any given time. Due to TDMA modulations, there are eight time slots in both the uplink and downlink directions. Although all eight can be used for sending traffic and signalling information, TS0 is used for signalling and the remaining seven are used for carrying user traffic. Hence, seven subscribers can talk at the same time by using one TRX. Increasing the number of TRXs would increase the capacity and the number of subscribers talking at the same time would also increase (and not in multiples of seven). The subscriber increase would be taking place in a series of $7 + 8 + 7 + 8 \ldots$ with each increasing TRX.

Another aspect to be kept in mind during detailed capacity planning is the antennae heights. As the heights of the antennae are reduced, the area covered would be less, leading to more base stations for a proper coverage in a given region. This would lead to the using of many more frequency channels which would in turn lead to problems in frequency planning. Also, if the antenna heights are increased, the coverage area would increase, thus leading to a less number of frequencies being used. This basically means that proper frequency planning not only directly impacts quality but also capacity and coverage in the network.

Frequency Planning and Spectrum Efficiency

The frequency spectrum is always a big issue with the operators, regulators and of course radio planning engineers. The area to be covered is usually much larger as compared to the number of channels available. Hence, a common technique is used which is called as 'frequency re-use'. The same frequency is used in different cells in the network in such a way that interference is least and at the same time provides maximum capacity and coverage to the network (as shown in Figure 2.10). One important factor in the frequency re-use is the distance. This is based on considering that the cells are hexagons. However, as cells are not 'practically' hexagons, hence the frequency re-use factor is not common in the network. It also varies between the BCCH and TCH layers. The BCCH TRX should be interference-free, hence the frequency re-use factor is higher on the BCCH TRX than the TCH. To keep interferences at a lower level, some factors if kept in mind prove to be quite useful. One of the ways is to control the transmitted

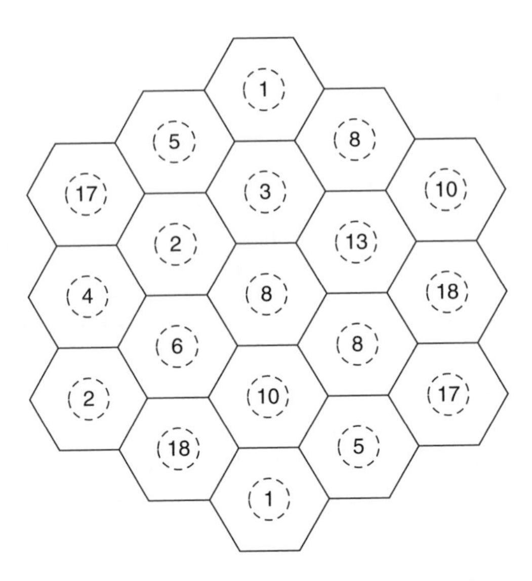

Figure 2.10 An example of frequency re-use.

power. Not only the power transmitted from the BTS can be controlled, but also the power transmitted by the mobile can be controlled as well. This is done because the power received at the BST should be just above the required threshold in order to reduce the interference to other mobiles. The software feature in the BSS is capable of controlling the power transmitted by the mobile station and the value of power transmitted to the BSS is carried out by the BCCH. Another place where the BSS controls the power in order to reduce interference is the 'handover'. This is a phenomenon when a mobile gets detached from one cell (at the cell edge) and gets attached to another cell. The main criteria for control for cell selection for the mobile to get attached are enough power and less interference. During the 'silent mode' (i.e. when a subscriber is not speaking), the BSS asks the mobile to reduce its power which in turn reduces interference. This is the DTX or discontinuous transmission feature in the mobile. Another very important technique used to reduce the interference is frequency hopping (FH). In this, the frequency of the signal is changed in every burst in such a way that the network experiences minimum interference levels. As every burst of frequency will fade in a different way and time, the 'de-correlation' between the bursts increases, resulting in increases in the coding signal (known as 'frequency diversity'). Also, some mobiles are affected by interference more than others as each mobile has one constant frequency. Due to FH, the interference spreads in the network, due to which the effect of interfering signals gets reduced (called as 'interference diversity'). There are two types of FH: base band and RF. In the former, calls are 'hopped' between the TRX and the number of frequencies are constant while in the latter, calls stay on one TRX only while the frequency changes with every frame. RF FH is more robust as it is non co-related to the number of TRXs.

Parameter Planning

Usually, parameter planning is considered to be a part of detail planning; however, it is mentioned separately here to signify the importance of the parameters of the radio network. Radio resource management and mobility management is a part of parameter planning. Let us now understand how the mobile station gets connected to the network. There are two states of mobile station in a network: 'idle' and 'dedicated'. When connected/logged into the network, the transition of the state request comes from the mobile station through the RACH (channels explained before in this chapter). FCCH and SCH help the mobile to get connected to the network. Once connected, the mobile station keeps getting information from the BTS through the BCCH. When a call is initiated, the PCH transfers the information requesting the allocation of the dedicated channel, where information of the dedicated channel request being granted comes from the AGH. As there is no fixed time when the mobile generates this request, hence other factors such as traffic increase, congestion, etc. come into play, a radio planning engineer needs to keep these aspects under control in order to keep a control on throughput. Let us now look at some important parameters used in radio network planning.

LAC (Location Area Code)

The nearby cells are combined into one logical region and a two-bit code is assigned. The mobile can roam freely within the area without informing the HLR or CLR. This parameter is significant as it reduces the amount of signalling that is needed for informing the HLR every time the mobile station changes the location area.

BSIC (Base Station Identity Code)

A six-digit code (with values between 0 and 7) consists of two parts, with the first three of the code stating the NCC (Network Colour Code) and the other three stating the BCC (Base Station Colour Code). The former one separates the various networks (i.e. it separates the various operators) and the latter one separates the base stations. The BSIC is used for cell identification in the 'dedicated mode'. All of the cells in the same area and having the same frequency should not be allocated to the same BCC codes.

RxLev

This is the received signal level of the BCCH channel.

RxLevAccessMin

This is the minimum signal level required by the mobile to access the cell (normally a signal level close to −100 dBm is required). RxLev is compared to RxLevAccessMin and if it is lower, then it keeps on searching for another BCCH. It is the limiting factor in the downlink direction.

MSTxPower and MSTxPowerMax

The MSTxPower parameter is related to the maximum power MS that can be used on network parameter settings. It is a limiting factor in the uplink direction and cell size. Based on the type and class, MSTxPower is the maximum transmission power of the mobile station.

In the idle mode, the cell selection is controlled by two parameters based on the criteria C1 and C2. C1 is known as the criterion for path loss which determines, in the idle mode, the selection of the cell, while C2 is known as the cell reselection of the criteria.

Some other important parameters include the timing advance (TA) and IMSIAttach/ Detach. The former one tells us how far the mobile is from the BTS while the latter one defines the location update (*attach* is activated when the mobile station is powered while *detach* is activated when the mobile is powered down).

Parameter Tuning

The parameters are tuned to make the network delivery of a better quality to the mobile subscribers. However, the parameters behaviour needs to be observed as well. This is done though two main methods: drive testing and statistics. While statistics give the idea about the real behaviour of the parameters by mobile subscribers, irrespective of the geographical locations, drive testing brings in simulation of the end users' opinion of the network quality. Statistics are extracted at the cell level regions and network level. Thus although statistics present the advantage of 'going deep' into analysis at the cell level, they are at a disadvantage of not giving any geographic approach. Here drive tests become handy as they give the inputs from the geographic level. Statistics are accumulated through the network management system, while the drive tests are carried out through the software loaded on a laptop and a mobile with a SIM card. Drive tests are performed to carry out both the outdoor and indoor coverage. Due to logistics constraints, drive tests cannot be performed on the whole network, thus, making both the statistics and drive tests results necessary for the optimization engineers. One or more handsets are mounted on a 'car kit' and connected via cables to the laptop. Each of these make continuous blank calls to the MSC and log files are recorded for analysis at the later stage.

The parameters that are monitored are known as Key Performance Indicators (KPIs). The latter are related to coverage and capacity. Some of the main KPIs are explained below.

BER

BER (Bit Error Rate) is an estimated number of bit errors in a number of bursts to which they correspond to values from 0 to 7 ('best-to-worst') of the RxQual (Receiver Quality). The RxQual is considered as a basic measurement. It simply reflects the average BER over a period of 0.5 s. The number of bit errors is accumulated in a BER sum for each SACCH multi-frame and the result is classified from 0 to 7 according to the BER–RXQUAL conversion table shown in Table 2.4.

DCR

DCR (Drop Call Rate) is the ratio between the drop in traffic channels during the conversation to the number of successful 'seizers' on the cells or areas.

Table 2.4 BER versus RXQUAL conversion table

RXQUAL	BER
0	BER < 0.2 %
1	0.2 % < BER < 0.4 %
2	0.4 % < BER < 0.8 %
3	0.8 % < BER < 1.6 %
4	1.6 % < BER < 3.2 %
5	3.2 % < BER < 6.4 %
6	6.4 % < BER < 12.8 %
7	12.8 % < BER

FER

FER (Frame Error Rate) represents the percentage of blocks with an incorrect CRC (Cyclic Redundancy Check). The range of the FER goes from 0 to 100 % with the former 'being better'. This is more stable than the BER and depends upon the codec as well.

CSR

CSR (Call Success Rate) is the measure of the ability of the cell/area/network to provide the TCH to the mobile from where the call is originating. Another parameter related to this is the CSF (Call-Set Up Failure Rate). There are a couple of ways on how these data are used by the operators: some start by considering the CSF after the SDCCH is assigned by not considering SDCCH blocking, while some 'modelize' the user's perception and consider all types of failures.

HSR

HSR (Handover Success Rate) gives the percentage of the successful outgoing handover attempts. A higher value of the 'handover failure' would result in higher values of the DCR.

SDCCH Drop

High SDCCH traffic is not the same as high TCH traffic. Due to bad links or any product-related problems, an 'SDCCH drop' can take place. A high SDCCH drop explains the high values of the CSF.

The tuning is based on a few inputs and these inputs come from the planning software, drive tests and measurements on the network management systems. The inputs from the drive tests and network management systems are used in the software to calculate new values for the parameters. The performance assessment is based on three factors: coverage, traffic and quality and the amount of traffic and blocking; resource availability and access measurements; handover measurements, receiver levels and power control measurements. Once the measurements and analysis are carried out, the process of optimizing the network begins by tuning the parameters.

2.3.3 Transmission Network Planning and Optimization

2.3.3.1 Dimensioning

The main aspects of dimensioning in a transmission network are to plan the capacity of the BTS–BSC and BSC–MSC transmission links. If the transmission link is a microwave link, then other aspects such as link budget, line-of-sight calculation, topologies planning, etc. are also considered.

One PCM or E1 consists of 31 timeslots (TSs) that are numbered from TS0 to TS31. The TS0 is used for management while TS1 to TS31 are used for traffic and signalling. Signalling is from 16 kbps, 32 kbps and 64k bps. Transmission planning engineers need to plan for two interfaces: the A_{bis} and A_{ter} interfaces. On the air interface the blocking permitted is generally 2 %, while on the A_{ter} interface the blocking is 0.1 % to 0.5 %. The A_{bis} interface should not have any blocking, that is zero blocking. Once the radio planners give the number of base stations and their locations, the transmission planner needs to come up with the list of links needed along with the capacity. Traffic calculations and the capacity of each BTS are given by

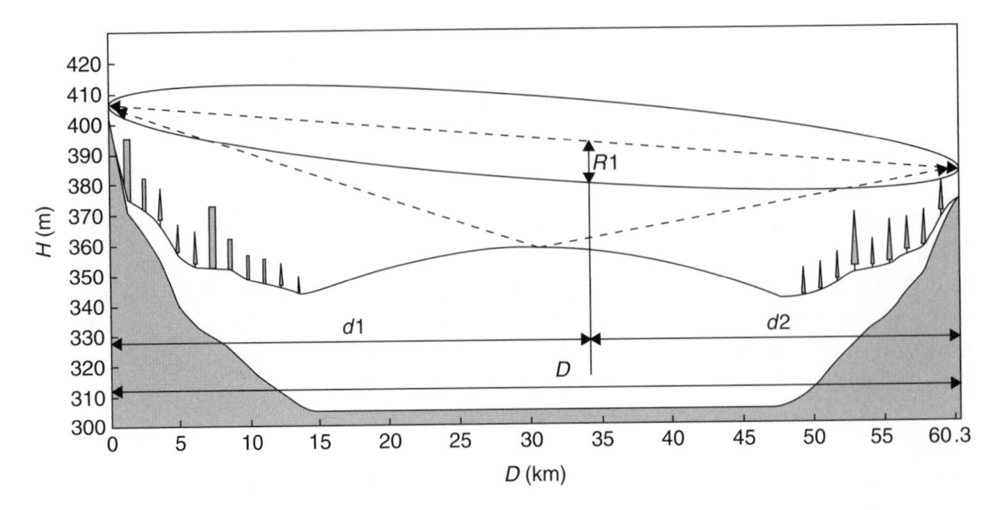

Figure 2.11 The first Fresnel zone.

the radio planners. Based on the topology, the microwave links and other transmission links, a related inventory is prepared. One E1 on the A_{bis} can carry approximately 96 subscribers while on the A_{ter} interface there are approximately 120 subscribers.

2.3.3.2 Nominal Planning

Site surveys form a very important part of the nominal planning process – after all, the radio sites need to be connected to the BSC and subsequently to the MSC. During the site surveys, the transmission planners requirements are quite opposite in terms of height of the building. A radio planner looks out for a small height building while the transmission planner looks out for the highest building so that the 'line of sight' can be achieved. The line of sight achieved means that the microwave link is clear of all obstructions in its path. This is done by calculating the radius of the Fresnel zone. The radius of the Fresnel zone, $R1$ (see Figure 2.11) should be clear of all obstacles in the path. The planners should make sure that there is some clearance between the radius of the Fresnel zone and the highest obstacle on the path. The radius of the first Fresnel zone is given as follows:

$$F1 = 12.75\sqrt{\frac{d1 \times d2}{f \times D}}$$

where

$F1$ is the radius of the first Fresnel zone;
f is the frequency of the transmitting signal (GHz);
D, $d1$ and $d2$ are the distances (km).

2.3.3.3 Link Budget Calculation

These calculations give an idea of the amount of power that would be received by the receiving antenna under 'live' conditions. An important constituent is the free space loss (FSL), that is the loss of signal strength in the free space. This is dependent on frequency and hop distance and is calculated as:

$$FSL\ (L_{fs}) = 92.5 + 20\ \log\ d + 20\ \log\ f$$

Another constituent is the antenna gain. This is dependent upon the frequency of operation and diameter (for parabolic antennae) and is calculated as:

$$G = 20\ \log\ (Da) + 20\ \log\ f + 17.5$$

The hop loss L_h can be calculated as:

$$L_h = L_{fs} - G_t - G_r + L_{ext} + L_{atm}$$

where

G_t is the gain of the transmitting antenna;
G_r is the gain of the receiving antenna;
L_{ex} is the extra attenuation (due to rain, etc.);
L_{atm} is the atmospheric losses due to water vapour and oxygen.

The received signal level P_{rx} can be given as:

$$P_{rx} = P_t - L_h$$

The difference between the received power and the receiver threshold (R_{xth}) is known as the 'fade margin' (FM):

$$FM = P_{rx} - R_{xth}$$

The received signal, however, is also dependent upon atmospheric conditions. We know that there are losses due to changes in the atmospheric conditions. Due to these changes, the signal may fade more than normal or even change its path. Also, due to the fact that the signal takes many paths to reach the receiver, the fading (also known as multi-path fading) may take place at the receiving end as well, in particular, if the signal is reflected from a surface like water or 'paddy fields'. Due to temperature changes during the 24 h period, the transmitted waves change the way they move in the atmosphere towards the receiving antenna. Sometimes, the temperature changes are so large that the transmitted signal gets caught in a duct-like formation in the atmosphere and the signal is not able to reach the receiving antenna. The atmosphere acts like a waveguide and the trapped signal moves into it towards the direction of the duct. To prevent multi-path fading problems, two antennae are installed on the same tower (usually one a few metres below the other), known, respectively, as the 'main' and 'diversity' antennae.

This is known as 'space diversity'. When ducts are formed, the angle of transmission of the transmitting antenna is changed or the diversity antenna is used, depending upon the type of duct formation. Another form of attenuation that takes place, especially above 10 GHz, is the rain attenuation. Above 15 GHz, this attenuation becomes more severe. The attenuation increases with the increase in distance, frequency and rain rate. It is calculated as follows:

$$A = \alpha \times R^{\beta}$$

where

A is the attenuation due to rain;
R is the rainfall rate;
α and β – these are constants (defined for spherical drops and are polarization-independent), with values given in ITU Recommendation 838).

One more fading that we should know about is 'k-fading'. This term derives from the 'k-factor'. The latter is defined as the ratio of the *effective* radius of the earth to the *actual* radius of the earth. This factor is dependent upon the curvature of the propagating microwave signal from the transmitting to the receiving antenna, which is in turn dependent upon the temperature, pressure and humidity. Now, as these parameters vary the signal starts to travel in directions that would lead to a 'faded' signal strength at the receiving antenna. Thus, during the planning phase, the link budget calculation should be done, not only for standard conditions of temperature and pressure (i.e. $k = 4/3$) but also for non-standard conditions such as $k = 0.6$, etc. A brief summary of the factors influencing the performance of the 'link' are described in the following.

Influence of Weather

Although the microwave beam is conventionally shown as a line, the actual method of propagation is as a wavefront where the important portion of the wavefront involves a sizeable transverse area. For free-space propagation, it is essential that all the potential obstructions along the path be removed from the beam centre-line by at least $0.6F1$ where $F1$ is the radius of the first Fresnel zone.

Influence of Rain and Fog

At microwave frequencies of up to 6 and 8 GHz, rain attenuation is not considered sufficient to warrant any special considerations. Under saturation conditions, a 30-mile path suffers a few dBs of attenuation at 6 GHz. Fog conditions can also be considered similarly. At microwave frequencies of 11 or 12 GHz, attenuation can be serious. The amount of the attenuation depends upon the rate of rainfall, the size of the drops and the length of exposure.

Influence of the Objects in Azimuth

The potential problems with 'off-path objects' are the reflections and these usually turn out to come from the buildings. The energy travelling in the longer path lags behind the main beam. The most serious cases are the ones of the multiple reflections. In this case the delay is so great so as to cause a delay in the distortion in the base band.

Atmospheric Absorption

This is due to the oxygen and water vapour which exists in the atmosphere. This effect is much less in the 2–8 GHz range and is usually neglected. In the higher bands, the effect is small but not negligible.

2.3.3.4 Detailed Planning

Some important aspects of detailed transmission planning are frequency, timeslot allocation, 2 Mbps synchronization and transmission management system planning.

Bad frequency planning would lead to interference and hence degradation of the system performance. A cellular operator would have only a few selected 'frequency spots' to be used. Frequency allocation is carried out in a way that no two same or adjacent frequencies are allocated to adjacent 'hops'. In addition, frequency allocation is carried out in such a way that an 'over-reach phenomenon (the signal of one 'hop' reaching the antennae of the third 'hop' in the chain) does not take place.

Time slot allocation planning involves the allocation of traffic channels for traffic, signalling, pilot bits and loop control bits on the PCM or 1E1 channel. There are a few ways of allocating time slots. One of the ways is to allocate timeslots to traffic for all BTSs/TRXs and then allocating the remaining ones to the signalling. Another technique is to allocate the traffic and signalling for one TRX together and then proceed to allocating the traffic and signalling timeslot for the next TRX. Some timeslots on the A_{bis} needs to be reserved for pilot bits (used for synchronization) and some for loop control function as well.

The 2 Mbps plans are about routing the traffic on the E1s. When BTSs are connected to the BSC directly ('star/point-to-point topology'), each E1 which goes from the BTS is connected to the ET ports on the BSC. However, 2 Mbps planning is really needed in the cases of chain or loop topology, especially in cases where one E1 needs to carry the traffic on more than one BTS. Although one E1 carries the traffic of more than one BTS, still only one ET port is allocated to it. To keep the network elements, SDH and PDH and other transmission equipment synchronized, synchronization planning is carried out. For this, the 'clock signals' from the standard sources are used. Some of these are:

- PRC (primary reference clock);
- Slave clock (synchronization supply unit (SSU));
- SEC (SDH equipment clock).

Of these, the PRC is the most accurate (10^{-11}) while the SDH clock signal, which is the internal clock of the equipment, is the least accurate (10^{-4}). The clock signal moves from the PRC or any other standard source to the MSC, from the MSC to the BSC and from the BSC to the BTS. A couple of important aspects to remember during the synchronization process are the amount of equipment in a chain to be synchronized, which should be in accordance with the ITU Recommendation G.813, and loop topology, where there should be no 'loop' of the timing signal. Synchronization should always be protected.

After all of the network designing, implementation and commissioning, there needs to be a system that can track the issues arising in the transmission network. This is carried out by a transmission network management system. Whenever there is a fault in the transmission network, an alarm is raised by the software monitoring the network. Many times, the software-related issues are rectified from the office itself and a site visit is not required. This management system is based on the 'master-slave protocols'. In most cases, the NMS acts as the 'master' while the network elements act as 'slaves'. Once the master is defined, the definition of the management bus and its transfer method is decided. The next step is to decide the parameters for each of the network elements that the master would control. Network elements can be managed by attaching the slave to the masters through a cable. Other techniques involve sending the management bits through the PCM signal from TS0 to TS31 and through the auxiliary data channels of the 'frame overhead' of for example the radio frames.

Parameter Tuning

As compared to a radio network, there are less numbers of parameters in a transmission network. The main monitoring and subsequent tuning happens for capacity and quality of the transmission network. Within a few years of inception of the mobile network, the subscriber base increased substantially. This poses two challenges to planning engineers: first, on how to utilize the current capacity effectively and secondly, on how to increase the capacity without disturbing the traffic. During the 'roll-out phase' of the network, the pressure of meeting deadlines and practical aspects such as sites not better ready in time, result in the BTS sites becoming connected to the BSC or other BTSs, in not the best possible way from the capacity utilization perspective. Thus, the E1 utilization can be done once the optimization process gets underway. Also, the new sites that are added, especially the ones in chain or loop topologies, can use the remaining E1 capacity. From the quality perspective, there are three main reasons of degradation: interference due to poor frequency planning, degradation in the microwave links due to atmospheric changes and degradation due to the synchronization problems. Unlike in radio planning, in transmission planning the data collection times can range from a few days to a few years. Based on this analysis, solutions such as change in frequency plans, movement of the antennae or addition of antennae (space diversity) and proper source and movement of the clock signal in the network can be carried out.

The types of diversity arrangement are as follows:

- frequency diversity;
- space diversity;
- hybrid diversity (special combination of the frequency and space diversity).

Frequency Diversity

In this, two frequencies are used on the same microwave link. This is 'full' and has simple equipment redundancy. Also, with two 'full end-to-end paths', full testing can be done without interruption of the services. However, the major disadvantage with frequency diversity is that double the amount of the frequency spectrum is required.

Space Diversity

In this, two antennae are used on the same tower to catch the receiving signal and provide good diversity protection against atmospheric conditions. This is quite a popular method of diversity. It provides a 'spectrum-efficient solution'. Though it can provide full equipment redundancy (with HSB) it does not give 'end-to-end' operation paths. It is also more expensive than frequency diversity because of the additional equipment required.

Hybrid Diversity
This provides the best of the frequency and space diversity but has the same disadvantage as the ordinary frequency diversity, that is it requires two RF frequencies to obtain one working channel.

2.3.4 Core Network Planning and Optimization

The number of elements are lowest in the core network but perhaps these are the most important ones, for example the MSC, HLR, etc. After all, elements like the MSC are very few in the network, sometimes only being present once. Hence, careful dimensioning and planning is needed for the number of elements in the core network and the related connectivity and signalling of these elements.

2.3.4.1 Dimensioning

The inputs include subscriber figures (estimated – now and for a few years later), the number of BTSs, BSCs and other network elements, traffic figures (estimated – now and for a few years later), traffic moving from one network to the external network and into one network from the external networks and the services that would be delivered to its subscribers (e.g. voice, SMS, MMS, etc.). Of these, traffic calculations are the most important ones constituting two parts: traffic generated in one network and moving within/ outside of one network and the traffic generated from other networks moving within one network. Another important aspect is the 'calling and moving interest'. The 'calling interest' indicates the distance over which calls are made, that is long distance or short distance, while 'moving interest' defines the subscriber behaviour of moving within the network. Once the number of subscribers have been determined, the number of switches can be calculated. The subscriber number directly affects the VLR (visitor location register). Thus:

$$\text{Number of Switches} = \text{Number of subscribers/VLR (or HLR) Capacity}$$

The location of the MSC is usually decided at the 'headquarters' or at an easily accessible place. In case there are many switches, these are located at a high subscriber density area to save transmission costs. All these MSCs are connected in a cyclic fashion so as to protect the traffic. Also, some of these are interconnected as well (as shown in Figure 2.12). Thus, more than one protection topology is used. The concept of a 'transit switch' is added, where the traffic is routed through this switch. For larger networks, more transit switches are used and the topology between the transit switches is usually fully 'meshed'. The next step is to develop routing plans, that is how the traffic should be routed so that no one route is over loaded and in case of failure, which path will the traffic take.

Another important aspect is the signalling (see Figure 2.13). Signalling networks can be of different types. 'Signalling Network Indicators' give information on the type of signalling network used. Signalling in the NSS network is SS7. Signalling can be transferred on the 64 kbps timeslots in the PCM.

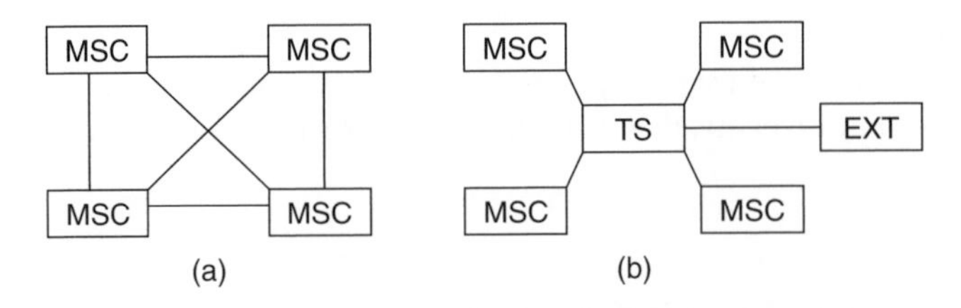

Figure 2.12 Switch network planning.

Signalling Point

Any point that is capable of sending or receiving the signalling is called the Signalling Point (SP). Signalling traffic passes through the STP or Signalling Transfer Point and finally reaches the location that it is meant for, called the Signalling End Point (SEP).

Signalling Link

The logical connection between two SPs is called the Signalling Link. SPs are connected using PCM links (also called PCM circuits). These links carry messages of higher layers.

In GSM networks, as SS7 is used for signalling, four different types of networks are possible, namely NA0, NA1, IN0 and IN1. The former two are for national networks while the latter codes are used for international networks. Signalling dimensioning involves: end-to-end traffic calculations, routing the traffic and the transmission needed. The call parameters related to successful calls and especially short calls (they generate large signalling traffic) is taken as the inputs. Based on all the inputs above, the dimensioning output consists of traffic flow calculations, element calculations (number of switches, etc.), routing plans (including transmission) and signalling plans.

2.3.4.2 Detail Planning

A detail plan will consist of:

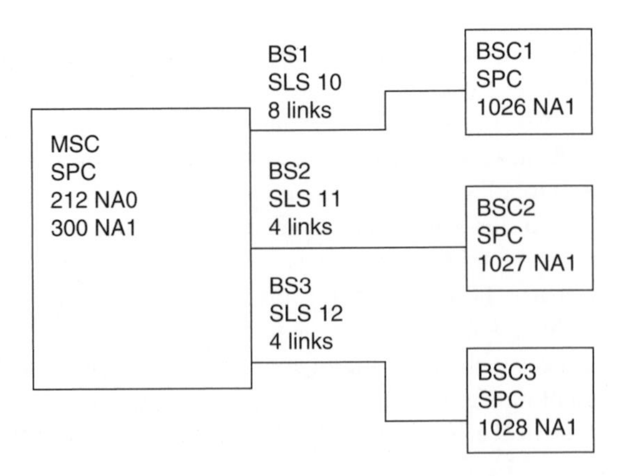

Figure 2.13 Signalling network planning.

Detailed Routing
Naming conventions are adopted and names created for 'Destinations' and 'Sub-destinations', 'Circuit Groups' and so on. Routing types are also fixed. Routing source data for each switch are created.

Digit Analysis
A standardized approach in programming the digit analysis is defined and source data for digit analysis are produced for each switch.

Detailed Signalling
Naming and numbering conventions are adopted and used to create signalling points in the network, signalling links and link sets, signalling routes and route sets.

Detailed Numbering
The number of groups used by each switch are defined as well as V5 numbering if used. Allocation of geographically determined number groups are promoted in order to help the operation of the network.

Detailed Charging
Charging zones and cases are defined for the network and the corresponding source data for each switch are generated. Collection and transfer of charging records are determined and corresponding source data generated.

DCN Settings
Parameters for logical and physical connections of the switches to DCN networks are generated so as to implement DCN planning at the switch level.

Synchronization Planning
General synchronization planning is detailed at each switch level: the physical connections are defined plus priority settings in each switch. The source data for synchronization are generated

In the analysis and dimensioning phases, the minor information related to routing is decided; however, in the detail planning phase, routing plans are fixed, including the naming convention. Detail signalling plans consist of defining the signalling end points (SEPs), signalling transfer points (STPs), signalling link numbers and signalling link sets. The numbering group used by each switch is finalized. This is usually done by taking into account aspects such as geographical locations in order to make it easier. Simultaneously, charging zones and their cases are defined along with determination of collection and charging record transfer. The main categories of the numbering plans include IMSI (International Mobile Subscriber Identity), MSISDN (Mobile Subscriber ISDN number), value-added services number (e.g. Virtual Private Network), roaming numbers, handover numbers, test numbers, emergency numbers, etc. Signalling principles remain the same as defined before and are defined at switch level along with the priority settings. This detail plan will consist of the following.

2.3.4.3 Parameter Tuning

The key performance indicators in the core network include traffic-related parameters, signalling performance-related parameters and measurements related to the HLR, VLR and other network elements in the core network. The parameters are collected from the network management system through various testing/measurement reports from the network. Traffic- and signaling-related data analysis would result in figures related to the traffic handled by switches and exchanges, the exact amount of traffic under each traffic class, subscriber calling-related measurements, configurations and the load on the signalling network. The final optimization plans would consist of switch and signalling optimization network plans. The changes/suggestion would include the following: in the case of congestion, an extra PCM to be used (including changes in topology) and/or traffic routing to be modified, a uniform distribution of signalling links and sets to be distributed uniformly across the network, thus making sure that load sharing is equal, and proposals for redundancy of signalling control units.

2.4 EGPRS Technology

Data services are possible on the GSM network via two means: SMS (the short message system) and GSM data service. The maximum data rate in these networks was limited to 9.6 kbps on one time slot. Although both these SMS and GSM data service have circuit switch connections of 9.6 kbps, the data capability of the former (SMS) is quite restricted. Several limitations in the GSM system, such as low bit rate or a long set-up time enabled the development of the next level of data services in these networks. The next step in the development was 'High Speed Circuit Switched Data' (HSCSD) that enabled the increase in data rates in GSM networks. In HSCSD, although the maximum radio interface bit rate for a 14.4 kbps channel coding is 115.2 kbps, the maximum achievable data rate is 64 bps due to limitations in the A-interface and core network. There are two types of configurations that exist in HSCSD networks: symmetric and asymmetric. The former one consist of co-allocated bi-directional full rate traffic channels while the latter consists of co-located uni- or bi-directional full-rate traffic channels.

GPRSs (General Packet Radio Services) are standardized switching data services that are based on the GSM network. This brings a new set of bearer services to the GSM network, thereby providing data transfer in packet mode in the network. The date rates in the GPRS network provide a ten-fold increase in the speed, that is from 9.6 kbps to 115 kbps (a maximum theoretical speed of 171.2 kbps is possible by using the 8 Air-interface time slots simultaneously). GPRS networks on top of GSM network would mean adding packet core network elements to a traditional GSM network. GPRS networks are known as 2.5G networks. Although GPRS provided packet switching capability to the conventional GSM network, high data speeds were not possible. The data speed was only incremental and did not suffice the ever-growing demand for higher speed data connection by the subscriber, thus leading to the birth of 'Enhanced Data Rates' for GSM evolution of the EDGE network. EDGE networks were capable of much higher data speeds but were slightly less capable than 3G networks and hence were also called 2.75G networks. Some enhancements to the GPRS network would see date rates increase three-fold. These networks can offer theoretically a speed of 384 kbps.

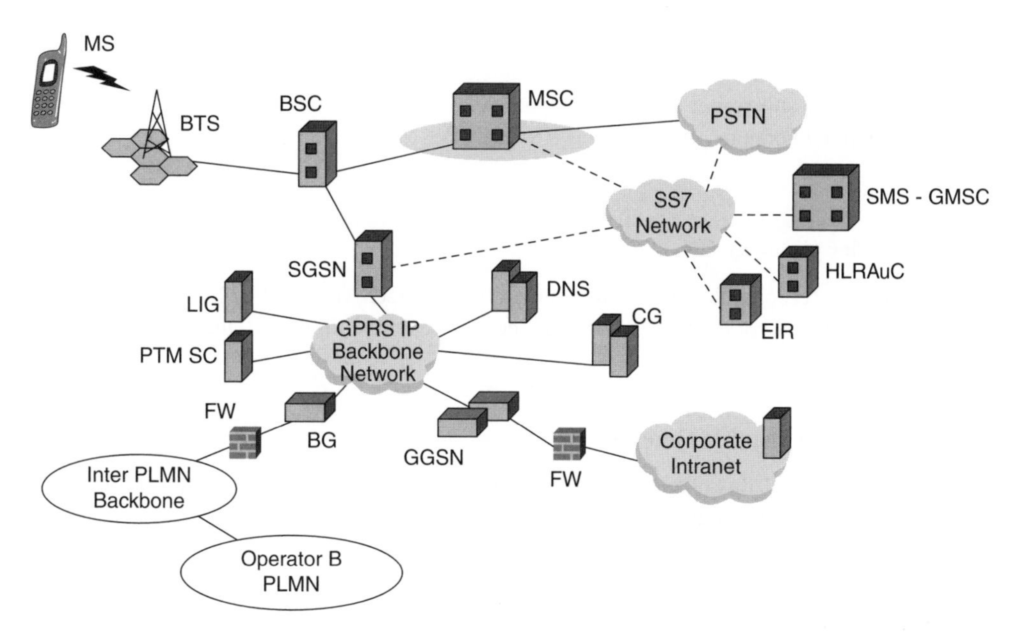

Figure 2.14 GPRS network.

EDGE and EGPRS are interchangeably used to refer to the same technology. The reader should make no distinction between the two terms used in this book.

2.4.1 EGPRS Network Elements

In simpler terms, addition of the packet core elements to a conventional GSM network would convert it into an EGPRS network. The new elements, as seen in Figure 2.14, are the Mobile Station (MS), Serving GPRS Support Node (SGSN), GPRS Gateway Support Node (GGSN), Border Gateway (BG), Legal Interception Gateway (LIG), Domain Name System (DNS) and Packet Control Unit (PCU).

2.4.1.1 Mobile Station (MS)

The main difference between the GSM and GPRS mobiles is the data capability. The GPRS mobile is capable of handling data at higher speeds. There are three classes of GPRS mobiles: class A, class B and class C. Class A mobiles can be used on either networks simultaneously, Class B can be connected on both networks but used only on one at a time while Class C can only be used on either one of the networks.

Serving GPRS Support Node (SGSN)
The SGSN connects the BSC and GGSN while serving as an access point of to the GPRS mobile station for the GPRS network. This element is responsible for functions such as authentication and mobility management.

GPRS Gateway Support Node (GGSN)

The GGSN connects the GPRS network to the external networks, for example Internet and X.25, and acts as a sub-router from the perspective of external networks. It also acts as a 'wall' to the external networks to protect the GPRS network.

2.4.1.2 Border Gateway (BG)

This element is responsible for secure connections between various PLMNs over the inter-PLMN backbone network. The BG contains security features, firewalls, etc.

2.4.1.3 Legal Interception Gateway (LIG)

As the name suggests, this element is required for delivering the intercepted user information to the legal agencies (law enforcement agencies).

2.4.1.4 Domain Name System (DNS)

This element is responsible for mapping the domain name to the IP address. In the GPRS network, it is used to map logical access point names (APNs) to the GGSN IP addresses.

2.4.1.5 Packet Control Unit (PCU)

This unit is physically located in the BSC and is responsible for the packet control related functions in the BSC.

2.4.2 Interfaces in the EGPRS Network

Apart from the interfaces discussed above in conventional GSM networks, there are a few more interfaces in the packet core network (see Figure 2.15).

2.4.2.1 G_b Interface

This is the interface between the BSC and SGSN. It carries the GPRS traffic and signalling between the BSS and GPRS packet core network.

2.4.2.2 G_n Interface

Within same PLMNs, the interface between two GSNs is the G_n interface. It provides the data and signalling interface intra-PLMN backbone.

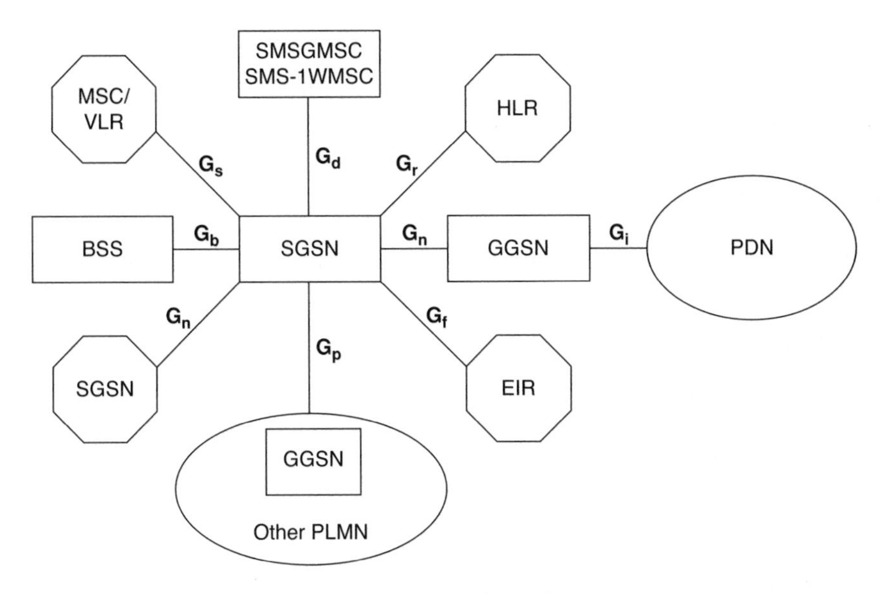

Figure 2.15 GPRS interfaces.

2.4.2.3 Gb Interface

This is the interface between the BSS and SGSN. It carries the traffic and signalling information between the BSS (of the GSM) and the GPRS network.

2.4.2.4 Gn Interface

This is the interface between the SGSN and SGSN/GGSN of the same network. This provides data and signalling for intra-system functioning,

2.4.2.5 Gd Interface

This is present between the SMS-GSMC/SMS-IWMSC and SGSN, thus providing a better use of the SMS services.

2.4.2.6 Gp Interface

This is the interface between the SGSN and the GGSN of another PLMN – an interface between the two GPRS networks.

2.4.2.7 Gs Interface

This is the interface between the SGSN and MSC/VLR. Location data handling and paging requests through the MSC are handled via this interface.

2.4.2.8 Gr Interface

This is the interface between the SGSN and HLR through which all the subscriber information can be assessed by the SGSN from the HLR.

2.4.2.9 Gf Interface

This interface gives the SGSN the equipment information that is present in the EIR.

2.4.2.10 Gi Interface

This is the interface between the GGSN and the external networks. This is not a standard interface, as the specifications would depend upon the type of interface that would be connected to the GPRS network.

2.4.3 Channels in the EGPRS Network

Moving onto the existing channels in the GSM network which are voice channels, there are some new channels in a GPRS network that are related to the packet functionality. These logical channels are allocated to the physical channel called the PDCH (packet data channel) (see Table 2.5).

Channel allocation in the EGPRS (EDGE) network is similar to that of the GPRS network. The PACCH is the only associated channel when physical resources are assigned while the BCCH, PCH, RACH and AGCH are signalling channels.

Table 2.5 EGPRS channels

Channel	Abbreviation	Function/Application
Packet Broadcast Control Channel *(DL)*	PBCCH	Broadcast system information specific to packet data
Packet Access Grant Channel *(DL)*	PAGCH	Notifies that mobile about resource assignment before actual packet transfer
Packet Notification Channel *(DL)*	PNCH	Used for sending information to multiple mobile stations
Packet Paging Channel *(DL)*	PPCH	Pages a mobile station before packet transfer process begins
Packet Random Access Channel *(UL)*	PRACH	Used by the mobile station for initialization of the uplink packet transfer.
Packet Common Control Channel	PCCCH	Contain logical channels for common control signalling
Packet Data Traffic Channel	PDTCH	Channel temporarily used for data transfer
Packet Associated Control Channel	PACCH	Used for signalling information transfer for a given mobile

Table 2.6 Coding schemes for the GPRS network

Coding Schemes	Code Rate	Data Rates (Kbps)	Data Rates (kbps) (excluding headers: RLC/MAC)
CS-1	1/2	9.05	8
CS-2	~2/3	13.4	12
CS-3	~3/4	15.6	14.4
CS-4	1	21.4	20

2.4.4 Coding Schemes

There are four coding schemes in the GPRS network (see Table 2.6), that is CS-1, CS-2, CS-3 and CS-4. In CS-1, a half-rate convolution coding is used for forward-error correction and has a data rate of 9.05 kbps. Both the CS-2 and CS-3 are similar to CS-1 except that 'puncturing' is used. This technique increases the data rates at the expense of redundancy. With no FEC in CS-4, the data rate is further increased. Since the EGPRS system is an enhancement to the GPRS system, the data rates increase to 59.6 kbps.

The first four coding schemes are still GSMK while the last five are 8-PSK (see Table 2.7). In GPRS systems, the modulation used is GMSK while in the EGPRS system the modulation used is 8-PSK. Apart from data rates system, a couple of other features (link adaptation and incremental redundancy) make the EGPRS systems much more attractive.

Link adaptation is used for maximizing the channel throughput with the lowest amount of delay in the changing climatic conditions. Thus, the link adaptation features try to maintain the quality of the signal in adverse conditions. Incremental redundancy (IR) improves the throughput and this is done by automatically adapting the total amount of transmitted redundancy to the radio channel conditions. This is carried out by using two techniques: ARQ (Automatic Repeat reQuest) and FEC (Forward Error Correction). The 'power control feature' in EGPRS systems is more complicated than in GSM systems due to addition of the data. Uplink power control is used to reduce interference (and increase mobile battery life) while the downlink power control reduces the power of the BTS and hence the interference in the network. TBF (Temporary Block Flow) is another concept in EGPRS networks. This is

Table 2.7 Coding schemes for the EGPRS network

MCS	Modulation	User Rate
1	GMSK	8.8 kbps
2	GMSK	11.2 kbps
3	GMSK	14.8 kbps
4	GMSK	17.6 kbps
5	8-PSK	22.4 kbps
6	8-PSK	29.6 kbps
7	8-PSK	44.8 kbps
8	8-PSK	54.4 kbps
9	8-PSK	59.2 kbps

defined as the temporary connection that is established for the data flow between the network and mobile station.

2.5 EGPRS Network Design and Optimization

The planning process in an EGPRS network is similar to that of GSM networks. However, some important changes are highlighted here. As mentioned before, the main planning tasks are coverage, capacity, frequency and parameter planning

The main task in coverage planning is to provide coverage planning for both uplink and downlink in a balanced way (using link budget calculations). This means in practice keeping the provision of sufficient Carrier-to-Noise (C/N) ratios across the coverage area for successful data transmission, on both the uplink and downlink. Each of the coding schemes, CS/MCS, defined in a EGPRS system is suited for a particular coverage area, that is as the coding scheme goes higher (more data being transmitted), the coverage area becomes smaller (as shown in Figure 2.16). One important aspect of the link budget is that the for CS-2, body loss is not used, hence giving a 3 dB advantage. The EGPRS system is an 'interference-limited system' rather than a 'frequency-limited system'. In the GPRS system, CS-3 and CS-4 implementation has many requirements; hence usually CS-1 and CS-2 are implemented.

For capacity planning, the fundamental change from the GSM system is that the EGPRS system carries three types of traffic: voice, CS-data and PS-data. CS traffic always has a higher priority than the PS traffic. Some of the PS traffic are 'delay-sensitive' and hence, dedicate timeslots are used to carry it. The EGPRS system has higher data rates, that is an average throughput per radio timeslot changes. However, the concept of dedicated and default (for

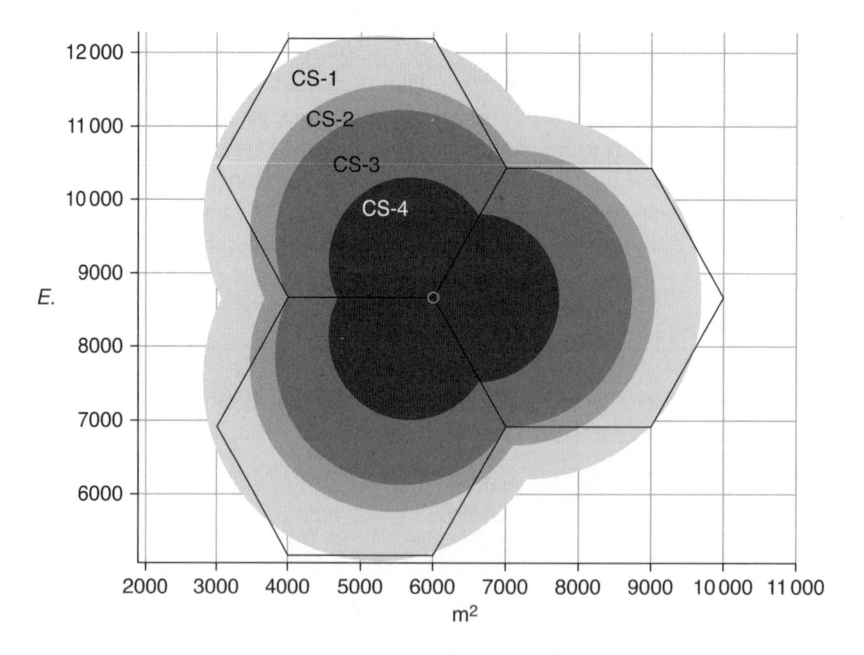

Figure 2.16 Coverage areas of different GPRS coding schemes.

voice and CS-data) territory remains the same for EGPRS system. The main parameters used when planning for capacity in EGPRS networks are as follows:

- peak circuit-switched traffic load (erlangs);
- peak packet-switched load (kbps);
- packet-switched load during overall traffic peak hour (kbps);
- circuit-switched traffic load during overall traffic peak hour (erlangs).

CS traffic calculations are similar to that of GSM networks using erlang B tables, blocking and C/I thresholds. For PS traffic, the dedicated territory is used and for CS and voice, the default territory is used. It is possible to upgrade and downgrade EGPRS territories in cases where CS traffic decreases and increases. The data throughput is one of the key indicators of a EGPRS network (after all, a EGPRS network is installed for catering to the data users). The number of users will define the throughput capacity of the network. As mentioned above, the more the number of data users, the less would be the coverage and a lesser number of users would be able to access higher capacity, that is higher data rates. Under 'busy conditions', several users may use the same timeslot. Here, the concept of the rate reduction factor (RRF) comes into play which can be quantified by using the effective reduction from the rate achievable at low loads for the given GPRS territory occupancy. This leads to Quality of Service (QoS) that is offered to EGPRS subscribers. During dimensioning for a given QoS, RRF can be used. This gives an estimate of how well an EGPRS territory can be loaded before the QoS offered to users is degraded to an unacceptable level.

Coverage and frequency planning go hand-in-hand as in the GSM network. The EGPRS system is an interference-limited system and not a noise-limited system. Frequency planning and power control are a couple of methods used to reduce interference in a EGPRS network. Frequency planning in a EGPRS network is similar to that of a GSM network. Power control is more important in the downlink direction and is accomplished by using the BCCH layer.

The transmission planning process in a EGPRS network is similar to that for a GSM network. However, due to addition of packet data, there are a few additions in planning that include dimensioning of dynamic Abis (this is a vendor-specific feature) and dimensioning of the packet control unit in the BSC. There are few timeslots on the Abis that are dedicated for the PS traffic. Due to this, the number of TSs for the Voice and CS data decreases and hence the number of voice users decreases. Usually, in the EGPRS network, especially sites that are hotspots, the capacity of E1 is increased from the usual 1E1 to 2E1s. The concept behind dynamic Abis dimensioning is to find the number of timeslots that can be assigned to a pool in such a manner that the Abis does not become a limitation for the air-interface throughput. The main inputs for this would be the number of radio timeslots required for PS traffic (i.e. dedicated and default territories in an EDGE radio network), capacity of radio timeslots, blocking that could take place on the Abis-interface, etc. Based on these inputs and the number of PCM timeslots available for the pool, the required number of PCM timeslots can be calculated. Equipment limitations and capability should, however, be kept into consideration.

From the core network planning perspective, PCU dimensioning, Gb-interface planning and SGSN dimensioning are the key aspects. Based on the capacity of the BSC, inputs such as the number of TRXs that can be supported, amount of the traffic handled and the number of traffic channels that can be handled would lead to the number of PCUs that would be needed to support the traffic. Gb-interface dimensioning would come up with the number of E1s that would be needed between the PCU and SGSN. The total number of subscribers, processing capacity of the SGSN and interface capacity would lead to the number of SGSNs needed in the network. Dedicated PCM links (also known as 'Frame Relay Links') could be used to send the packet data from the BSC to the SGSN. The main result of the frame relay dimensioning is to find the number of timeslots that are needed to send the packet data traffic from the BSC to the SGSN.

2.5.1 Parameter Tuning

The key parameters in the EGPRS systems are the same as in the GSM systems; however, with the addition of a PS core network, packet data-related parameters are new additions to the list.

2.5.1.1 GPRS Enabled

This parameter defines whether the cell/radio is allowed to handle GPRS traffic.

2.5.1.2 Default GPRS Capacity

This represents timeslots that are always allocated to the GPRS territory unless preempted by CSW traffic.

2.5.1.3 Dedicated GPRS Capacity

This represents the number of TCHs which can be 'dedicatedly reserved' for GPRS use and is thereby removed for circuit-switched traffic.

2.5.1.4 Additional GPRS Capacity

This represents the additional timeslots, over and above the default territory, that may be used by GPRS traffic, if circuit-switched traffic permits (see Figure 2.17).

2.5.1.5 Maximum GPRS Capacity

This represents the maximum number of timeslots in a given cell that can be used up for GPRS traffic.

2.5.1.6 GPRS Traffic Preference

This decides if GPRS traffic should be assigned to the BCCH carrier prior to other TCH carriers within the same cell.

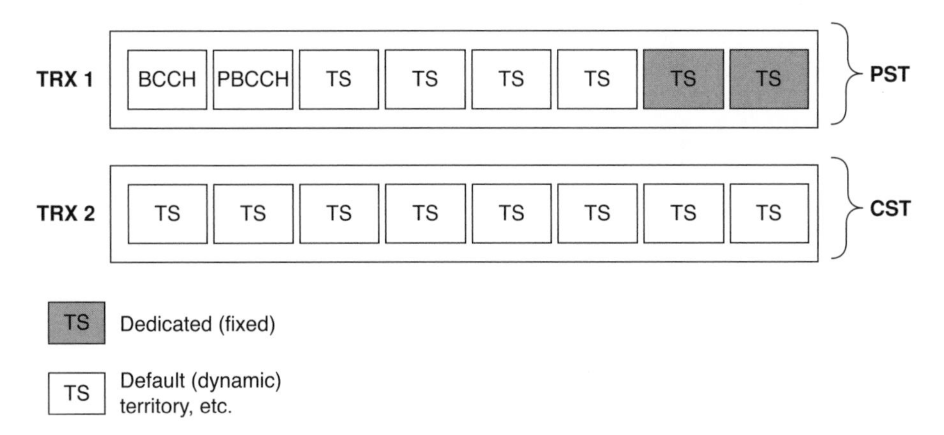

Figure 2.17 GPRS timeslot allocation to CS and PS traffic.

2.5.1.7 Territory Update Guard Time

This determines a timer value, that is the 'guard time' between two subsequent territory updates.

2.5.1.8 Intracell Handover for GPRS Territory Upgrade

During the GPRS territory upgrade procedure, CS calls in the timeslots to be included in the territory are handed over to other available timeslots in the same cell.

There are a few more parameters in the EGPRS transmission and core network. The major ones are related to the concepts of EDGE territory, Dynamic Abis, BLER (Block Error Rate), MCS schemes, BCCH usage, etc. These are the parameters that directly influence the actual throughput and latency (and ultimately end-user application throughput). The parameters related to TBF (Temporary Block Flow) include the TBF set-up success rate, TBF drops and TBF blocking. Based on this parameter monitoring, tuning is carried out in order to bring the network up to the highest quality standards.

Coverage in the GSM, as well as in the GPRS networks, is dependent on the *C/I* ratios. With the EGPRS network by utilizing the resources of the GSM network, the interference would be high, thus degrading the *C/I* ratio. Degradation of the *C/I* ratio would in turn mean a reduction in the coverage areas. Moreover, the addition of the GPRS network would decrease the voice quality of the network.

In urban areas, there is always a problem of frequencies. The channel allocation to the EGPRS network in the initial phase is always based upon the GSM channel allocation, that is voice (CS) subscribers taking a priority. Thus, capacity plans and the channel allocations have to be studied and optimized again during this process.

Another aspect is the speed of the data services. Although the theoretical speed of a EGPRS network should be about 171 kbps, the practical values are still around 40–60 kbps. A speed of 171 kbps is possible only when all of the eight timeslots are utilized, which is not the case as usually 5 TS can be used for data. Moreover, the *C/I* requirements may not be satisfied for coding schemes beyond CS-1, thus reducing the data processing scheme.

3

UMTS

3.1 The 3G Evolution – UMTS

In Chapter 2, we saw the evolution of the systems from 1G to 2G or rather to 2.5G. Second generation (2G) systems were developed with transparency and compatibility in mind; these are semi-global system and definitely far better and developed than the first generation analogue systems. One of the most famous technologies was the GSM system which we have already seen in the previous chapter. The next generation or third generation (3G) technology has been developed to fulfil the dream for a truly global system – based on footprints of already successful GSM systems in terms of the system being fully specified and the interface being standard and open. However, due to drastic changes, third generation systems are much more superior to the second generation systems. The bit rates go up to 2 Mbps (with HSPA up to 42 Mbps). The data services offered are both in real time and non-real time with variable data rates (bandwidth on demand). The quality of service and spectrum efficiency is high and at the same time, multiplexing of various services with varying quality requirements on one channel is possible.

3GPP or the 3rd Generation Partnership Project is a collaboration agreement between various telecommunication organizations, such as ARIB, CCSA, ETSI, ATIS, TTA and TTC, established in 1998. The main job of 3GPP was to come up with the technical specifications that would be globally applicable for 3G networks. The 3GPP specifications were to be based on GSM specifications. 3GPP consists of Project Co-ordination Groups (PCGs) and Technical Specification Groups (TSGs), as shown in Figure 3.1.

The specifications of 3G are called 'releases' and each of these releases incorporate many documents/standards. Release 1 was called 'Release 98', 'Release 2000' was called 'Release 4' and after that it was 'Release 5', 'Release 6' and 'Release 7'. 3GPP's releases beyond 'Release 7' are called LTE (Long Term Evolution) systems – these are different systems with different air-interfaces. The first release is specified in 'Release 8' but there are specifications for WCDMA in 'Release 8' (covered later in this book). 'Release 98' concerned the pre-3G, that is GSM network. It was only in 'Release 99' that the first UMTS networks were specified. The main feature of 'Release 99' was the creation of UTRA (Universal Terrestrial Radio Access). Other features included LCS, CAMEL, narrowband AMR, etc. 'Release 4' came up with features such as bearer-independent CS network architecture, streaming, multimedia messaging,

Cellular Technologies for Emerging Markets: 2G, 3G and Beyond Ajay R. Mishra

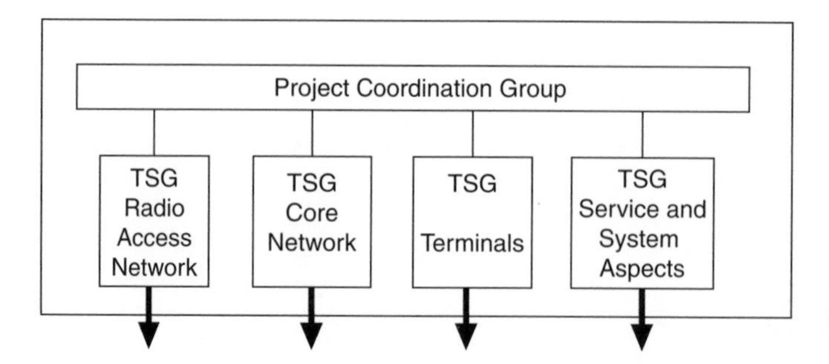

Figure 3.1 3GPP structure [www.3gpp.org].

GERAN, etc. 'Release 5' introduced IMSs (IP-based Multimedia Services), Wideband AMR, GTT (Global Text Telephony), HSDPA, etc. 'Release 6' introduced HSUPA, Push-to-Talk over cellular (enhancements to IMS), etc. 'Release 7' concerned HSPA+, real-time applications such as VoIP, etc.

3GPP has five main standardization areas: terminal, radio, core, services/systems and GERAN. The terminal groups are responsible for standardization work related to terminals such as USIM to the mobile interface, messaging, conformance testing, service capability protocols, etc. The radio group is responsible for the specifications related to air-interfaces (I_{ub}, I_{ur}, I_u), BTS radio performance, conformance testing, layers 1, 2 and 3 specifications, etc. The core group has the specification responsibility related to packet data, inter-working function between core and external network, signalling, etc. The services and systems groups look into specifications development related to services, evolutions, charging, accounting, architecture, etc.

UMTS (Universal Mobile Terrestrial System) is one of the 3G technologies. Because of the nature of 3G technology and its evolution from GSM, UMTS is also called 3GSM. The catalogue of 3G within the ITU is called the IMT-2000 or International Mobile Telecommunications for 2000 MHz. Several different proposals for the air-interface were incorporated. WCDMA (used in Europe, Asia, Japan and Korea) uses the frequency bands allocated by WARC-92. In North America, these bands were already used and hence 3G services are implemented, replacing part of the spectrum, a technique called 're-farming'. TD-CDMA contains the time division duplex components of UMTS and Chinese TD-SCDMA (integrated with UMTS-TDD). UWC-136 system has been included in the IMT-2000 group – this is TDMA-based and an enhancement of IS-36 and GSM. For low-mobility applications, DECT (Digital Enhanced Cordless Telecommunication) has been adopted. This is based on FD TDMA. UMTS uses WCDMA as its air-interface technology. The frequency band of 175 MHz has been reserved for 3G systems (for GSM, 220 MHz is reserved in Europe). The range for 3G systems extends from 1880–1980 MHz, 2010–2025 MHz and 2110–2170 MHz, along with two 30 MHz segments for satellite-based systems. The UMTS system uses a channel bandwidth of 5 MHz. There are seven unpaired channels within 1900–1920 MHz and 2010–2025 MHz, implying that one 5 MHz has to implement in both the uplink and downlink directions. Between 1920 to 1980 MHz and 2110 to 2170 MHz, there are twelve paired frequency bands available (5 MHz each). However, in Japan 1900–1920 MHz is occupied for the Personal Hand Phone Service (PHS) and hence not available for 3G; similarly this is the case in the USA. This means that Europeans and Japanese systems would not be operational

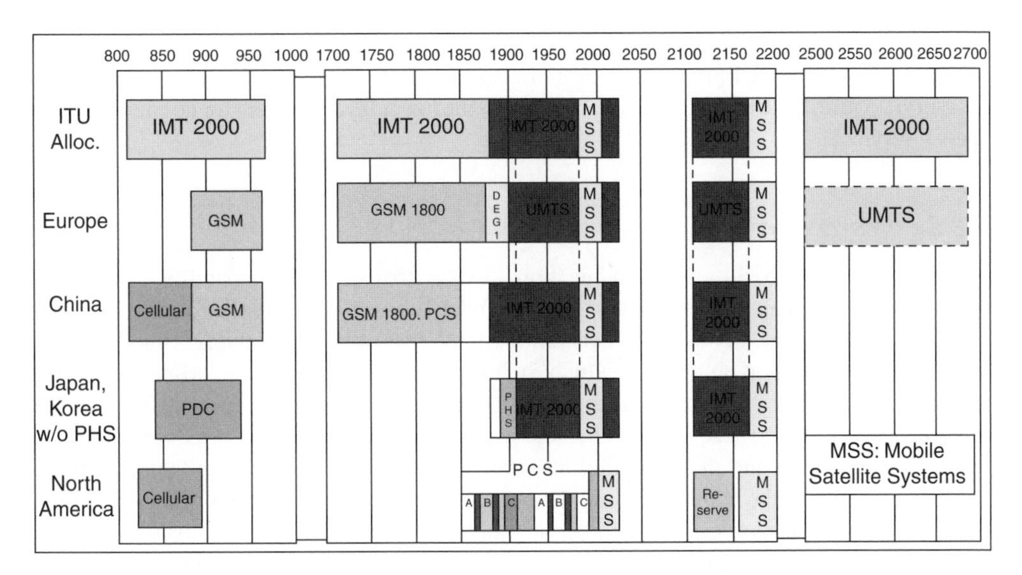

Figure 3.2 WRC-2000 IMT-2000 frequencies [www.itu/newsarchive/wrc2000/presskit/IMT-2000.html].

in the USA. The Federal Communications Commission has allocated 1915–1920 MHz with 1995–2000 MHz and 2020–2025 MHz with 2175–2180 MHz for advanced wireless system use. However, government and regulatory bodies are working on spectrum issues. Frequency allocation according to WRC 2000 is shown in Figure 3.2.

UMTS frequencies can be summarized as follows:

- **1920–1980** and **2110–2170** MHz: (FDD, W-CDMA), paired uplink and downlink, channel spacing is 5 MHz. Operator requires 3–4 channels (2 × 15 MHz or 2 × 20 MHz) to build a high-speed, high-capacity network.
- **1900–1920** and **2010–2025** MHz: (TDD, TD/CDMA), unpaired, channel spacing is 5 MHz. Transmission and receiving are not separated in frequency.
- **1980–2010** and **2170–2200** MHz: for satellite uplink and downlink.

3.2 UMTS Services and Applications

A modern telecommunication network such as UMTS can provide a wide variety of services. The service concepts and definitions of UMTS are mostly taken from GSM. The service parameters are often fixed in GSM, while they can be dynamically renegotiated in UMTS. The services provided by UMTS can be divided into four main classes (3GPP TS 23.107).

3.2.1 Teleservices

A teleservice is a type of telecommunication service that provides the complete end-to-end capability for communication between mobile users in accordance with standardized protocols.

The user has no direct responsibility for the end-point applications. Teleservices make use of the whole OSI model protocol stack and also include the terminal equipment functions. Teleservices utilize the bearer services provided by the lower layers (except for circuit-switched speech services).

3.2.2 Bearer Services

Bearer services are basic telecommunication services that offer the capability of the pure transmission of signals between access points. These services can be either circuit-switched or packet-switched. Bearer services concern the three lowest layers of the OSI model. They are end-to-end transport services in which the user is responsible for the end-point entities. A bearer service is defined using a set of characteristics that make it different from all the other bearer services. These service characteristics define such things as the traffic type, the traffic characteristics and the supported bit rates. UMTS allows for the negotiation of these parameters between the application and the network. There is a negotiation routine in which the application requests a certain bearer service, and the network checks the available resources and then grants the requested service or suggests a lower level of service. The application at the user side either accepts or rejects the network's suggestion. It also renegotiates the properties of a bearer service during an active connection in UMTS. This makes the UMTS bearer service much more flexible and allows the network resources to be much better utilized.

3.2.3 Supplementary Services

A supplementary service (SS) complements and enhances bearer services and teleservices. They cannot exist without these basic services, that is there are no stand-alone supplementary services. Supplementary services reside in the switch and may supplement several basic telecommunication services. Also, one basic telecommunication service may simultaneously use several SSs. The latter case requires that the interactions between active SSs be carefully specified.

3.2.4 Service Capabilities

Service capabilities are a set of building blocks that can be used to implement value-added services. As the value-added services are not standardized, it is possible to implement them in a way that produces unique services. The latter are more likely to attract and hold subscribers than constant 'price wars' among operators with identical services. Service capabilities are accessible to applications via a standardized application interface.

According to the UMTS Forum, the 3G data services are divided into three categories: content, connectivity and mobility. These are sub-divided into six categories: mobile Internet access, location-based services, multimedia messaging, mobile Intranet, customized 'infotainment' and 'rich voice'. UMTS offers teleservices, for example speech and SMS and bearer services. The characteristics of the bearer services can be negotiated before the sessions or during on-going sessions for both point-to-point and point-to-multipoint communications. These bearer services have different parameters for Quality of Service (QoS) such as BER (Bit Error Rate), Maximum Transfer Delay, etc. QoS is defined end-to-end, that is one user equipment to

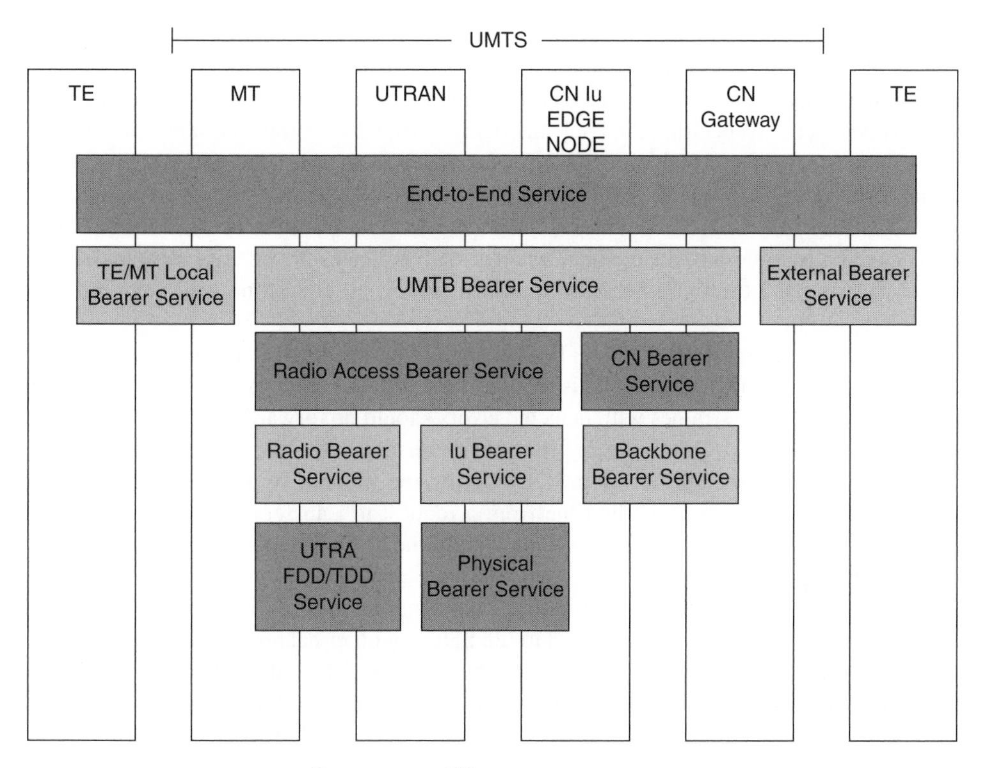

Figure 3.3 QoS architecture. © 2009. 3GPP™ TSs and TRs are the property of ARIB, ATIS, CCSA, ETSI, TTA AND TTC who jointly own the copyright in them. They are subject to further modifications and are therefore provided to you "as is" for information purposes only. Further use is strictly prohibited.

another user equipment. As shown in Figure 3.3, each bearer service offers individual services by using the services offered by layers below and at the same time including all aspects (control signalling, user plane transport, QoS management, etc.) to enable the provisions of contracted QoS.

3.3 UMTS Bearer Service QoS Parameters

Traffic type. This parameter is intended to describe the characteristics of the source, for example background, interactive, streaming, conversational.

Maximum bit rate. This is the maximum number of bits delivered by UMTS and to UMTS at an SAP within a period of time, divided by the duration of the period. The maximum bit rate is the upper limit a user or application can accept or provide. Practically, the maximum bit rate can be used to make code reservations in the downlink of the radio interface.

Guaranteed bit rate. This is the guaranteed number of bits delivered by the UMTS at an SAP within a period of time divided by the duration of the period. This describes the bit rate that the UMTS bearer service should guarantee to the user or applications.

Delivery order (Y/N?). This indicates whether the UMTS bearer should provide, in sequence, SDU delivery or not. This is derived from the user protocol (PDP type) and specifies if out-of-sequence SDUs are acceptable of not.

Maximum SDU size. The maximum SDU size allowed (for admission control and policing).

SDU format information. This is the list of possible exact sizes of the SDUs, in bits. This information is needed by UTRAN to be able to operate in the transparent RLC protocol mode which is beneficial to spectral efficiency and delay when RLC re-transmission is not used.

SDU error ratio. The fraction of SDUs lost or detected as erroneous and which is needed for conforming traffic.

Residual bit error ratio. This indicates the undetected bit error ratio in the delivered SDUs. If no detection is requested, the residual bit error ratio indicates the bit error ratio in delivered SDUs. This is used to configure radio interface protocols, algorithms and error detecting coding.

Delivery of erroneous SDUs (Y/N?). This indicates whether the SDUs detected as erroneous should be delivered or discarded. This is used to decide whether or not error detection is needed and whether the frames with detected errors should be forwarded or not.

Transfer delay (ms). This indicates the maximum delay for the 95th percentile of the distribution of delay for all delivered SDUs during the lifetime of a bearer service, where delay for an SDU is defined as the time from a request to transfer an SDU at one SAP to its delivery at the other SAP. This is used to specify the delay tolerated by the application. It allows UTRAN to set transport formats and ARQ parameters.

Traffic handling priority. This specifies the relative importance for handling of all the SDUs belonging to the UMTS bearer compared to the SDUs of other bearers. Within the interactive class, there is a definite need to differentiate between bearer qualities. This is handled by using the traffic handling priority attribute, to allow UMTS to schedule traffic accordingly. By definition, priority is an alternative to absolute guarantees, and thus these two attribute types cannot be used together for a single bearer.

Allocation/retention priority. This specifies the relative importance compared to other UMTS bearers for allocation and retention of the UMTS bearer. The allocation/retention priority attribute is a subscription attribute which is not negotiated from the mobile terminal. This priority is used for differentiating between bearers when performing allocation and retention of a bearer

3.4 QoS Classes

These are also called traffic classes (see Table 3.1). The main differentiation between the various traffic classes is the 'delay sensitivity'. There are four QoS classes defined in the UMTS standards (3GPP TS 23.107):

- conversational;
- streaming;
- interactive;
- background.

3.4.1 Conversational Class

As the name suggests, this is for applications related to 'real time conversations', for example speech, VoIP, video conferencing, etc. The constraints in terms of delay are given by human

Table 3.1 UTMS QoS classes. © 2009. 3GPP™ TSs and TRs are the property of ARIB, ATIS, CCSA, ETSI, TTA AND TTC who jointly own the copyright in them. They are subject to further modifications and are therefore provided to you "as is" for information purposes only. Further use is strictly prohibited

Traffic class	Conversational class conversational RT	Streaming class streaming RT	Interactive class Interactive best effort	Background Background best effort
Fundamental characteristics	Preserve time relation (variation) between information entities of the stream Conversational pattern (stringent and low delay)	Preserve time relation (variation) between information entities of the stream	Request response pattern Preserve payload content	Destination is not expecting the data within a certain time Preserve payload content
Example of the application	Voice	Streaming video	Web browsing	Background download of emails

interactivity. This class is characterized by a very low tolerance level in terms of delay and any failure to provide a low enough transfer delay would lead to a degraded quality of service. The fundamental QoS characteristics are as follows: preserve time relation (variation) between information entities of streams and conversational pattern (stringent and low delay).

3.4.2 Streaming Class

Downloading of the real time video/audio stream falls under the streaming class. It is one way data flow (unidirectional) and is aimed at 'love' (human) destinations. It is characterized by the time variations between information entities (i.e. samples, packets) within which a flow is preserved, though it does not have any requirements on low transfer delay.

3.4.3 Interactive Class

Examples like web browsing, remote terminal emulation, database retrieval, etc. best describe this class. In this class, the human interaction is to a machine located at a distance. Telemachines are also typical example of this class of service. The round trip delay time is a key attribute of this class. The fundamental QoSs of this class are the request response pattern and the preserve payload content.

3.4.4 Background Class

In this class, the destination is not expecting the data within a certain time frame, that is delay is not a concern. Computer-delivering e-mails, SMSs, downloading of databases, etc. are typical examples of this class. The fundamental QoSs for this service are preserving payload content and not expecting data within a certain time frame.

3.5 WCDMA Concepts

3.5.1 Spreading and De-Spreading

As discussed before, WCDMA has higher bit rates of 2 MHz. This means that a larger bandwidth is required to support such higher bit rates and in this case, it is 5MHz. The user information bits are spread over a large bandwidth by multiplying this user information with 'quasi-random' bits called 'chips'. As the actual information is transmitted on a wider bandwidth the system become tolerant towards the narrowband interfering signals, in the process solving the problems of the number of subscribers that could simultaneously be logged to the network in the TDMA and FDMA systems. The spread spectrum technique thus allows efficient use of the spectrum and allows multiple numbers of users to use the system simultaneously – in the same frequency band. One of the most common techniques used is the 'direct sequence' CDMA or DS-CDMA. In this technique, the bipolar user data bit stream is multiplied by the user-specific bipolar code sequence. This signal (i.e. multiplying signal) is called the 'code'. The rate at which the data spreads is called the chip rate. A BPSK (Binary Phase Shift Keying) modulated signal is usually used as the original signal and the second modulation is performed by multiplying the original signal with a sequence of bits (of a wideband signal). This would spread the original signal, making it a wider bandwidth signal. Each bit of the original signal is called the 'symbol' while that of the multiplying signal is referred to as 'chips'. The ratio of chip rate to symbol rate is called the 'spreading factor' (SF). It is also defined as the factor by which the spectrum spreads. The multiplied signal (this is different for every user) is transmitted over a bandwidth that is larger than what would have been required to transmit the original data. Every user in the system uses this process and at the same time leaves a user-specific code that would allow the transmitted signal to be re-constructed (de-spreading) at the receiving end, as shown in Figure 3.4. All the users in the network transmit at the same frequency, thereby increasing interference. When the receiver receives the sum of several user signals, it is multiplied by the spreading code while the respective user and the effect of other signals are removed. The output generated from the received signal and the spreading code is integrated periodically for the duration of the bit and the original signal is sampled at the end of the period.

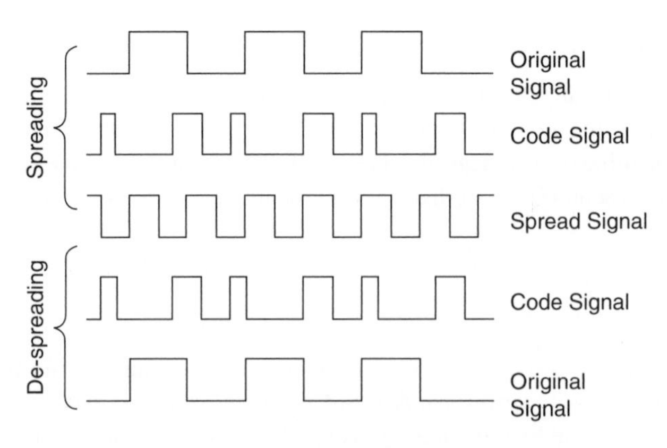

Figure 3.4 Spreading/de-spreading phenomena.

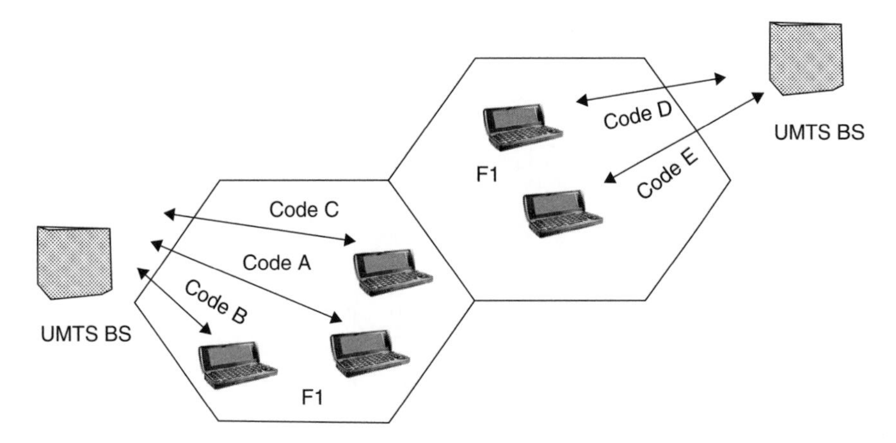

Figure 3.5 Code assignment to subscribers.

3.5.2 Code Channels

Unlike the GSM/ EDGE systems where the users are separated by frequency/time, in UMTS systems, the users are separated by codes. Each user will have a unique code wherein the signal from other users is seen as interference. This makes it an interference-limited system.

The spreading codes (codes A, B, etc., shown in Figure 3.5) are orthogonal to each other. Although there is no co-relation between the codes, in some cases it can be so low that the interference does not takes place between the users (these are called 'quasi-orthogonal' codes). The set of codes that are orthogonal/'quasi-orthogonal' to one another is called the 'code family'. For mobile applications, large code families are used. The chip sequence is dependent on the spreading factor (defined earlier) as the chip rate is higher by the spreading factor than the bit rate. The orthogonal codes used are of variable spreading factors and are called as OVSF codes or Orthogonal Variable Spreading Factor Codes. These codes can be created using a code tree (as shown in Figure 3.6). Each node of the code tree has two branches with a double length code and the same spreading factor. A code with an SF of N is created from the code with an SF of N/2. Thus, at the kth level, a set of 2^k spreading codes with a length of 2^k chips is available. There is one limitation though – two codes of different levels of the code tree are orthogonal to one another only if one of the two codes is not the 'mother code' of the other one. Due to this, the number of simultaneous codes depends on the bit rate and spreading factor.

Apart from spreading, there are two other codes that should be known: scrambling and channelization codes (spreading is done with the channelisation codes and scrambling with the scrambling codes). The scrambling codes are used to separate the terminals and base stations. During transmission, the spread signals are aggregated and then scrambled. The process is done through chip-by-chip multiplications of the signal by a scrambling code that has same rate as the spread chip streams – thereby not changing the signal bandwidth but making them separable from each other. Due to this, the entire family of codes is available to each transmitter. This helps in a way that two transmitters can use the same spreading codes as the differentiation can be done by scrambling codes. Channelization codes are used for separation from a single source. These are based on the OVSF technique, allowing the SF

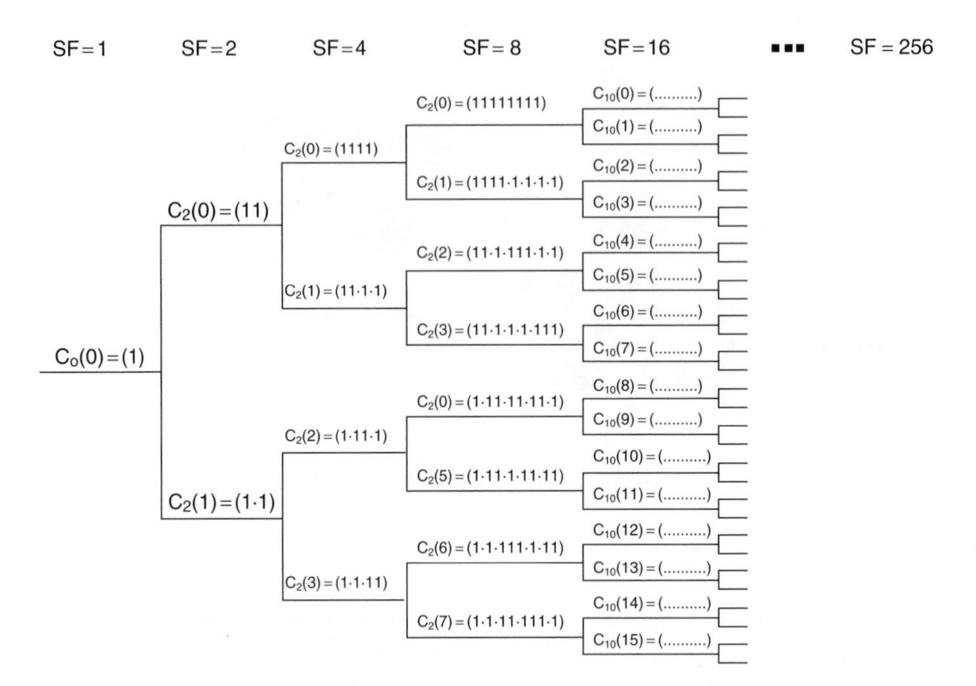

Figure 3.6 OVSF code tree.

to be changed while maintaining the orthogonality between different length spreading codes. Both the spreading and uplink scrambling codes are allocated by the system, while in downlink allocating the scrambling codes of the cells is a part of the network planning process.

3.5.3 Processing Gain

The amount by which the power density of the carrier signal is increased in the receiver is called the processing gain and is dependent on the spreading factor. Due to the processing gain, the system becomes more robust against interference. During the de-spreading phase, the sum of the user and interfering signals is multiplied by the spreading code of the user which helps in re-constructing the original signal. The bandwidth of the original signal is narrower than the spread signal by an amount of the SF. However, the interfering signal spreads in the receiver, thereby decreasing its power density by an amount of SF. This makes these systems more robust against the narrowband interfering signals.

3.5.4 Cell Breathing

The load factor directly corresponds to the traffic supported by the cell. Due to changes in the traffic load, the cell coverage area/ range changes. This is called 'cell breathing'. For the receiver to be guaranteed a sufficient quality of service, the service-specific C/I is maintained. This is possible if the transmitter power is sufficient so that even in the case of attenuation,

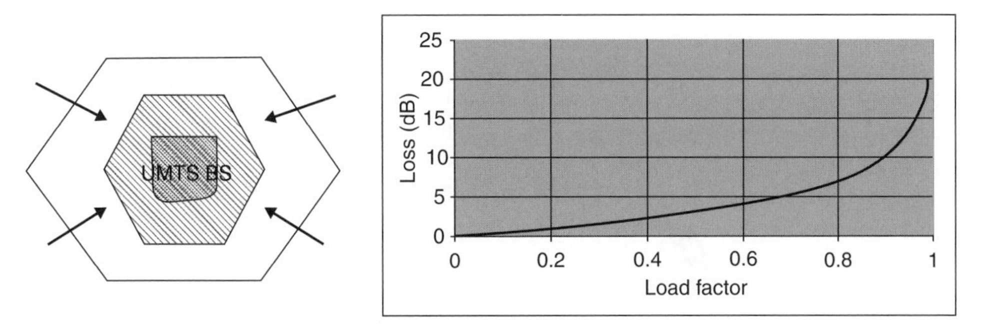

Figure 3.7 Cell breathing.

the C/I can be guaranteed. But as the traffic grows in the network, interference increases. This higher interference power would mean a decrease in received signal power and hence the coverage area. However, this traffic load does not increase or decreases suddenly; it is usually a gradual increase/decrease, as shown in Figure 3.7. Thus, the coverage also increases or decreases gradually giving a 'feeling' that the cell is breathing. The intracell interference is responsible for cell breathing. This interference is dependent on the receiver quality and orthrogonality of the spreading codes.

3.5.5 Handover

There are a few types of handovers. Hard handover is defined as the handover between two frequencies and is used in GSM. In UMTS, when a UE communicates to sectors from different base stations, it is called a soft handover (shown in Figure 3.8). A point to remember here is that frequency does not change in the soft handover. In the hard handover, connection is switched 'hard', that is at a particular time while in the soft handover, there is no fixed switchover point and the transfer from one base station to another is 'soft'. Another kind of handover is the

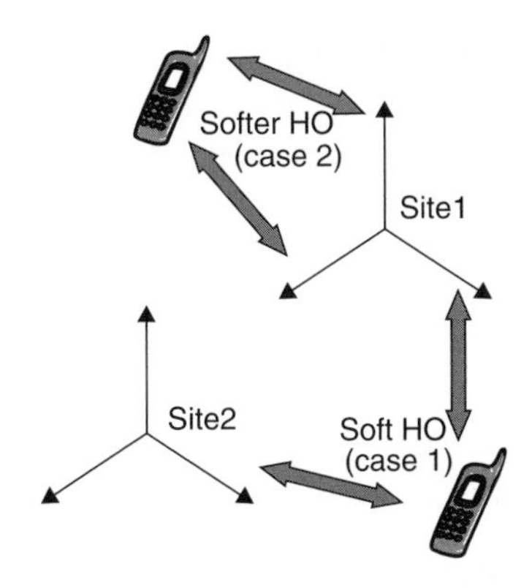

Figure 3.8 Handover in the WCDMA radio network.

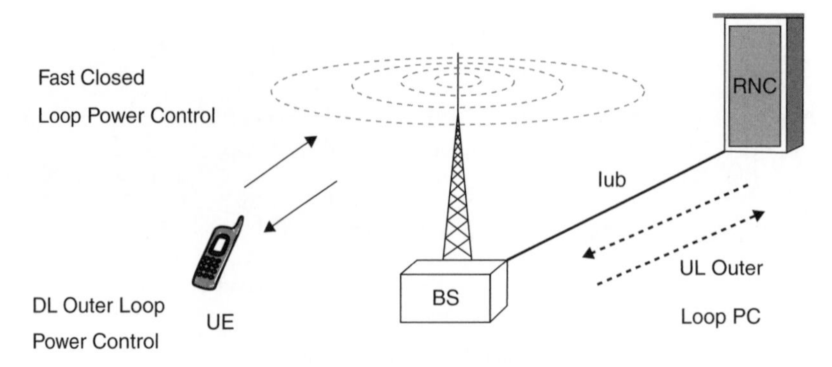

Figure 3.9 Power control.

'softer' handover. In here, the transmission also takes place between the UE and the different sectors of same base station. The mobile takes two different paths (interfaces) for communicating with the base station; hence two different codes are used so that the mobile can recognize the different paths. There is not much difference in the downlink direction; however there is much difference in the uplink direction for the two handovers. In the case of the soft handover, the received signal is routed to the RNC from both the base stations so that the best RNC can take a decision on keeping the signal which is giving a better quality. As this decision-making process takes time (about 10–80 ms), double air-interface capacity is utilized by a single mobile for one call. Hence, radio engineers have to take into account this process when doing the capacity planning for the radio network. About 30 % of the links witness the soft handover phenomenon against 10 % for a softer handover. The measurements generally include received signal code power (signal power received on one code), signal power received within the channel bandwidth and Ec/No (ratio of received signal power to the received signal power indicator).

3.5.6 Power Control

The power control mechanism in the UMTS (shown in Figure 3.9) is quite important as it prevents 'near–far' effects. In WCDMA networks, as the frequency re-use factor is unity, it increases the importance of the fast and accurate power control mechanism. The closed loop power control feature is used in these networks in the absence of which the mobile closer to the base station will easily 'over-shout' the other mobiles in the cell, causing a blocking effect. This is done by measuring the interference levels. Another type of power control feature used is the 'slow power control' or 'outer loop power control' which maintains set quality.

3.5.7 Channels in WCDMA

There are three different types of channels: *logical*, *transport* and *physical*. Logical channels are used with layer 2 for data transfer for higher layers. Transport channels are used for data transfer between layer 2 and layer 1, while physical channels are used for data transfer for

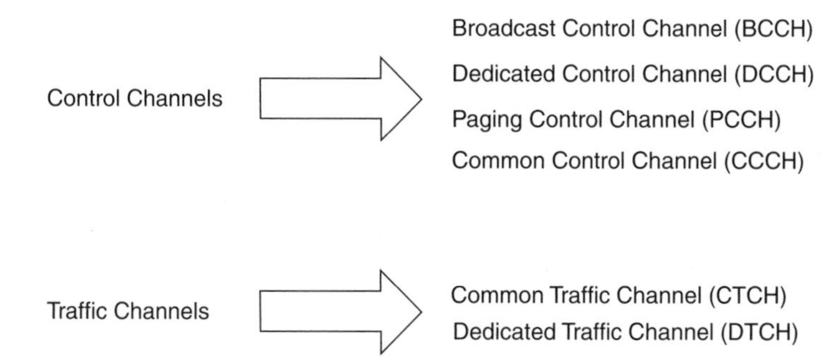

Figure 3.10 Logical channels (control and traffic).

layer 2 and higher layers. Some of the key channels in each of the categories are described in the following.

Logical channels consist of control and traffic channels. The logical channels are shown in Figure 3.10.

Transport channels, like the logical channels, are of two types: dedicated transport channels and common transport channels, based on the applications of the channels. The transport channels are described in Figure 3.11.

There are two kinds of *physical channels*: dedicated and common channels. The transport channels are mapped onto the physical channels (shown in Figure 3.12).

3.5.8 Rate Matching

The connection between the user terminal and the network is called the 'bearer'. The bearer is not fixed as in the 2G networks. In the 3G networks, the bearer is negotiated and network checks are available, plus resources and user subscriptions before allowing the bearer establishments. The end-to-end services that the user uses define the base band data rate (in kbps), which will use the negotiated bearer services. The base band data should then be protected from the errors as they transmit from the terminal to the network radio terminal. It further adds significant overhead to the original data, for example convolution coding may double or triple the original base band bit rate. After error protection, the BB data rate is matched to the bearer data rate in

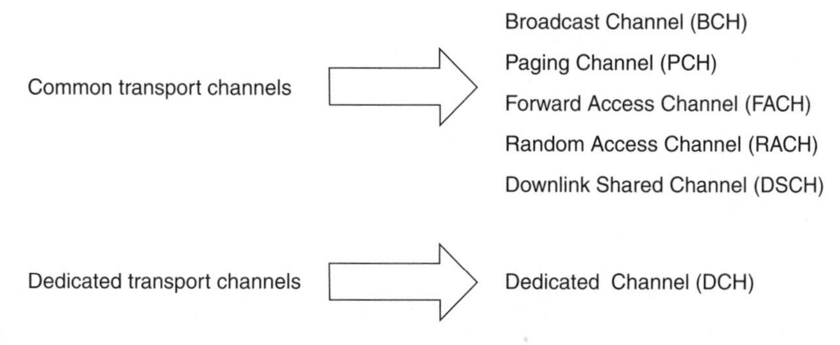

Figure 3.11 Transport channels.

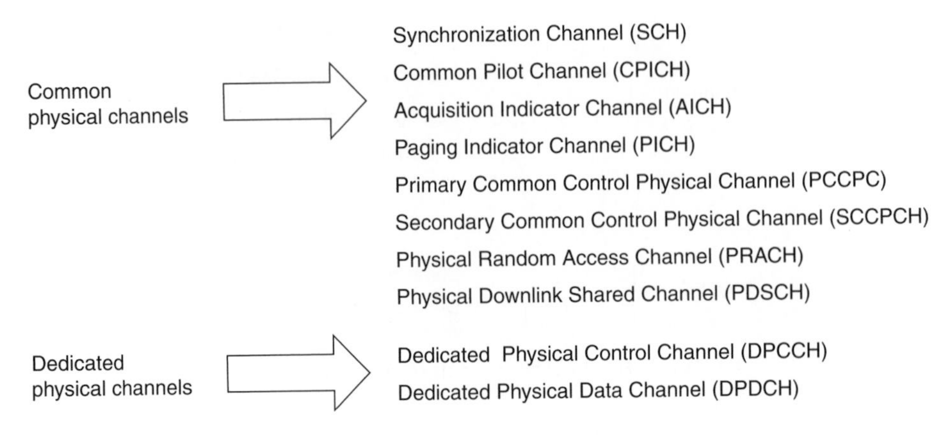

Figure 3.12 Physical channels.

the WCDMA air-interface. This is called 'rate matching'. The latter is carried out in order to match the bits into the frames.

Difference between the Uplink and Downlink direction: the channel bit rate is different in both of the directions, that is the uplink and downlink directions, although the system uses the same spreading factor because of the difference in the modulation methods, channel coding, interleaving and rate matching.

3.6 ATM

In ATM, the transfer mode is the one in which the information is organized into cells: it is asynchronous in the sense that the recurrence of cells containing information from an individual user is not necessarily periodic. Some of the basic characteristics that are offered by the networks are to handle various types of traffic (voice, data – real-time and non-real-time) and give reliability to users at lower costs. The ATM (Asynchronous Transfer Mode) technique has these characteristics and is hence chosen as the transmission technique of UMTS networks. Frame relay and ATM are almost similar techniques, but as the frames in frame relay are of variable length and hence variable delay, they are not suited for UMTS applications, while the ATM is based on the 'cell' – small and fixed packets size making it ideal for UMTS applications.

3.6.1 ATM Cell

An ATM cell consists of a 53 byte octet (octet header and 48 octet payload) as shown in Figure 3.13. ATM technology is independent of the physical medium and is scalable, that is multiplexing of the circuits leading to faster circuits is possible. Thus, ATM is a switching and multiplexing technique capable of giving a desired quality of service for different applications.

The ATM cells (each cell being 53 bytes, with a 5 byte header and 48 byte payload) are multiplexed to form virtual channels that are subsequently multiplexed to form various paths. These paths are not permanent and are hence called 'virtual'. These virtual paths are carried

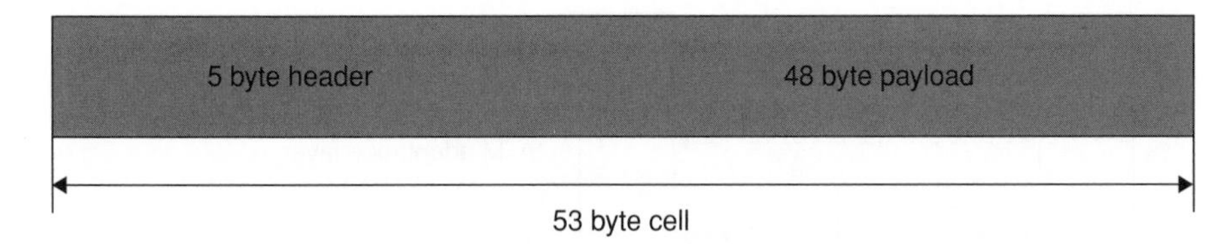

Figure 3.13 ATM cell structure.

by the transmission media and their number may vary depending upon the capacity of the transmission media.

3.6.2 Virtual Channels and Virtual Paths

As mentioned before, many virtual channels form one virtual path (shown in Figure 3.14). Let's now have a look into aspects related to virtual channels and virtual paths.

3.6.2.1 Virtual Channel

This is a sequence of the ATM cells belonging to a particular type of service or destination which comprises the virtual channels.

3.6.2.2 Virtual Path

This is the number of VCs sharing one single link that 'bundles' different VCs into one path, called the 'virtual path'. Also, to simplify the routing and switching of cells belonging to a particular destination and a particular type of service, VCs are combined into a single VP.

3.6.2.3 Virtual Channel Identifier

To identify a particular type of VC, the header of an ATM cell has a virtual channel identifier.

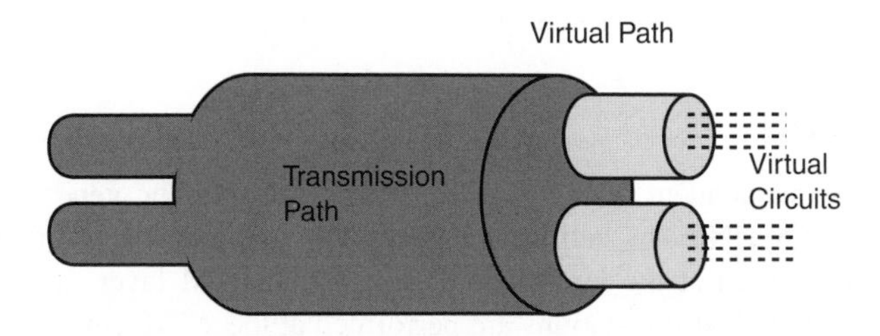

Figure 3.14 Virtual circuits and virtual paths.

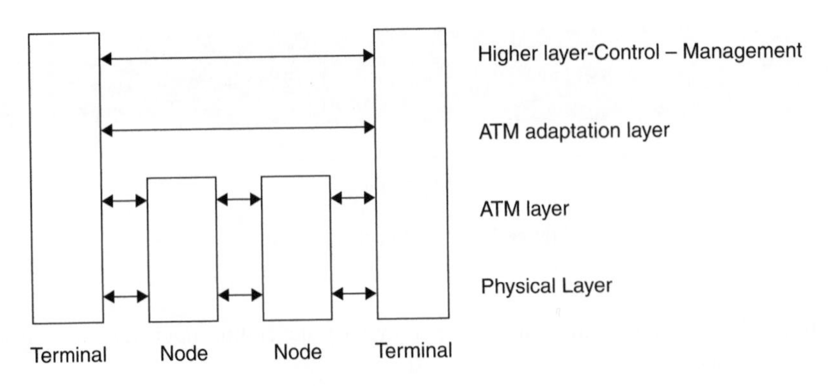

Figure 3.15 ATM protocol structure.

3.6.2.4 Virtual Path Identifier

To identify a particular type of VP, the header of the ATM cell has a virtual path identifier (VPI).

3.6.3 Protocol Reference Model

The protocol reference model is shown in Figure 3.15.

3.6.3.1 Physical Layer

There are two sub-layers in the physical layer: TC (Transmission Convergence Layer) and PM (Physical Medium Layer).

The TC layer is in charge of maintaining the cell rate by insertion and extraction of idle cells in order to adapt the rate of ATM cells to the payload capacity of the transmission system. HEC generation and verification perform error-free transportation of the headers. Cell delineation is the mechanism that enables the receiver to recover the cell boundaries. It is based on the identification of the header fields of the consecutive cells. If the HEC value is correct for a certain number of consecutive cells, then the cell delineation mechanism is achieved. To protect the cell delineation mechanism from malicious attack, the information is scrambled before the transmission.

The PM layer includes only the physical medium-dependent function. Its specification then depends upon the medium used. It provides the bit transmission capability, including bit alignment. The physical medium can be fibre, co-axial cable, etc.

3.6.3.2 ATM Layer

The ATM layer is independent of the physical medium layer. The generic flow control is defined at the B-ISDN UNI. The cell header generation and extraction function, except for the HEC value, is applied at the termination point of the ATM layer. It includes both PT and CLP bits. Cell VPI/VCI translations are performed at the ATM switching modes and/or cross-connect nodes are to be encountered on a route from the transmitter to the receiver. The cell mux/demux is based on the VCI/VPI values.

3.6.3.3 ATM Adaptation Layer (AAL)

With so many services requiring greater speeds, it would be suitable if the ATM provides them a common platform to be transferred. But in spite of using different versions of ATM layers for different services, another layer is introduced above the ATM layer that provides mapping of different types of applications to the ATM layer. This layer is service-dependent. Thus, the use of the ATM makes it mandatory for the need for an adaptation layer to support other information-based protocols not based on the ATM.

AAL is sub-divided into the two logical sub-layers: SAR and CS. The SAR (segmentation and re-assembly) sub-layer performs the segmentation and re-assembly of information in the ATM cells. Segmentation means to segment higher layer PDUs into a suitable size for the information field of the ATM cell (48 octet) and receiving size the re-assembly of the particular information fields into higher layer PDUs. The CS (Convergence Layer) performs other service adaptation. This layer is service-dependent. AAL functions are service-dependent. Connections falling into a particular class have similar QoS requirements. These classes can also have their own buffers. Real-time classes will have shorter buffers than the delay tolerant classes.

The various services provided by the AAL (ATM Adaptation Layer) are as follows:

- handling of transmission errors;
- segmentation and re-assembly;
- handling of lost cell conditions;
- flow control and timing control.

Different types and quality of services would require different AAL protocols to support the service. To minimize this wide variety of protocols, the ITU-T has defined four classes of service that cover a broad range of applications. This classification is based on three categories, which have a timing relation between the source and the destination, the bit rate and finally the connection mode. Based on these properties, different AAL protocols have been devised which come under the four classes defined. Table 3.2 gives details of this service classification as given by the ITU-T.

AAL Type 1

AAL type1, a connection-oriented service, is suitable for handling circuit emulation services, such as voice and video conferencing. AAL1 requires timing synchronization between the

Table 3.2 AAL protocol classification as given by ITU-T. Reproduced from the International Telecommunication Union (ITU)

	Class A	Class B	Class C	Class D
Timing relation between source and destination	Required		Not Required	
Bit rate	Constant	Variable		
Connection mode	Connection Oriented			connection-less
AAL Protocol	Type 1	Type 2	Type 3/4, Type 5	Type 3/4

source and destination. For this reason, AAL1 depends on a medium, such as SONET, that supports clocking.

AAL Type 2

The type 2 deals with the variable bit rate information and is intended for analogue applications, such as video and audio that require timing information but do not require a constant bit rate.

AAL Type 3/4

AAL3/4 supports both connection-oriented and connection-less data. It was designed for network service providers and is closely aligned with the Switched Multi Megabit Data Service (SMDS). AAL3/4 is used to transmit SMDS packets over an ATM network.

AAL Type 5

AAL5 is the primary AAL for data and supports both connection-oriented and connection-less data. It is used to transfer most non-SMDS data, such as classical IP over ATM and LAN Emulation (LANE). AAL5 is also known as the simple and efficient adaptation layer (SEAL) because the SAR sub-layer simply accepts the CS-PDU and segments it into 48-octet SAR PDUs without adding any additional fields.

3.6.4 Performance of the ATM (QoS Parameters)

The parameters are defined by the ATM Forum and are shown in Table 3.3.

Constant bit rate (CBR). The user traffic is continuous and steady under CBR. Supported by AAL1, this class has tight delay and delay variations bounds. Voice and video conferencing are typical examples of CBR traffic.

Real-time variable bit rate (rt-VBR). Supported by AAL3/4 and AAL5, the nature of traffic is 'bursty' and insensitive to smaller delay variations.

Non-real-time variable bit rate (nrt-VBR). This is similar to rt-VBR, except that the traffic is 'non-real time'. This class is used for frame relay inter-working. Both rt-VBR and nrt-VBR have a variable bandwidth.

Unspecified bit rate (UBR). This is the best effort transmission. The user may make a request for a maximum traffic rate, but the network provides the best rate possible. As in the above two cases, AAL3/4 and AAL5 support this type of traffic.

Available bit rate (ABR). Under this, there is no guarantee of bandwidth. Also there is no timing relationship between the source and destination.

Table 3.3 ATM QoS parameters

Service category	Traffic parameters				Qos parameters		
CBR	PCR	CDVT			max CTD	ptp CDV	CLR
rt-VBR	PCR	CDVT	SCR	MBS	max CTD	ptp CDV	CLR
nrt-VBR	PCR	CDVT	SCR	MBS	CLR	—	—
UBR	PCR(opt)	—	—	—	—	—	—
ABR	PCR	MCR	—	—	—	—	—

Sustainable cell rate (SCR). This is the upper bound on the average rate of the conforming cells of an ATM connection, over time scales which are long relative to those for which the PCR is defined.

Maximum burst size (MBS). This is the maximum number of cells that can be transmitted on peak cell rate.

Minimum cell rate (MCR). This is the rate negotiated between the end-systems and the network(s), such that the actual cell rate sent by the end-system on the ABR connection need never be less than MCR.

The performance of the ATM layers can be measured by monitoring certain parameters. These parameters can be divided into two groups: negotiable and non-negotiable.

Parameters that are negotiated are:

- peak-to-peak cell delay variation (CDV);
- maximum cell transfer delay (maximum CTD);
- mean cell transfer delay (mean CTD);
- cell loss ratio (CLR).

Cell delay variation. The CDV parameter describes variability in the pattern of cell arrival (entry or exit) events at an MP (Measurement Point) with reference to the negotiated peak cell rate $1/T$ (see Recommendation I.371); it includes the variability present at the cell source (customer equipment) and the cumulative effects of variability introduced (or removed) in all connection portions between the cell source and the specified MP. It can be related to cell conformance at the MP, and to network queues. It can also be related to the buffering procedures that might be used in AAL 1 to compensate for cell delay variation.

Cell transfer delay. Cell transfer delay (CTD) is the time, $t2 - t1$, between the occurrence of two corresponding cell transfer events, CRE1 at time $t1$ and CRE2 at time $t2$, where $t2 > t1$ and $t2 - t1 \leq T\text{max}$. The value of Tmax is for further study, but should be larger than the largest practically conceivable cell transfer delay.

Mean cell transfer delay. Mean cell transfer delay is the arithmetic average of a specified number of cell transfer delays.

Cell loss ratio. Cell loss ratio (CLR) is the ratio of total lost cells to total transmitted cells in a population of interest. Lost cells and transmitted cells in severely 'errored' cell blocks are excluded from the calculation of cell loss ratio.

QoS parameters that are not negotiated are as follows:

- cell error ratio (CER);
- severely errored cell block ratio (SECBR);
- cell mis-insertion rate (CMR).

Cell error ratio. The cell error ratio (CER) is the ratio of total errored cells to the total of successfully transferred cells, plus tagged cells, plus errored cells in a population of interest.

Severely errored cell block ratio. The severely errored cell block ratio (SECBR) is the ratio of total severely errored cell blocks to total cell blocks in a population of interest.

Cell 'mis-insertion' rate. The cell mis-insertion rate (CMR) is the total number of mis-inserted cells observed during a specified time interval divided by the time interval duration2 (equivalently, the number of mis-inserted cells per connection second).

Availability ratio. The availability ratio (AR) applies to ATM semi-permanent connection portions. The service AR is defined as the proportion of time that the connection portion is in the available state over an observation period. The service AR is calculated by dividing the total service available time during the observation period by the duration of the observation period.

Mean time between outage. The service MTBO is defined as the average duration of a time interval during which the portion is available from the service perspective. Consecutive intervals of available time during which the user attempts to transmit cells are concatenated. The network MTBO is defined as the average duration of a continuous time interval during which the portion is available from the network perspective.

Some other aspects related to ATM are described in the following.

Connection admission control (CAC). CAC algorithms determine if a new connection is to be accepted or rejected. It accepts a connection request only if sufficient resources are available and if it does not affects the QoS of the existing circuit.

CAC considers the following factors for a new connection request:

- traffic parameters of new connections and QoS requirements;
- existing traffic contract and connection;
- BW, allocated and unallocated;
- over-booking parameter.

Conformance monitoring and performance. This has two major ATM traffic function mechanisms. These are traffic shaping and policing. Policing is a Usage Parameter Function (UPF) mechanism which ensures that during the connection, the network uses the traffic contract defined for the connection to check that it stays within there contracted services. If there are non-conforming cells, then the network take appropriate action on them, for example setting the CLP bit of the non-conforming cells, thus making the cells eligible for discarding. This is a 'discard' that is done to prevent any non-conforming cells in affecting the QoS of the conforming cells of the connections. Traffic shaping functions modify the traffic flow and changes the characteristics of the user cells streams to achieve improved network efficiency and to get the lowest cell loss. Traffic shaping (properties as shown in the following list) may take place at either the egress or the ingress sides of the network:

- Modifies the traffic flow for an ATM that was not able to keep the traffic contract, ingress side.
- Ensures traffic conforms to the traffic contract on the egress of the switch.
- Avoids overflow for the subsequent ATM with a small buffer.
- Constrains data burst at the egress of the switch.
- Limits peak rate.
- Constrains 'jitter'.

Conformance of each cell of the ATM is evaluated through the generic cell rate algorithm.

Queuing. To maintain the network optimum performance, the ATM switch performs a series of cell treatment mechanisms, such as queuing, buffering, cell servicing and congestion control, thus maintaining the desired QoS. Queuing occurs in the following cases:

- When the two cells arrive at the same time and are going to the same destination, then queuing will occur.
- Cells of higher bit rates pass through a virtual connection with lower bandwidths and thus congestion takes place.

Buffering of cells occurs when two or more conforming cells are destined to the same output at the same time. Cells servicing, such as dropping of cells, occurs when the non-conforming cells, with the CLP bit set to one, arrives which causes congestion.

Circuit emulation service (CES). This is a technique for carrying CBR traffic (Circuit Traffic) over the ATM network, as the ATM is a packet rather than a circuit-oriented transmission technology. It must emulate circuit characteristics in order to support the CBR- and TDM-based traffic. CES user AAL1 is order to encapsulate 64 kbps TDM TS into the ATM cells. There are two possible CES encapsulations:

- Un-structured: intended to emulate pt-to-pt, JT1 or E1.
- Structured: intended to emulate pt.-to-pt, fractional JT1 or E1 circuit.

Inverse-multiplexing ATM (IMA). This technique involves inverse multiplexing and de-multiplexing of the ATM cells in a cyclic fashion among links grouped together to form higher bandwidth logical links whose total rate is approximately the same as the sum of the individual link rates. This is called the IMA group. The aim is to support broadband ATM traffic to be transported efficiently through the existing PDH transmission network.

The IMA group is terminated at each end of the IMA virtual link. The receiving end reconstructs the ATM cell stream after accounting for the link differential delays. The links can be added or removed to/from the operational IMA group to dynamically adapt the bandwidth to changing needs. In the case of link failures, the group stays operational with a reduced bandwidth, using the remaining links.

Fractional links. The user of fractional E1 or JT1 enables the adding of full or partial E1/JT1 channels with either 2G or 3G traffic or both. This method is used for maximizing the usage of available transmission capacity by sharing the same physical transmission links between the two types of traffic.

3.6.5 Planning of ATM Networks

Call sessions assigned by the virtual circuit between the end terminals may have a broad spectrum of holding times and the bit rate requirements. A group of call sessions between the two points in a network can be aggregated and transported over a pre-assigned path and hence beyond the call layer; connections also have to be managed at the path layer. The bit stream is also 'bursty' in nature. Thus, to achieve the transport level dimensioning, virtual path level dimensioning has to be achieved in terms of fixed or effective bandwidths, through the fixed bandwidths allocated for all the VPs on a particular transmission line, which may be shared by the dynamic VP bundles. An example of ATM planning is shown in Figure 3.16.

For these to happen, various switching mechanisms have to be employed. ATM cells, which last a few microseconds, are switched by the ATM switches. 'Burst'/packets are switched by 'burst switching' while 'non-bursty' switching of a path can be done during 'call set up'.

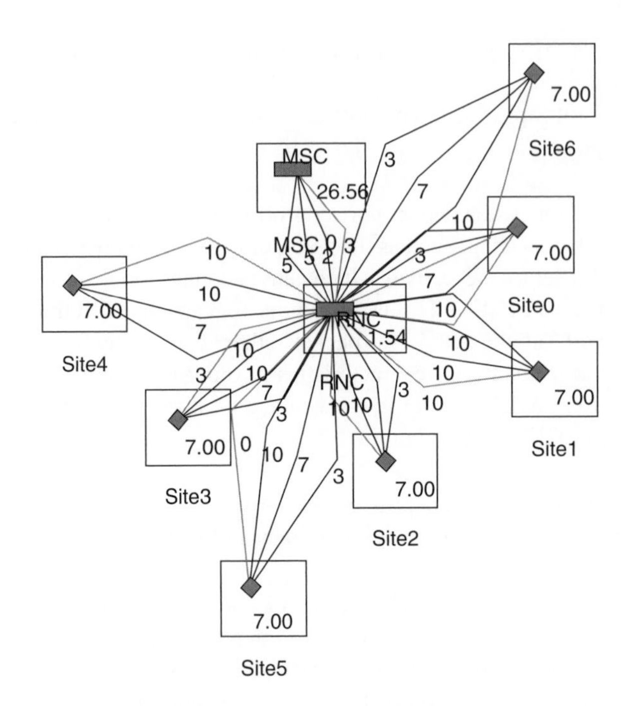

Figure 3.16 ATM planning.

Call sessions that last for minutes are switched by assigning a virtual circuit established on a sequence of paths. These paths, that may lasts for hours or days, are switched and configured by either the ATM or facility cross-connect. An example of an ATM network is shown in Figure 3.16.

ATM planning consists of defining/finding the various number of aspects such as those listed below (based on the parameters/ definitions mentioned above):

- Defining the type of traffic.
- Defining the VP and VC connections.
- The number of VPs and VCs, that is VPIs and VCIs.
- Defining the VP and VC cross-connections.
- Defining the physical, VP and VC connection parameters.

3.7 Protocol Stack

The protocol structure for UTRAN is based on the OSI model, as shown in Figure 3.17. There are two main layers: the radio network and transport network and two planes, the user plane and control plane. The user-related information, such as coded voice, etc. (including data streams and data bearers) is transported through the user plane. The control plane is used for the UMTS specific control signaling. Both the application protocol (such as RANAP, RNSAP and NBAP) and the signaling bearer for transporting the application protocol messages are included. The transport control plane is used for control signalling in the transport layer

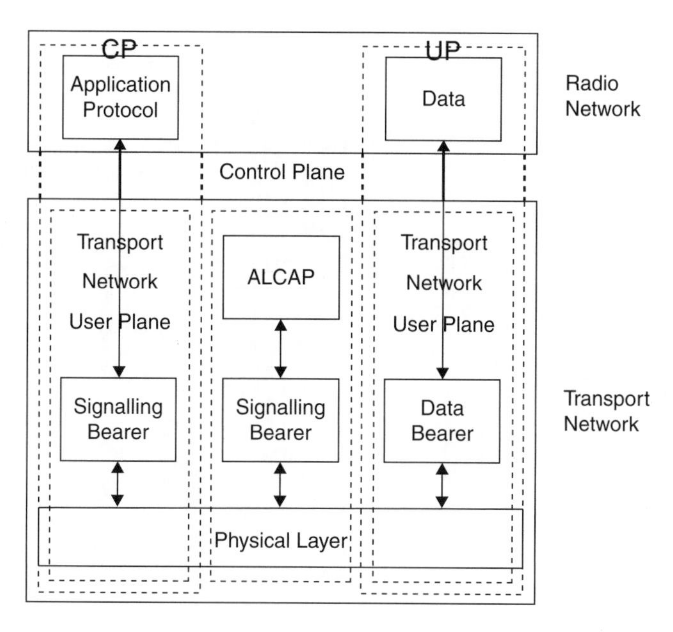

Figure 3.17 General protocol structure for UTRAN.

(as shown in Figure 3.17). It includes ALCAP and the signalling bearer needed for it. Due to the transport plane, the application protocol in the radio plane is independent of the technology selected for the data bearer in the user plane. The transport user plane contains the data bearer in the user plane and the signalling bearer for application protocol.

The RANAP (Radio Access Network Application Protocol) is the signalling protocol defined between the RAN and CN. The main function of RANAP is to control the resources on the Iu interface, providing both the control and dedicated control services.

The RNSAP (Radio Network Sub-system Application Protocol) is the one providing the signalling on the Iur interface. As the Iur interface is present between the two RNCs (one is the SRNC and the other is the DRNC), it is managed by these two RNCs. The RNSAP is responsible for the management of the bearer signalling on the Iur interface apart from the transport and traffic management functions.

The NBAP (Node B Application Protocol) is the one maintaining the control plane signalling on the Iub interface. This protocol is responsible for communication between the node B (WCDMA base station) and the UE.

The ALCAP (Access Link Control Application Part) is required for the setting-up of the data bearer and signalling in the transport network control plane. The presence of ALCAP is necessary as the user plane and the control plane can be separated and independent from each other. The ALCAP may or may not exist depending upon the type of data bearer.

3.8 WCDMA Network Architecture – Radio and Core

A simplified block diagram of a WCDMA network is shown in Figure 3.18. Let's have a look now at the radio and core network elements of this network.

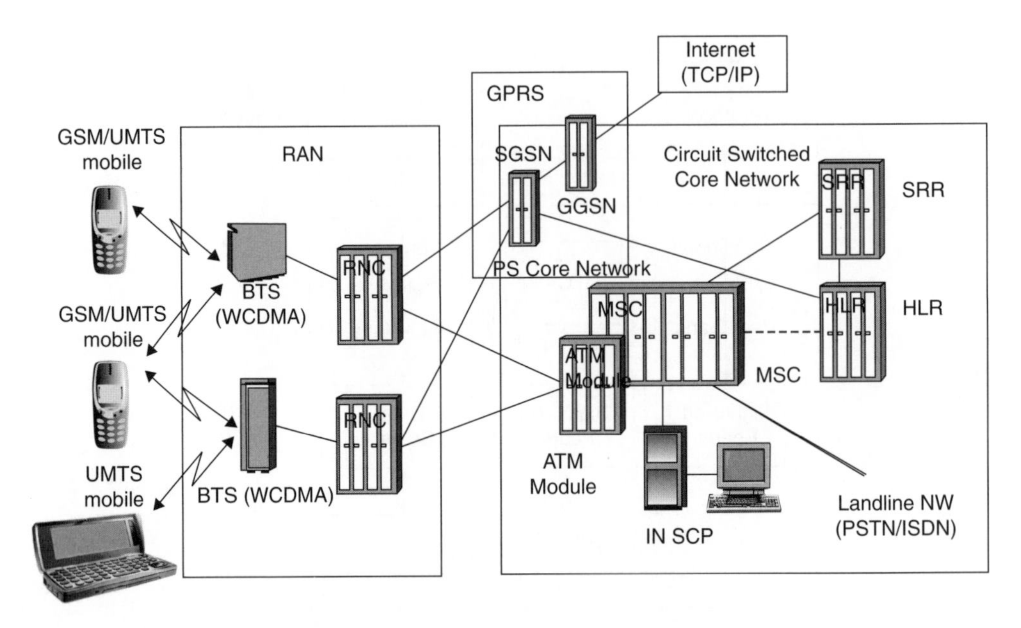

Figure 3.18 ULTRAN network architecture.

3.8.1 Radio Network

3.8.1.1 UMTS Terminal

The UMTS terminal, also called the User Equipment (UE) is responsible for the processing of the radio signal. The UE is able is support radio standards and contains the USIM which is similar to the SIM card in GSM mobiles. The UE is capable of functions such as power control, error correction, modulation, etc. The UE works with other network elements such as Node B, RNC, etc. in call connection set up/release, handover execution, power control measurements, bearer negotiations, mobility management, service requests, etc.

3.8.1.2 Node B or BS

The base station for the UMTS is called Node B (shown in Figure 3.19). It is also called the BS (base station) and does functions similar to the base station of the GSM network. There are three types of UMTS base stations: FDD, TDD and dual-mode base stations. The base station is connected to the RNC through ATM links. The ATM 'cross-connect' (AXC) is responsible for cross-connection at the ATM level. It also acts as an interface between the Application Manager and the Interface Units. The AXC can be a standalone unit also. The inner loop power control function is performed by the base station. It also performs functions such as softer handover, and measures connection quality and strength.

Application Manager (AM). This unit is responsible for operation and management functions and carrier control. From a transmission network planning perspective, the AM takes the same place as the TRX in GSM transmission network planning. The number of AMs is dependent

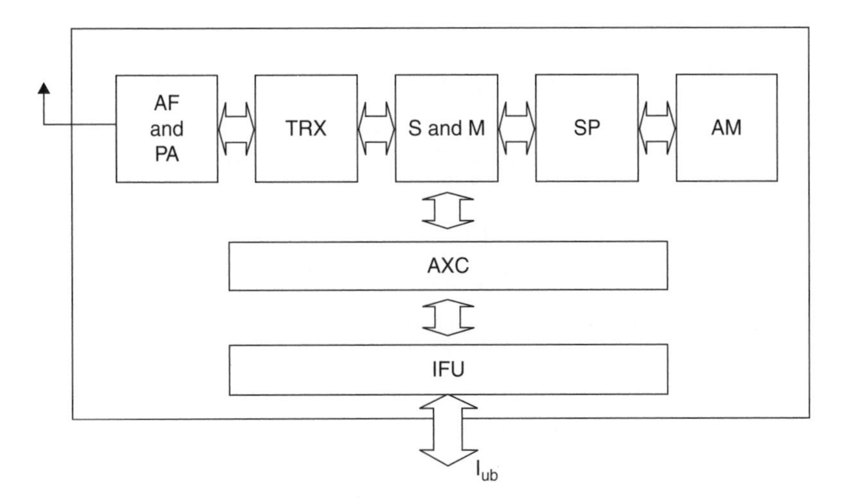

Figure 3.19 Simplified diagram of the 3G base station.

upon the traffic that is carried by the base station and is in no way dependent upon the number of cells or number of TRXs.

Signal Processors (SPs). The major function of this unit is to perform coding, decoding and code channel processing (in both the receiving and transmitting directions).

Transceivers (TRXs). Transceivers (TRXs) contain one transmitter and two receivers. A TRX consists of two frequency bands that are separated by a frequency of 190 MHz.

Antenna Filter (AF) and Power Amplifier (PA). These combine and isolate the transmitted and received signals. They also amplify the received signals.

ATM Cross Connect (AXC). This is responsible for cross-connection at the ATM level. It also acts as an interface between the AM and the interface units. The AXC can also be a standalone unit.

Interface Unit (IFU). This is the interface between the base station and Iub interface. This may vary depending upon the type of transmission, that is E1 /T1/JT1 or PDH/SDH, etc.

3.8.1.3 RNC

The Radio Network Controller (RNC) is the 'brain' of the network (shown in Figure 3.20). It is similar to the base station controller in the GSM network. Many protocols between the UE and RAN are implemented in the RNC. The RNC does the resource management functions in the network, such as call admission control, handover management, power control, channel allocation, packet scheduling, code allocation, encryptions, ATM switching, protocol conversion, etc. The RNC plays a dual role in WCDMA radio networks. It may also be known as SRNC (Serving RNC) or DRNC (Drifting RNC). From one mobile, if the RNC terminates both the data and related signalling then it is called the 'serving RNC'. If the cell that is used by this UE is controlled by any other RNC, other than the SRNC, then it is called the DRNC. There is another RNC, called the Controlling RNC (CRNC). This is responsible for configuration of Node B.

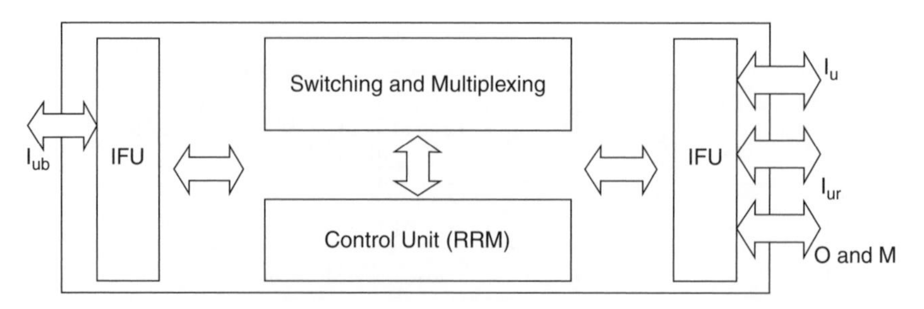

Figure 3.20 Simplified diagram of the RNC.

Control Unit. Radio resource management is considered to be the most important function of the RNC. All three network planning entities, radio, transmission and core, are affected by it. The control unit is responsible for the RRM functions such as the handovers, admission control, power control, load control, etc. It is also responsible for the control mechanisms for packet scheduling and location-based services.

Switching Unit. The ATM forms the backbone of the WCDMA networks. In RNC, the functionalities of the switching unit include providing the required support for the ATM traffic, ATM cell switching, AAL2 switching and multiplexing the traffic.

Interface Unit (IFU). There are mainly two kinds of interface units: PDH and or SDH. On either of the interface, that is on the Iub/Iur/Iu, PDH or SDH interface can be configured

3.8.2 Core Network

The core network is divided into circuit-switched and packet-switched domains. Some of the circuit-switched elements are Mobile Services Switching Centres (MSCs), the Visitor Location Register (VLR) and Gateway MSC. The packet-switched elements are the Serving GPRS Support Node (SGSN) and Gateway GPRS Support Node (GGSN). Other network elements, like the Equipment Identity Register (EIR), Home Location Register (HLR), Visitor Location Register (VLR) and Authentication Centre (AUC), are shared by both these domains.

The mobile switching centre (MSC) is the centre-piece of the circuit-switched core network. The same MSC can be used to serve both the GSM-BSS and the UTRAN connections. In addition to the radio access networks, it has interfaces to the fixed PSTN network, other MSCs, the packet-switched network (SGSN) and various core network registers (HLR, EIR, AuC). Physically, the VLR is implemented in connection with the MSC, so the interface between them (the B interface) exists only logically.

The functions of an MSC include paging, coordination of call setup from all MSs in the MSC's jurisdiction, dynamic allocation of resources, location registration, interworking functions (IWFs) with other type of networks, handover management (especially the complex inter-MSC handovers), billing of subscribers (not the actual billing, but collecting the data for the billing centre), encryption parameter management, signalling exchange between different interfaces, frequency allocation management in the whole MSC area and echo canceller operation and control. The MSC terminates the MM and CM protocols of the air-interface protocol stack, so the MSC has to manage these protocols, or delegate some responsibilities to other core network elements.

The Asynchronous Transfer Mode (ATM) is defined for UMTS core transmission. The ATM Adaptation Layer type 2 (AAL2) handles circuit-switched connection and packet-connection protocol AAL5 is designed for data delivery.

3.9 Network Planning in 3G

The network planning process in UMTS networks is similar to that of the GSM networks (as seen in Chapter 2) except for the fact that the number of parameters increases. Also, transmission planning changes quite drastically as compared to the one in the GSM network. The optimization process becomes more challenging as quality targets change, becoming more severe, due to the different quality of service classes that need to be provided to the subscriber.

3.9.1 Dimensioning

The process of dimensioning delivers a lower bound on the number of network elements and its configurations required to provide given services for a number of customers in a service area with certain quality and within a specific period of time. During dimensioning, a vendor selection is performed which significantly influences the dimensioning output, since the vendor's system performance and feature sets influence the parameters in the dimensioning. The number of required network nodes may differ between different vendors or the offered service can be obtained with different quality. Besides the performance criterion, the vendor's solution scalability and system development 'road map' are important. Both are required to flexibly extend network capacity, coverage or quality and adjust the network to future needs and progress.

Dimensioning delivers principal information for system planning, deployment and optimization: topology of RAN, coverage, capacity and quality requirements, usually given within certain thresholds or key performance indicators (KPIs). The dimensioning stage may be supported by field trials and initial propagation model adjustments for a better fine-tuning of the dimensioning parameters. The dimensioning output can be fed back into the network definition to fine-tune the business case.

The process consists of two main phases: coverage and capacity dimensioning. The coverage dimensioning phase is strictly connected with rollout targets. Having the rollout targets defined, to obtain a total number of coverage sites, the cell range and area should be calculated. From the total path loss, taken from a link budget using an appropriate propagation model and assuming a particular site configuration, the desired cell range and area can be calculated (different per clutter type).The total area can be divided into fractions of a particular clutter type, from which the number of sites per clutter type is obtained.

An alternate way is to have an approximate figure by calculating the number of Node Bs, considering these to be equal to the number of existing 2G sites within a desired area and then multiplying by the ratio between 2G and UMTS cell areas. Having the number of sites available, the next phase is to compare the network capacity with a calculated traffic demand. If the capacity is too small, the site capacity configuration should be changed or else additional sites should be added.

The UTRAN transmission network consists of two elements: the transmission links between the Node Bs and the RNC and the RNCs themselves. The transmission links can use microwave point-to-point links, a LMDS network, fibre optics networks, etc. From a generic approach, the capacity configurations can be differentiated on a per clutter type basis. Thus, different transmission link capacity types (e.g. microwave links) can be directly assigned to them. For the remaining part of the UTRAN transmission network (RNC-related), the number and configuration of the RNCs has to be known. The limiting factors in RNC dimensioning are Iub traffic capacity (in Mbps, in number of channels and in number of carriers per RNC), interface capacities (e.g. STM-1 and E1) and maximum number of cells per RNC.

The number and capacity of transmission links depend on the number of network elements that have to be connected, as well as the overall core network traffic. The number of ATM switches may be equal to the number of MSCs (large switches), plus the number of RNCs (small/medium ones). The number of other transmission nodes (e.g. SDH add-drop multiplexers) depends on the particular network topology. The core network consists of MSS Server, Media Gateway, SGSN, GGSN, HLRi/AuC, Charging Gateway, Border Gateway, Lawful Interception Gateway, Firewall, DNS/DHCP and EIR.

Load control mechanism is taken into account during the dimensioning as well. This is responsible for controlling the load of the network. The main difference between the admission control (this is responsible for letting users in to the network – preventing overload situation) and the load control is that load control comes into action after overload happens in the network.

The process of dimensioning in WCDMA is similar to that of the GSM networks. However, with parameters the dimensioning of WCDMA networks becomes more complex. The dimensioning outputs for WCDMA would include coverage/cell size calculations, capacity calculations, link budget calculations and estimates for the access transmission network. The inputs for dimensioning from capacity and coverage perspectives would include subscriber figures (present and forecasted), traffic density, spectrum availability and geographical area – size and topography to be covered. As compared to 2G/GSM, the quality criterion in WCDMA is quite strict. The inputs would include information about MS class, indoor coverage, redundancy, and blocking and location probability. Unlike in GSM radio network dimensioning, the traffic forecast figure alone would be good enough; however, in WCDMA, the subscriber figure alone is not sufficient. Some figures, such as the type of services and quality of service that would be used by a subscriber, would be needed. The link budget is needed for path loss and cell size calculations. There are some new parameters in the link budget calculations as compared to GSM. An example of a link budget is shown in Table 3.4. Let's have a quick look at the parameters in the WCDMA link budget calculations.

3.9.1.1 Noise Power of Receiver

This is defined as the minimum baseband signal strength of the receiver. To calculate it, the receiver noise density has to be scaled to the WCDMA carrier bandwidth. The receiver noise density is the summation of the thermal noise density (10log kT, where k is the Boltzmann constant and T is the temperature, in Kelvin) and the receiver noise figure (equipment-specific values assumed to be 3 dB at the base station and 8 dB at the mobile station).

$$\text{Receiver noise power (dBm)} = \text{Receiver Noise Figure} + kTB$$

Table 3.4 An example of a WCDMA link budget

Link Budgets		voice		LCD		UDD		UDD		UDD	
Data rate (kb/s):		12.2	12.2	64	64	64	64	128	128	384	384
Load:		10 %	10 %	10 %	10 %	10 %	10 %	10 %	10 %	10 %	10 %
RECEIVING END		Uplink Node B	Downlink UE	Uplink Node B	Downlink UE	Uplink Node B	Downlink UE	Uplink Node B	Downlink UE	Uplink Node B	Downlink UE
Thermal Noise Density	dBm/Hz	−174	−174	−174	−174	−174	−174	−174	−174	−174	−174
BTS Receiver Noise Figure	dB	3.00	8.00	3.00	8.00	3.00	8.00	3.00	8.00	3.00	8.00
BTS Receiver Noise Density	dBm/Hz	−171.00	−166.00	−171.00	−166.00	−171.00	−166.00	−171.00	−166.00	−171.00	−166.00
BTS Noise Power (NoW)	dBm	**−105.16**	**−100.16**	**−105.16**	**−105.16**	**−105.16**	**−100.16**	**−105.16**	**−100.16**	**−105.16**	**−105.16**
Required Eb/No	dB	4.00	6.50	2.00	5.50	2.00	5.50	1.50	5.00	1.00	4.50
Soft handover MDC gain	dB	0.00	1.20	0.00	1.20	0.00	1.20	0.00	1.20	0.00	1.20
Processing gain	dB	24.98	24.98	17.78	17.78	17.78	17.78	14.77	14.77	10.00	10.00
Interference margin (NR)	dB	0.46	0.46	0.46	0.46	0.46	0.46	0.46	0.46	0.46	0.46
Required BTS Ec/Io [q]	dB	**−20.52**	**−19.22**	**−15.32**	**−15.02**	**−15.32**	**−13.02**	**−12.81**	**−10.51**	**−8.54**	**−8.24**
Required Signal Power [S]	dBm	**−125.68**	**−119.38**	**−120.48**	**−113.18**	**−120.48**	**−113.18**	**−117.97**	**−110.67**	**−113.70**	**−106.40**
Cable loss	dB	−0.50	0.50	−0.50	0.50	−0.50	0.50	−0.50	0.50	−0.50	0.50
Body loss	dB	0.00	5.00	0.00	0.00	0.00	0.00	0.00	0.00	0.00	0.00
Antenna gain RX	dBi	18.00	0.00	18.00	0.00	18.00	0.00	18.00	0.00	18.00	0.00
Soft handover gain	dB	2.00	2.00	2.00	2.00	2.00	2.00	2.00	2.00	2.00	2.00
Power control headroom	dB	3.00	0.00	3.00	0.00	3.00	0.00	3.00	0.00	3.00	0.00
Sensitivity	dBm	**−143.18**	**−115.88**	**−137.98**	**−114.68**	**−137.98**	**−114.68**	**−135.47**	**−112.17**	**−131.20**	**−107.90**
TRANSMITTING END		UE	Node B	UE	Node B	UE	Node B	UE	Node B	UE	Node B
Power per connection	dBm	21.00	28.30	21.00	29.30	21.00	29.30	21.00	29.30	21.00	21.30
Maximum Power per connection	dBm	21.00	40.00	21.00	40.00	21.00	40.00	21.00	40.00	21.00	40.00
Cable loss	dB	0.00	3.00	0.00	3.00	0.00	3.00	0.00	3.00	0.00	0.00
Body loss	dB	5.00	0.00	0.00	0.00	0.00	0.00	0.00	0.00	0.00	0.00
Antenna gain TX	dBi	0.00	18.00	0.00	18.00	0.00	18.00	0.00	18.00	0.00	18.00
Peak EIRP	dBm	**16.00**	**43.30**	**21.00**	**44.30**	**21.00**	**44.30**	**21.00**	**44.30**	**21.00**	**44.30**
Maximum Isotropic loss	dB	**159.18**	**170.88**	**158.98**	**169.68**	**158.98**	**169.68**	**156.47**	**167.17**	**152.20**	**162.90**
Isotropic path loss to the cell border			**159.18**		**158.98**		**158.98**		**156.47**		**152.20**

3.9.1.2 E_b/N_o

This is the ratio of average bit energy and spectral density and is dependent on the type of service, radio channel and UE speed.

$$\frac{E_b}{N_o} = \frac{P_{rx}}{I} \times \frac{W}{R} \text{dB}$$

where:

P_{rx} is the received power;
I is the power received from a signal other than 'its own';
R is the bit rate;
W is the bandwidth.

E_b/N_o is larger than S/N by an amount of spreading factor. Communication is possible in the network even when the S/N for a connection at receiver is much small than unity.

3.9.1.3 Macro Diversity Gain (MDG)

This is the result of soft handover and is calculated by taking the difference of received signal branched and UE speeds (UL is 0 dB and DL is 1 dB). It is calculated by averaging over all connections.

3.9.1.4 Soft Handover Gain

This is gain against shadow fading. Based on the minimal transmit power of the mobile station, the best BTS is chosen. This is done through the handover algorithm (based on power) against the hard handover algorithm based on geometrical distance (average figure: 2 dB)

3.9.1.5 Processing Gain

The amount by which the power density of the signal is increased in the receiver is called the processing gain (*PG*) and corresponds to the spreading factor (*W*) and subsequently to the number of chips transmitted per bit (*R*).

$$PG = W/R$$

3.9.1.6 Interference Margin

This states the decrease in receiver sensitivity due to increases in the subscriber load. It is calculated from the UL and DL loading values (η).

$$\text{Interference margin (IM)} = -10\log_{10}(1-\eta)\text{ dB}$$

3.9.1.7 Ec/I_0

In order to meet the baseband E_b/N_o criteria, the required Ec/I_0 is calculated as follows:

$$\text{Required } Ec/I_0 = E_b/N_o - MDC - PG + IM$$

3.9.1.8 Required Signal Power

This is the summation of the required Ec/I_0 and receiver noise power.

3.9.1.9 Power Control Headroom

The margin against the fast fading is called the power control headroom. This parameter is important for slow moving mobiles and is needed because at the cell edge mobiles do not have enough power to follow the dip in fast fading.

3.9.1.10 Isotropic Power

This is defined as the minimum power needed to fulfil the E_b/N_o for a given service:

$$\text{Isotropic power (IP)} = \text{required signal power} + \text{loss (cable and body)} - \text{antenna gain}$$
$$-\text{SHO gain} - \text{power control headroom}$$

3.9.1.11 Body Loss

This is used for speech-related services as the mobile is closest to the body then, while for data services, it is 0 dB.

3.9.1.12 Peak EIRP

This is the maximum transmitted power after the antenna.

3.9.2 Load Factor

As the WCDMA is an interference-limited system, the amount of traffic supported by the base station needs to be calculated and is done through the load factor. We understand the phenomenon of cell breathing, wherein the coverage area of the cell decreases with the increase in number of subscribers and vice versa. For the calculations, the energy per user bit is calculated (E_b/N_o). Both the uplink and downlink load calculations are carried out. The UL load affects the noise level at the base station. During dimensioning, the estimate values range from 30 % to 70 %. The downlink load is always higher than the uplink load. In the UL, there is one transmitter per connection; however, in the DL, all connections share one transmitter. Some aspects that affect the load factor are traffic, interference, transmit power, speed of UE, SHO, services offered, etc. In the case of congestion, load control decreases the load to within limits. This can be done by uplink and downlink fast load controls.

The system planning typically begins with a field trial of available technology solutions. This will help to verify the system behaviour in a real environment. In parallel, the propagation model calibration is performed and the grid is defined in the service area. Next, the site selection together with site configuration takes place. This is strongly supported by an advanced planning and optimization tool. With regard to UMTS planning, trade-off between coverage and capacity cannot be described by a static coverage prediction only. Finally, the network data base's parameter fill is determined for the system. The more realistic number of network elements and its cost of deployment are fed back to adjust the business case.

Most of the radio network-planning activities are related to the cell layer that provides a wide area coverage, that is the network 'footprint' into a market. It is thus typically related to the macro-cell layer deployment. Before any planning activity can commence, the key parameters of the radio network layout need to be defined to find an optimum between coverage and capacity with a minimum investment. These parameters are set at priority with certain assumptions; for instance, the site-to-site distance is calculated using a link budget, where parameters are averaged and the antenna height is fixed in advance.

The process of searching an adequate Node B candidate site is initialized, resulting in a site location. The assignment of Node Bs should follow an ideal geometry grid, where the neighbouring grid nodes are elements of equilateral triangles. The corresponding cell shapes are hexagons or diamonds, depending on the antenna direction and antenna horizontal beamwidth. In a practical scenario, the actual Node B placement, configuration and its antennae tilt arrangements will hence differ from the ideal grid and the initial settings will be adjusted according to the local environment and traffic requirements.

The site-to-site distance determines the network density and, together with the antenna height, the achieved coverage and capacity. If the antenna height is increased but the site-to-site distance remains constant, then the coverage probability increases and so does the area coverage per cell; hence, more potential traffic can be served. On the other hand, the interference range and strength to the surrounding cells rises as well. If the site-to-site distance increases but the antenna height remains constant, then the coverage probability per cell decreases; thus less potential traffic can be served. However, the interference between cells also decreases, thereby leaving some headroom for a potential capacity increase. The aim is to find the best trade-off between both parameters for coverage and capacity through iterative planning activities with a reference network and assumed site configuration.

Sectorization is a commonly used technique to enhance capacity or coverage by equipping the site with sectors; it is traditionally provided by means of directional high-gain antennae. Each of the sectors creates a new cell. This is different from cell splitting, where the sectors may still belong to one logical cell. In the case of cell splitting, the enhancement leads only to a coverage extension in the downlink and uplink; it could also provide additional capacity improvements in the uplink, but due to power splitting leads to a decrease of the downlink capacity.

The one sector solution may also be realized by a sectorized antenna to cover, for example a tunnel. Two sectors may be employed for line coverage at the macro layer (e.g. sites along a motorway) or for coverage and capacity of dense areas at the micro layer. Three sectors can be used to provide regular footprints of the network service at the macro layer with low or medium loads, and six sectors to provide additional capacity and improved coverage to regular macro layers.

The performance of a sectorized site is mainly given by the horizontal (azimuth) characteristics of the used antennae and, obviously, the number of sectors. The more the sectors are used, the narrower the beamwidth of the antennae has to be; an optimum antenna beamwidth can be found for each configuration. A narrower antenna provides usually a higher directional gain and hence improves the link budget in both directions. In addition, less overlapping between neighbouring sites is provided; thus less interference is received from other cells and less soft handover overhead is present. The improved directional link gain offers an improved coverage in the main direction; however it requires that the edges of the service area are to be covered by the adjacent sectors. The directional gain can be used to increase the site-to-site distance and hence reduce the network density. Thus, the greater allowable path loss, together with less interference and less soft handover overhead, increases the network capacity.

Another key parameter influencing the performance of UTRAN is the direction of the antennae (sectors). The dependency of the sector orientation on the Node Bs' relative arrangement points out that proper adjustment controls the soft handover overhead and interference leakage. Also, the spacing between sectors ought to be as identical as possible, as this would allow an equal distribution of softer handover areas among all sectors.

The antenna azimuth of a physically placed Node B should provide a clear best server coverage scenario, leading to dominant sectors and pilots in the serving area. The areas with insufficient pilot dominancy and extensive cell overlapping should be avoided. Two basic measures can be used to minimize the lack of dominant pilots: pointing sectors into an area in between the sectors of surrounding sites and preventing adjacent site sectors to 'point' at each other. In parallel, the sector orientation should be adjusted to traffic requirements, so as to evenly spread the expected load among all sectors. The sector separation should be kept regularly. The physical antenna mounting should leave sufficient freedom for further optimization processes, so that some future orientation (and tilt) changes can be performed.

The antenna tilting technique is a very powerful measure to increase the network performance. The main aim of tilting is to increase local area cell dominance and reduce the cell overlapping, hence the inter cell interference. It is achieved by concentrating the radiation pattern of an antenna towards the anticipated cell serving area, thereby increasing the cell isolation to neighbouring cells. Basically there are two possibilities to tilt an antenna: mechanically or electrically. In the mechanical tilt, the antenna body moves with reference to the ground plane. The mechanical tilt affects predominantly the main direction and the tilt effect decreases from the main direction towards the side lobes in the azimuth. The gain reduction is a function of the azimuth and the down-tilt angle: the horizontal half power beam-width increases with an increasing mechanical down-tilt angle. The mechanical tilt can be carried out by adjustable brackets. The electrical method is realized by adaptation of the relative phases between all elements of the antenna array (dipoles), allowing the antenna pattern to be tilted evenly in all horizontal directions. Contrary to the mechanical approach, there is no dependency between the azimuth and the tilt angle. Consequently, the gain reduction is independent of the azimuth and the horizontal half-power beamwidth remains unchanged as a function of the down-tilt angle. The shape of the vertical pattern is modified in dependency of the tilt angle. The phase adjustment is performed by means of different feeder lengths for each dipole, which results in a fixed electrical tilt, or by a phase shifter, which results in an adjustable electrical tilt.

Hierarchical cell structures (HCSs) are a powerful network configuration technique used to balance network traffic and hence increase the overall network performance. The main aim of HCSs is to provide a set of overlaying cells in the same geographic area, which facilitates traffic to be assigned to various layers according to some criteria. Interference is a paramount issue in such arrangements, thereby requiring stringent interference avoidance or cancellation mechanisms. Interference can be avoided (or at least minimized) if the cell layers are separated in frequency, time or space.

As mentioned before, the frequency re-use factor is one, that is the frequency re-use factor is one. This causes cell coupling in terms of interference and hence transmission powers, even between quite distant areas of the same network. This property requires careful simulations, when it comes to the network planning of a real network with the use of network planning tools. The influence of the number of cells assumed for simulations is clearly visible for both parameter values as well as its dependency on the cell load. Thus, for reliable system level simulations used in network planning, the size of the simulated area should be extended by two additional tiers of surrounding sites. Such a structure is generally considered for simulations where only the inner cells are used for performance analysis, while the outer cells just represent sources of interference which exist in the real network. An example of a WCDMA link budget is shown in Table 3.4.

3.9.3 Dimensioning in the Transmission and Core Networks

What needs to be dimensioned in the transmission network? These include interface capacity (required and spare) and the number of RNCs needed, Iub, Iu and Iur interface capacity along with the spare capacity requirements. For capacity calculations, the traffic generated by each BS connected to the RNC and overhead (plus signalling) are needed. Factors affecting the overheads are the soft handover, re-transmission, protocol stack and signalling. As we have understood from before, the UE is connected to more than one BS (or sector) during handover, thereby increasing the capacity requirements. For dimensioning purpose, the soft handover overhead is almost 40 %. The protocol overheads consist of RLC, FP, AAL2 and ATM Protocols and are based on the type of service for which dimensioning is carried out. Signalling overheads are used for all services and some estimates are around 10 % for Iub, 1 % for Iu and 2–4 % for Iur, which can be used as the overhead interface capacities. For RNC dimensioning (similar to BSC dimensioning in GSM), the capacity of the Iub interface, number of carriers, number and type of interfaces (Iu, Iucs, Iups, Iur) and the (number of) base station capacities are calculated.

3.9.3.1 Parameter Planning

As compared to the GSM, the number of parameters increase tremendously in the UMTS transmission network. Site planning, microwave link planning and synchronization planning remain similar to that of GSM networks. However, now we have AXC and RNC parameter planning as new items. There parameters need to be planned and configured. Also, AAL2 addressing schemes, digit analysis and AAL2 routing form a part of network planning.

Parameters planning for AXC configuration include that of the PDH, SDH, IMA, AAL2, IP, CES, etc. planning. For the RNC, parameters related to the SDH, PDH, Gateway Tunneling Protocol Units, signalling units, data combining and diversity units, connection configuration, ATM, etc. are defined and configured. An example of SDH parameter planning is shown in

Table 3.5 An example of SDH parameter setting in AXC

Parameter	Range, Value
Laser mode	On/off
Type	STM-0/STM-1
Loop on	Enabled/disabled

Table 3.5. In this, parameters that are related to laser-mode operations (on or off) or the type of SDH equipment that is being used (STM-0 or STM-1), etc. need to be configured.

On the core network side, the planning is not based on generation terminology but rather in terms of releases, for example 'Release 99', 'Release 4', 'Release 5', etc. planning. More details related to these are given in Chapter 6 later in this book. Some key tasks will include ET allocation numbers and types, points of interconnection (POI), the number of PCMs towards each POI, network element naming and numbering, signalling point codes for each network element, signalling point names, signalling links and link sets (names and numbering), signalling route and route sets (names and numbering), emergency and service call routing, etc.

3.9.4 Radio Resource Management

The radio resource management (RRM) is responsible for the efficient utilization of the radio resources on the air-interface. It is used to optimize the used capacity, maintain planned coverage and guarantee the quality of service. RRM can be broadly divided into handover control, power control, admission control, load control and packet scheduling. Power control and handover control are connection-based functions while the rest are network-based functions. Power control maintains the radio link level quality by adjusting the uplink and downlink power with the basic idea to get the quality requirements with minimum transmission powers to achieve low interferences in the radio network. The basic functions are open loop, fast closed loop and outer loop power control. The fundamental of power control is to achieve the minimum signal to interference ratio that is required to achieve the desired link quality. The active state mobility of the UE is controlled by the handover control by maintaining the radio link quality and minimizing the radio network interference. The load control updates the load information of the cells that are controlled by the RNC. This information is used by the admission control and packet scheduling for controlling the radio resources. The admission control feature decides if a radio access bearer should be admitted to the network or not, thereby helping in maintaining the stability and achieve higher traffic capacity in the network. For the NRT, the scheduling (including allocating and changing the bit rates of the bearer) of the radio resources is carried out by the packet scheduler.

3.10 Network Optimization

The 3G traffic classes (conversational, interactive, streaming, background), QoS provisioning mechanisms and the possibility for customer differentiation, together with the joint management and traffic sharing between 2G and 3G networks, provide a challenging 'playground' on one hand for vendors, and on the other hand for service providers and network operators. To be able to fully utilize the resources and to focus on the service provision rather than trouble shooting, advanced analysis methods for optimization are required.

Optimization theory as a whole concentrates on methods which are able to find the optimum and do this as quickly as possible. The majority of real-life optimization problems, including UMTS radio network optimization, cannot be treated in a direct and simple way by means of analytical optimization, since the mathematical model of the network is far too complicated.

Network optimization can initially be seen as an involving task as a large number of variables are available for tuning the impacting different aspects of the network performance. Even after careful network planning, the first step of optimization should concentrate on planning criteria. This is necessary as radio propagation is affected by so many factors (buildings, terrain, vegetation, etc.) that propagation models are never fully accurate. Optimization takes into account any difference between the predicted and actual coverages, both in terms of received signal (RSCP) and the quality of the received signal (E_c/N_o).

The same qualitative metrics defined for planning should be considered: cell overlap, cell transition and coverage containment of each cell. Assuming that the user equipment is used to measure the RF condition in parallel with a pilot scanner, reselection parameters can be estimated by considering the dynamics introduced by the mobility testing. During network planning, dynamic conditions cannot be considered, as network planning tools are static by nature. Thus, simulation is used in the planning tool to achieve the dynamic effect instead of the static network. Once the RF conditions are known, dynamic simulation can be used to estimate the handover parameters, even before placing any calls on the network. However dynamic simulations would be very 'heavy' in terms of processing power and therefore planning tools which are static simulators are used. Dynamic simulators are good for R and D because of the detailed information required about algorithms, etc. is a must.

Following the system launching, there are mainly three sources of system performance verification: customer complaints, network and drive test statistics. This is used to assess prevailing network deficiencies and system trends. The first will be used to perform typically *ad hoc* optimization actions and the second to identify any future bottlenecks or drawbacks. The optimization process comprises network performance analysis and, based on analytical results, making decisions about some parameter settings. This process is repeated until the achieved results are acceptable, that is the network performance is good enough.

The overall end-to-end quality target is defined and for each service type the quality criterion is determined. The thresholds are then set for each related key performance indicator (KPI). Network performance data can be gathered from Network Management Systems (NMSs), drive tests, protocol analysers and/or customer complaints. Network reporting tools provide statistical and pre-analysed information about the quality. Based on the network configuration and status of the network, the quality in detail is analysed and individual corrections are done iteratively by solving the individual parameters affecting the reported quality. Tuning of the individual parameters or parameter sets is carried out in an iterative loop until the quality is met.

The selection of the data for performance analysis consists of two aspects; the data are selected based on functional area (or a subset of that), that is accessibility, reliability, traffic performance and distributions, and the purpose of the analysis. For getting an overall performance evaluation of the network the selection of the counters and other performance indicators is different from those one would choose for optimization or trouble shooting cases. The optimization case is more focused and thus more problem-specific indicators are required. In addition, uplink and downlink are often analysed separately. After the optimization has been performed and the changes implemented in the network, it is essential to check the function of the optimization target, but equally important it is to derive the overall performance distribution and compare it to the pre-optimization case.

Service optimization is needed to refine the parameter settings (reselection, access and handover). Because the same basic processes are used for all types of services, it is best to set the parameters while performing the simpler and best understood of all services: *voice*. This is fully justified when the call flow differences for the different services are considered, either for access or for handover; the main difference between voice and other services is the resource availability. Testing with voice service greatly simplifies the testing procedure and during analysis limits the number of parameters, or variable, to tune. During this effort, parameter setting will be the main effort. Different set of parameters are likely to be tried to achieve the best possible trade-offs: coverage versus capacity, call access (Mobile Originated and Mobile Terminated) reliability versus call set-up latency, call retention versus Active Set size, etc. The selection of the set of parameters to leave on the network will directly depend on the achieved performance and the operator priority KPIs.

Once the performance targets are reached for voice, optimizing advanced services such as video-telephony and packet-switched (PS) data service will concentrate on a limited set of parameters: power assignment, quality target (BLER target) and any bearer-specific parameters. During optimization of the PS data service the importance of good RF optimization will be apparent when channel switching is considered. Channel switching is a generic terms referring to the capability of the network to change the PS data bearer to a different data rate (rate switching) or a different state (type switching). Channel switching is intended to adapt the bearer to the user needs and to limit the resource utilization.

Saving resources will be achieved by reducing the data rate when the RF conditions degrade. By reducing the data rate, the spreading gain increases, resulting in a lower required power to sustain the link. Once the basic services are optimized, that is, the call delivery and call retention performance targets are met; the optimization can focus on service continuity, through inter-system changes, and application-specific optimization. Inter-system changes, either reselection or handover, should be optimized only after the basic WCDMA optimization is completed to ensure that the WCDMA coverage boundary is stable.

Application optimization can be seen as a final touch of service optimization and is typically limited to the PS domain. The system parameters are optimized, not to get the highest throughput or the lowest delay but to increase the subscriber experience while using a given application. Considering video-streaming, the main issue for this application-based optimization might be different applications that may have conflicting requirements. The different applications and their impacts on the network should be prioritized. Irrespective of the application considered, the main controls available to the optimization engineer are the RLC parameters, target quality and channel switching parameters. The art in this process is to improve the end-user perceived quality, while improving the cell or system capacity.

Due to the fact that in the future the operation of cellular networks will be strongly service-driven, advanced methods for network analysis are required. Compared to the current situation with the provision of voice and simple 'best-effort' data services only, the change in the operators' tasks is enormous. Effective analysis of the networks is currently challenging enough, due to the fact that the amount of data the network elements produces is very high. The operators' task is to filter the relevant information to a level that can be easily handled.

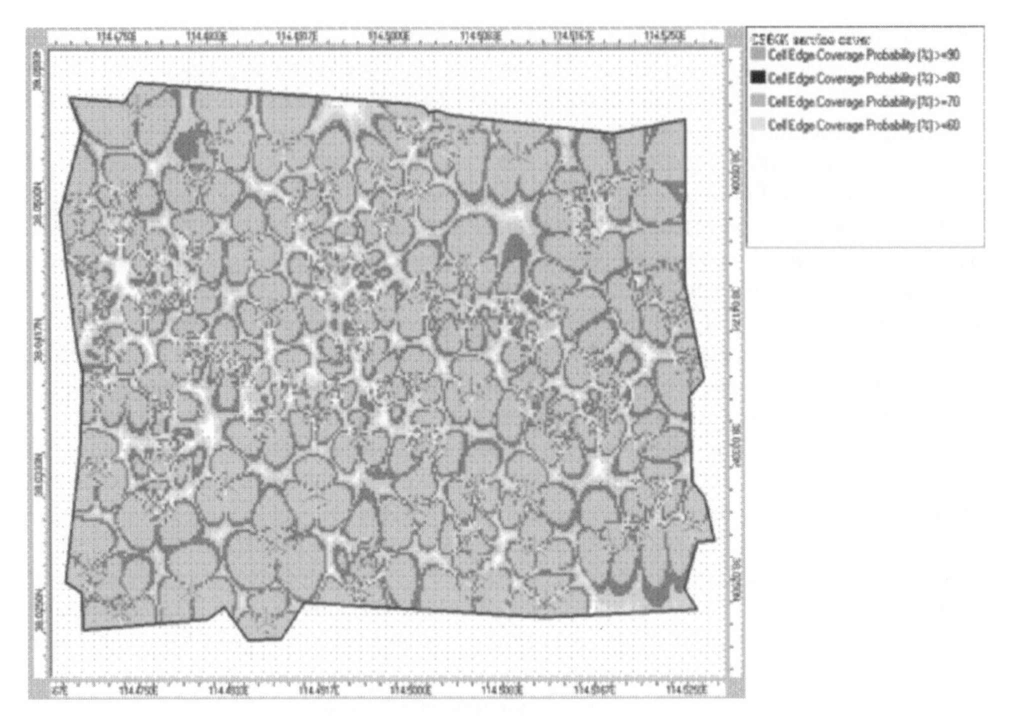

Figure 3.21 Coverage plot (cell edge).

The data set must include all the essential parts needed to conclude the service quality. The whole process from setting up to releasing the call must be included. Performance of handover and power control has an impact on the quality and the end-user experiences.

3.10.1 Coverage and Capacity Enhancements

We have seen before that coverage and capacity are interrelated (refer to the cell-breathing concept). With the coverage increase, the average transmission power of the base station increases. In addition, the capacity decreases (both the UL and DL capacities are important and the UL capacity decreases if the average path loss increases per user). An increase in coverage means an increase in link performance. E_b/N_o decrease would result in coverage increase as for a lower E_b/N_o, less power is needed, thereby covering more area. Uplink coverage can be improved by decreasing the interference margin or by reducing the base station noise figure or by increasing the antenna gain.

Also increasing the multipath diversity also improved E_b/N_o and improved protection against fast fading, etc. Other ways to improve the coverage is through antenna tilts. An example of the coverage plot at the cell edge is shown in Figure 3.21. Along with link budget calculations, load factor is used for understanding the capacity calculations. The load factor is dependent upon E_b/N_o, processing gain, interference, activity factor, etc. The orthogonality factor and SHO are a couple of other factors associated with the load factor in the downlink direction.

To improve the capacity, apart from increasing the number of cells/carriers, etc., orthogonality should be maintained. Due to the multipath aspects, some orthogonality is lost, thus increasing the interference. Transmit diversity also increases the capacity (it can also be used to increase coverage). If multipath diversity is less, then the downlink transmit diversity increases the capacity to quite an extent. Lower bit rates would also increase the capacity. This is possible by using the Adaptive Mean Rate (AMR) codes; AMR is the speech codec scheme that is used in UMTS. It should be noted that the AMR increases the capacity in terms of number of users but not in terms of data. For capacity and coverage optimization (leading to better network quality), it is very much necessary to optimize the handover control features in these networks. One important parameter to mention at this stage is the transmitting power of the CPICH. This parameter affects the coverage and it should be set as low as possible. The optimum value of this parameter would determine the coverage and capacity, that is should a new user be permitted in the cell? This parameter also affects packet scheduling. If the value of the CPICH was not optimum, then it would lead to either network being under-utilized or even huge interferences (if the number of users is more than required), thus degrading the quality of the network. These parameters also affect the call success rate and drop call rates. As in GSM networks, these two factors directly determine the network quality. Packet scheduling is one of the most important aspects when controlling the congestion in the network. This handles the non-real time packet data deciding the timings of the packet initiation and the rate at which it should be delivered. As seen in the beginning of this chapter, there are four traffic classes that have been defined and each of these classes has different applications to take care of. The NRT packet data is 'bursty' in nature, containing one or more data calls. Packet scheduling is carried out for both uplink and downlink for non-real time bearers. The packet can be scheduled by using the time division or code division techniques or both. Packet scheduling and load control (inclusive of admission control) work in tandem. A higher load would lead to a higher interference, which would mean less calls being admitted in the network. This would affect the bit rates assigned to the NRT packet data. Thus, for packet scheduling, that is less delay and higher bit rates for NRT data, load control would be an important parameter to analyse and optimize. The transmitted power in the downlink and interference power in the uplink would be important parameters for optimization. When the thresholds of these two parameters are 'crossed', preventive measures to control the load are initiated. From the perspective of packet scheduling, this is important because assigning of the higher bit rates or lower bit rates is dependent upon the load control and AC.

4

CDMA

4.1 Introduction to CDMA

CDMA stands for Code Division Multiple Access – a cellular technology which was adopted by the Telecommunications Industry Association in 1993. This technology was originally known as IS-95 and is currently known as 'CDMAOne'. The first commercial deployment was in the year 1995 and by 1998 there were 16 million subscribers which rose to 35 million by 2001. Currently, there are more than 450 million subscribers of CDMA in the world.

CDMAOne is a complete set of wireless technologies based on TIA/EIA-95 CDMA standards including both the IS-95A (Interim Standard of the US Telecommunications Industry Association) and IS-95B and related services such as cellular, fixed wireless (e.g. WLL), PCS, etc. The IS-95A standard, which is the basis of commercial second generation CDMA systems, was first released in 1993 with a revision in 1995. The IS-95B standard combined the IS-95A, ANSI-J-STD-008 and SB-74 standards in a single document. IS-95A contains the structure of wideband 1.25 MHz CDMA channels, power control, hand-offs, call processing, registration techniques, etc. IS-95B defines the standards for 1.8–2.0 GHz CDMA PCS systems. As the IS-95B systems are capable of offering 64 kbps packet-switched data in addition to voice services, it is called the 2.5G technology. . While IS-95A provides CS data connections at 14.4 kbps, IS-95B provides 64 kbps packet-switched data. Hutchinson was the first operator to deploy IS-95A while the IS-95B was first deployed in 1999 in Korea. Thus, the first generation of CDMA technology was characterized by:

- Direct sequence spread modulation (helps in spreading the original information bandwidth).
- One code per user is deployed, that is use differentiated by codes and not frequency.
- RAKE receiver used to overcome multi-path interference.
- Enhanced power control technique.
- Multi-carrier parallel transmission resulting in less frequency-selective fading.
- Multi-user detection schemes support in detecting multi-user signals.

Cellular Technologies for Emerging Markets: 2G, 3G and Beyond Ajay R. Mishra

There are many advantages of CDMA technology-based mobile networks, a few of which are as follows:

- Multifold increase in capacity of AMPS systems and GSM systems (8–10 times of former and 4–5 times of latter).
- One single frequency usage makes network planning (frequency planning) much easier.
- Enhanced call quality, privacy and coverage characteristics.
- Bandwidth on demand feature.

One main aspect to remember here is that whereas the GSM standard is the specification for the whole network, the CDMA standard is for the air-interface only.

Improvements were made in IS-95 leading to the development of IS-2000 or CDMA 2000, a set of standards that meet some criteria laid out in IMT-2000 specifications for 3G networks. IS-2000 uses the same 1.25 MHz carrier shared channel as the IS-95 standard. Another name for IS-2000 is 1xRTT ('1 times Radio Transmission Technology'). A three-fold increase in the carrier bandwidth to 3.75 MHz would increase the data rates, although this has not been commercially deployed. Another set of CDMA-based standards known as Evolution-Data Optimized (1xEV-DO or IS-856) have been created to provide higher packet data transmission rates required by the IMT-2000. CDMA systems are usually confused with the WCDMA systems (described in Chapter 3); however, both these systems are incompatible with each other.

4.2 CDMA: Code Division Multiple Access

There are two fundamentals sources behind the working of radio systems: time and frequency. When the time is divided, keeping the frequency constant (i.e. one frequency), it is called Time Division Multiple Access, while when the division is by frequency on the same time domain, it is called Frequency Division Multiple Access. However, in Code Division Multiple Access, the same frequency is available to all for all of the time. In this technique, the subscribers are separated (or identified) by the use of codes, as shown in Figure 4.1.

CDMA is a form of 'spread-spectrum'. The core principle behind a spread spectrum is the use of noise-like carrier waves having a bandwidth much more than is required for

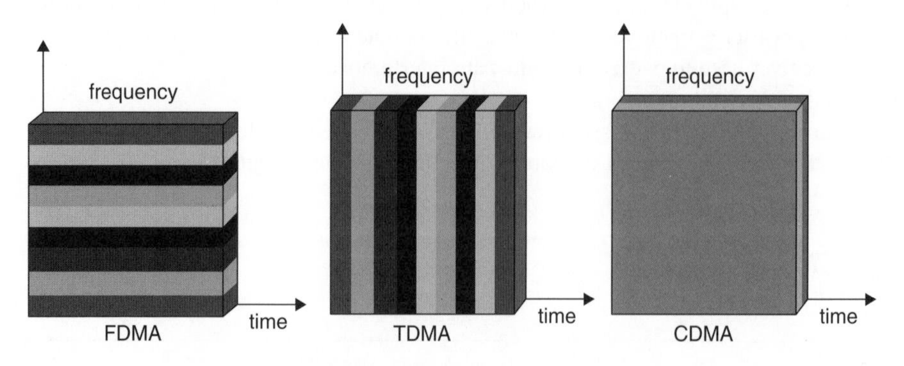

Figure 4.1 Multiple access techniques.

point-to-point communication systems for similar data rates. There are three key elements in spread spectrum communications:

- The signal occupies a higher bandwidth than what is required to send the information, thereby protecting the signal from interference, etc.
- This bandwidth is spread by means of codes that are data-independent.
- The receiver synchronizes the code to recover the data.

These (above) three steps are similar to what we have seen in Chapter 3. The signal is protected by a code called a 'pseudo-random code' or 'pseudo-noise'. There are three ways to spread the bandwidth of the signal: Frequency hopping (FH), time hopping (TH) and direct sequence (DS). In FH, the signal is rapidly switched 'pseudo-randomly' between different frequencies within the hopping bandwidth. In TH, the signal is transmitted 'pseudo-randomly' in short bursts while in DS the coding of (digital) data is done at a higher frequency. The same 'pseudo-random code' is generated both at the transmitter and receiver for the latter to extract the data from the received signal. In frequency hopping CDMA, a multi-tone oscillator generates frequencies that are discrete and the frequency pattern is chosen by the subscriber in a sequence that is orthogonal to each other. In the time hopping technique, the transmitter should provide a high dynamic range and high switching speed. Due to the high costs associated with the TH system, they are less widely implemented as compared to DS and FH systems.

A combination of the DS, FH or TH systems generates hybrid systems, as shown in Figure 4.2. Also, a combination of multi-carrier and multi-tone creates a hybrid system. In MC-CDMA, instead of applying to the spreading sequence in the time domain, it is applied in the frequency domain by mapping the chip of the spreading sequence to the OFDM sub-carrier. MT-CDMA employs the time domain spreading and multi-carrier transmission used in MC-DS-CDMA (in this, parallel transmission of DS-CDMA signals are sent using the OFDM structure). A DS-CDMA and rake receiver combination combats with the multi-path fading effects. In indoor environments, due to low time dispersion, high power consumption and difficulties in implementation, a combination of the techniques DS-CDMA, MC-CDMA, MC DS-CDMA, MT-CDMA, etc. is used to counter these effects.

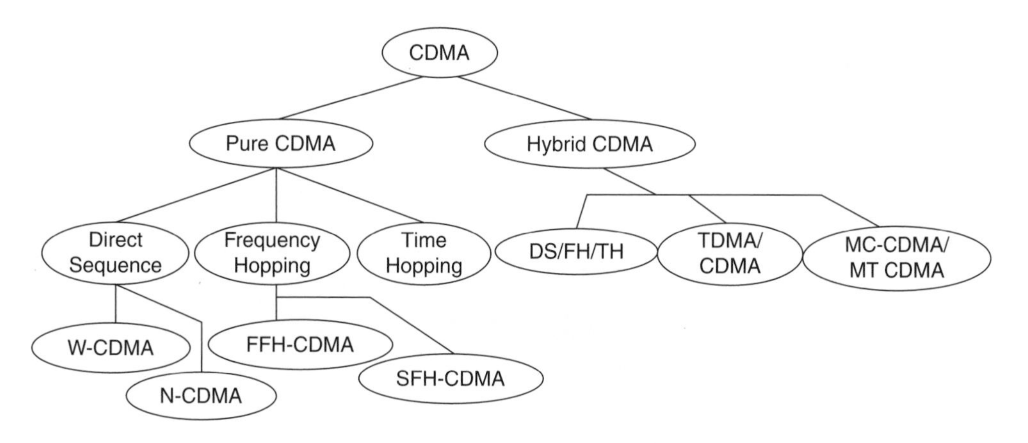

Figure 4.2 CDMA technologies.

4.3 Spread Spectrum Technique

The spread spectrum technique uses noise-like wide band signals that are hard to detect (as they are 'noise-like'). These signals are made to be much wider than the information they are carrying so that they are noise-like signals. The codes used for spreading, called spreading codes are also called 'pseudo-random codes' as they are not real Gaussian noise. The power transmitted by spread spectrum transmitters is similar to that of narrow band transmitters. In fact as these are wider signals, they transmit at lower spectral density. One important characteristic of the signals is that both the spread and narrow band signals can occupy the same band with almost no interference. How does this spreading happen? As shown in Figure 4.3, for each channel and successive connection, a pseudo-random code is generated which is modulated by the information data. This basically results in spreading of the information data. The resulting signal modulates a carrier which is amplified and broadcasted. In the reception side, the carrier signal is received and amplified. Then it is mixed with a local signal to get the original spread digital data. This is then mixed with the pseudo-random codes and the original information data are received.

4.3.1 Direct Sequence CDMA

Each of the CDMA technologies (direct sequence, frequency and time hopping) is based on a corresponding spread spectrum technique. In the direct sequence technology, the data modulated signal is modulated a second time (digital phase modulation) using a wide spread signal (pseudo random sequence) as shown in Figure 4.3. This is known as a direct sequence spread spectrum technique. As compared to the other techniques, this is relatively simple and does not require a high speed, fast settling frequency synthesizer. The spreading modulations of DS could be amplitude, frequency or BPSK modulations. It is usually BPSK modulation as it is possible to implement it with lower costs. As the bandwidth of the spreading signal is much larger than the original data, the transmitted signal bandwidth is dominated by the

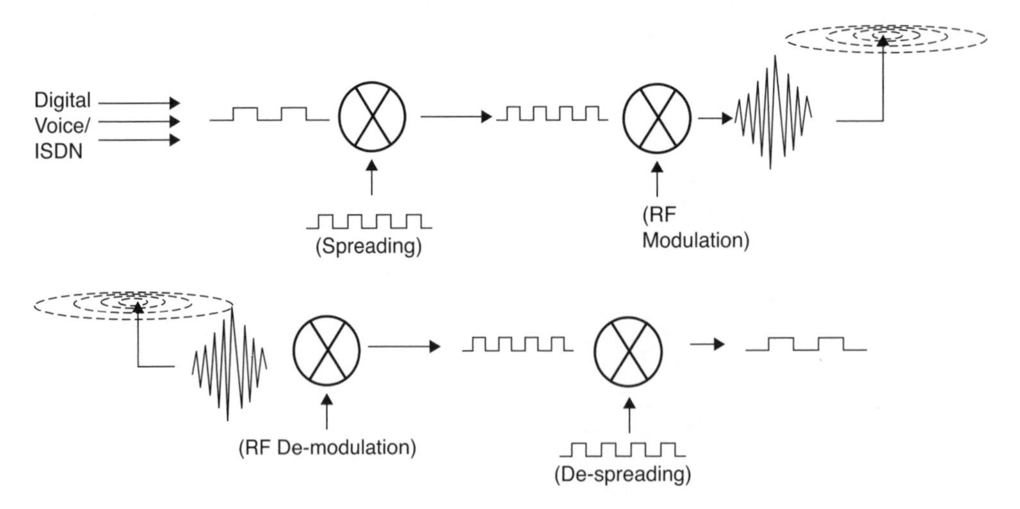

Figure 4.3 Spreading and de-spreading.

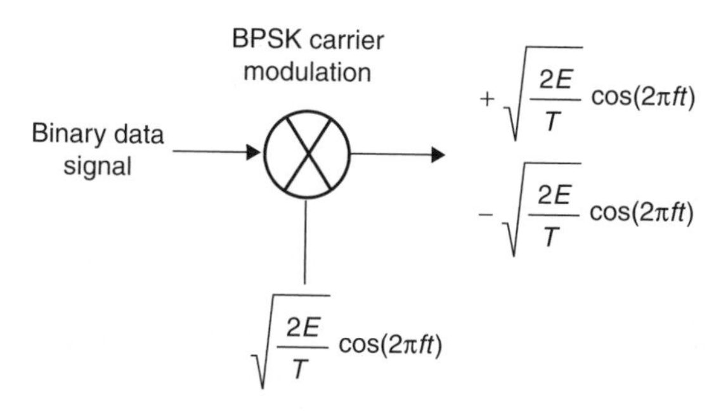

Figure 4.4 BPSK modulator.

spreading signal. Each bit of spreading sequence or code is called a 'chip'. Another important term to understand is processing gain. This is the ratio of the spread signal bandwidth to the un-spread signal's narrow bandwidth. In many cases, it is also defined as the ratio of the chip rate to the original data information. The processing gain cannot be increased to increase the performance of the system. The main reason is the increase in system complexity as the chip rate needs to be increased for increasing the processing gain, thereby increasing the sampling rate and hence the load at the CPU (or DSP chip). As the increase is not linear, a two-fold increase in the chip rate would lead to more than double the load at the CPU. There are a couple of direct sequence spreading techniques that we will look into briefly here. They are Binary Phase Shift Keying and Quadrature Phase Shift Keying. BPSK modulation is quite simple to implement as the BPSK modulator is a simple multiplier, as shown in Figure 4.4.

In BPSK, when the transmitted binary data is +1 it is sent in a positive co-sinusoidal form and when it is –1, it is negative co-sinusoidal. When the binary data is +1, the phase is zero degrees and when the binary data is –1, the phase is 180 degrees. The amplitude is thus between +1 and – 1 and the phase is between zero and 180 degrees. The demodulator decides on the type of bit (+1 or –1), depending upon the phase of the bit by judging whether it is closer to zero or 180 degrees. The input to the modulator is the data signal (+1 or –1) which is then multiplied by $\sqrt{2E/T} \times \cos(2\pi ft)$ giving a corresponding modulated signal. BPSK can transfer one bit of information (+1 or –1) per symbol bit. QPSK is able to transfer two bits per symbol, hence increasing the bandwidth efficiency by two. There are four symbols that are used to send these bits, 0, 1, 2 and 3. Mapping between the transmitted symbol and represented bits is shown in Table 4.1. The data bits are sent to a de-multiplexer whose output will give a bit stream, separating the odd and even bits. The odd bits are multiplied by a quadrature carrier and the even ones are multiplied by an in-phase carrier, and both of these are combined to form the QPSK signal.

Table 4.1 BPSK modulation

Transmitted signal	Bits
0	+1, +1
1	−1, +1
2	−1, −1
3	+1, −1

The BER performance of the BPSK DS-SS system is the same as that of the QPSK DS-SS system, although the bandwidth of the QPSK system is double that of the BPSK system.

4.3.2 Frequency Hopping CDMA

In frequency hopping CDMA the spreading of the spectrum takes place by switching the carrier frequency periodically from one to another. This technique is also called code controlled multi-frequency FSK modulation. This is quite similar to the FSK technique except that the number of frequencies is quite large. This frequency is selected from a group of frequencies that are separated by the width of the data modulation bandwidth. The spreading code in the FH systems controls the appearance sequence of the carrier frequencies, that is it does not modulate the signal directly. On the receiving side, the signal is mixed with the local oscillator that is synchronously hopping with the received signal. The FH system has a frequency synthesizer and sequence generator according to which the synthesizer generates the frequencies. As it is difficult to develop a fast-settling synthesizer with a large number of frequencies, FH systems are difficult and costly to implement. The processing gain in an FH system is defined in two ways: one for contiguous frequencies (i.e. all carrier frequencies are evenly spaced) and the other for non-contiguous frequencies. For the contiguous case, the processing gain is similar to that of a direct sequence spread spectrum. However, for the non-contiguous system, the total number of available frequencies is the processing gain for the system. FH CDMA systems, due to their complexity and costs, have been used in only a very few systems, for example 'Bluetooth'. They use fast frequency hopping, allowing a larger synchronization error and hence are an application of the FH CDMA technology.

4.3.3 Time Hopping CDMA

In the time hopping spread spectrum technique, a code sequence is used to key the transmitter on and off where on/off follows a pseudo-random code sequence. This leads to a code sequence that is used at the receiver to decode the transmitted signal. The pulse generator used in the TS-SS technique should be able to produce very narrow impulses having widths or the order of nanoseconds and high timing accuracy. Such types of generators are difficult to produce, thereby making the implementation of TS CDMA systems almost impossible.

4.4 Codes in CDMA System

IS-95 is an asymmetric system, that is both the forward and reverse links have different link structures and different codes to channelize the individual users; the forward link uses Walsh codes and the reverse link uses PN codes.

4.4.1 Walsh Codes

Walsh codes are used in CDMA systems to separate individual users which are occupying the same frequency band. These are a set of 64 binary sequences that are orthogonal to each other. These codes are generated by a 'Hadamard matrix'. In this, higher-order matrices are

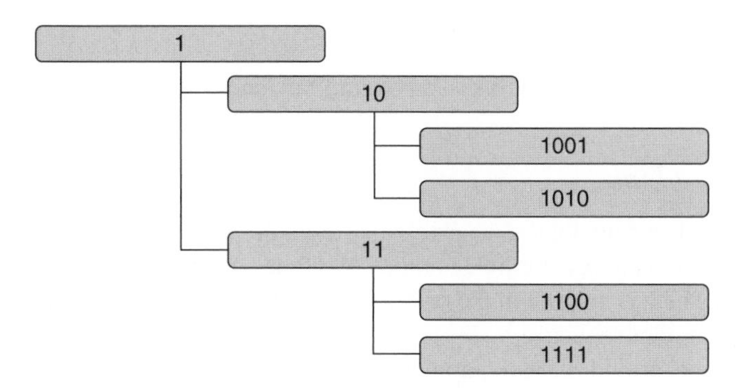

Figure 4.5 Walsh code.

generated by lower order matrices using recursion. In the first matrix, H_{2N}, the bar element contains the inverted elements *Hbar* and H_N. For deriving four orthogonal Walsh sequences, w_0, w_1, w_2 and w_3, a Hadamard matrix of the order of four is generated. In the orthogonal sequence, the correlation should be very small or zero, the difference between +1 and –1 should be either 1 or zero and the scale dot product of each code should be zero. The pilot channel in the forward link is identified by the Walsh sequence $w_{0.}$ An example of a Walsh code is shown in Figure 4.5.

The forward link in IS-95 uses a set of 64 orthogonal Walsh sequences. Plus, as w_0 is not used to transmit any baseband information, the number of channels in the forward link is 63. Walsh codes are used for channelization as well. If there are two subscribers, each sending a separate message, a separate Walsh code is used for each. The chip rate of a Walsh code is four times that of the message bit rate (i.e. a processing gain of four). Orthogonality during all stages of transmission holds the key to channelization of the Walsh codes.

4.4.2 PN Codes

IS-95 uses a 16-bit short PN code in the reverse link (or uplink). This is used to identify the different base stations or the cells where different base stations are located. This is done by assigning an offset of this code to the base stations under a common time reference. On the reverse link, these codes for used for signal robustness by mobile stations. For spreading modulations as signature codes for identifying base stations, 42-bit long PN codes are used. The length of the code is 2^{42}–1 chips. The PN codes are generated from the linear feedback registers. The maximal length of the code is dependent on the number of registers used:

$$L = 2^N - 1$$

where N is the number of registers used. If the number of registers is 3, then the length of the code is 7. The PN codes also satisfy the same conditions as that of the Walsh codes, that is the cross correlation needs to be small or zero, the code sequence needs to have equal numbers of 1 and –1 or the difference to be one and the scaled dot product. PN codes are used for channelization as well. The PN code rate is 1.2288 MHz.

4.5 Link Structure

In CDMA systems, there are two links – a forward link and a reverse link.

4.5.1 Forward Link

The forward link has four logical channels that include pilot, synchronization, paging and traffic channels while the reverse link contains two types of channels that include access and traffic channels. There is no pilot in the reverse link as it is not possible for the user equipment to transfer its own pilot sequence.

The pilot channel provides user equipment with phase and timing sequences. As mentioned before the pilot channel is identified by the Walsh sequence w_0. This contains no baseband information. The baseband stream is spread by the Walsh sequence and is then multiplied by the PN sequence. These pilot channels are transmitted by the sectors of the base stations. The baseband information is carried by the synchronization channel and it is responsible for synchronization of the user equipment with the base station. The synchronization channel has a data rate of 1.2 kbps. This information is 'error-protected' and interleaved. Each short PN sequence is synchronized by the synchronization channel frame. The user equipment, which reaches alignment with the synchronization channel, is able to read the synchronization channel message. The synchronization channel contains a 15 bit SID (system identification number) and a 16-bit network identification number (NID). This also contains other bits such as the pilot bit, system time and long code generator. The paging channel also carries baseband information such as access information, surrounding cells data, etc., apart from performing a paging function. The data rate of the paging channel is higher than the synchronization cells, i.e. to 4.8 kbps or 9.6 kbps. However, within a system, all paging channels have the same rate. Each CDMA carrier contains a maximum of seven paging channels. Pilot, synchronization and paging channels are broadcast channels while the traffic channels actually carry the user traffic and control channels. Apart from this, signalling messages are also sent on the traffic channels. The data rates are 1.2 kbps, 2.4 kbps, 4.8 kbps and 9.6 kbps. Speech signals are coded into frames of 20 ms containing 24, 48, 96 and 192 bits per frame. Except for 9.6 kbps, in the rest all data rates, symbols, etc. are repeated. This is done to reduce the interference by reducing the transmitted per repeated signal. The next step is to interleave the data to prevent against fading and then the signal is scrambled. For good communications, the coded data is multiplexed with the power control sub-channel to adjust the transmitted power of the user equipment so that the received E_b/N_o is within the acceptable limits. This process of replacing the code symbols with the power control bits does introduce errors which are corrected by the one-half rate code. The process to remove the code symbols is called 'puncturing'. This is then followed by spreading using Walsh codes. The data are then further spread by short PN sequences to provide the second layer of isolation, distinguishing between different transmitting sectors.

4.5.2 Reverse Link

In the reverse link, instead of the Walsh codes, long PN sequences are used for distinguishing users from each other. As mentioned before there are two channels: access and traffic. The access channel is used by the user equipment to access the network. One base station supports up to 32 access channels and the user equipment is distributed 'pseudo-randomly' in the cell in such a way that these access channel resources are loaded equally. The data rate is fixed at 4.8 kbps (96 bit frame presented every 20 ms). A one-third rate convolution encoder is used to pass the data to make the error coding more robust. After this, the data go through symbol repetition and block interleaving. This is followed by a 64-ary orthogonal modulator, with 64 Walsh codes being used to modulate groups of six symbols. Walsh functions are used so that the base station can identify between different symbols. However, one should remember that the Walsh codes used on reverse links are totally different to the ones used on the forward link. After this, the modulation codes are spread by the long PN sequence (2^{42}–1). After this, the data are passed through the baseband filters (one stream is delayed by half a chip). The outputs are used for modulating the quadrature carriers and the resultants are summed up. The access channel message contains 8 bit message lengths, an 8–842 bits message body and a 30 bit CRC. These messages are of two types: request and response messages. Response messages are transmitted in response to the message from the base stations while the request message is generated by the user equipment autonomously. The traffic channel carries speech, user and control data between the user equipment and base station. The format of the traffic channel is similar to that of the access channel except with the addition of a data burst randomizer which utilizes the voice activity factor. Here the Walsh codes are used to carry information while in the forward links they are used to distinguish between various users. Also, the user equipment transmits only one copy of the code symbols, discarding the repetition which is quite different to the way things are done on the forward link (to maintain constant energy per data bit, the base station reduces the transmitted power). Thus, the repeated symbols are removed in a random manner by a process called 'burst randomization' and is carried out by a 'data burst randomizer'. After this, the data are spread by the long code generator and the resulting data streams are used to modulate the phase of the radio carrier, similar to that of the access channels. The forward and reverse link channel parameters are given in Table 4.2.

4.6 Radio Resource Management

4.6.1 Call Processing

In this feature, we will look into how the call is set up in a CDMA system. There are a few steps for this, including cell selection, idle, access and traffic modes.

4.6.1.1 Cell Selection

This is the process in which the mobile finds and selects cell/base stations and networks that it will identify itself with. The mobile starts the selection process with finding the most appropriate network based on the subscriber requirements. Once the network is found, the mobile tries to find the primary carrier frequency followed by the finding of a pilot channel (as described before). The next step involves demodulating the corresponding synchronization channel. For synchronization with the network, the synchronization channel contains various parameters, including a 15-bit system identification (SID) needed to identify the network, a system time parameter which is used to fully synchronize the mobile to the network and a paging channel data rate which indicates the rate used by the paging channel (4.8 kbps and 9.6 kbps). After this the mobile gets connected to the network, the next stage, that is the idle stage starts.

Table 4.2 Forward (a) and reverse (b) link channel parameters

(a)

	Synchronization	Paging		Traffic			
Data Rate (bps)	1200	4800	9600	1200	2400	4800	9600
Code Repetition	2	2	1	8	4	2	1
Modulation Symbol rate (syms/s)	4800	19 200	19 200	19 200	19 200	19 200	19 200
PN (cpb)	256	64	64	64	64	64	64
PN Chips/Mod. Symbol	1024	256	128	1024	512	256	128

(b)

	Access	Traffic			
Date Rate (bps)	4800	1200	2400	4800	9600
Code Rate	1/3	1/3	1/3	1/3	1/3
Symbol Rate before repetition	14 400	3600	7200	14 400	28 800
Symbol Repetition	2	8	4	2	1
Symbol rate after repetition	28 800	28 800	28 800	28 800	28 800
PN (cpb)	256	128	128	128	128
PN Chips/Mod. Symbol	256	256	256	256	256
Transmit Duty Cycle	1	1/8	1/4	1/2	1
Code Symbol/Mod Symbol	6	6	6	6	6

4.6.1.2 Idle Mode

After getting connected to the network, the mobile monitors the paging and pilot channels. The paging channel is monitored to detect incoming calls while the pilot channels are monitored to have the mobile connected to the right base station. Transmission of the paging channel is done in slots of 80 ms each. When the mobile monitors the paging channel continuously, it is said to be in a 'non-slotted mode', while when it monitors the paging channels in the assigned paging channel slots it is said to be in the 'power saving slotted mode'. In the start, the mobile is not aware of the arrangement of the paging channel at the base station and hence gets tuned to the paging channel present on the Walsh code sequence. After this, it only gets allocated to the right paging channel. This paging channel contains information related to the network as well as the following: system parameter messages (contains the number of paging channels at the base station, registration channels and handover threshold), access parameter messages (parameters related to the system access), channel list messages (list of CDMA frequencies at the base station) and neighbour list message (details to the neighbouring base stations such as a pilot PN offset). Even when the mobile is in the idle state, handoffs do take place when the mobile moves from one cell area to another. This is called 'idle handoff'. This is done by maintaining three sets of pilot PN offsets – active set, neighbour set and remaining set. Also, three search windows are also used to search the pilots in the respective sets. There is no soft handoff in the idle state as the mobile monitors the paging channel only on one base station.

4.6.1.3 Access Process

In this process the mobile gets access to the network (initiated by either a paging channel or by subscriber call) using access channels. When the mobile sends a message on the access channel which, when this is accepted by the base station, it is responded to by a paging channel message called a 'configuration message'. The configuration message on the paging channel also contains the page message (contains IMSI), page-slotted message and general page message. A page response message is sent by the mobile so that the base station 'originates' the call following which the base station sends a channel-assignment message. After this the mobile tries to receive the traffic channels transmitted by the base station at a rate of 1.2 kbps. On the reverse link, the mobile then transmits the traffic channels at 9.6 kbps. When the base station receives the traffic channels from the mobile, it sends the acknowledgment order to the mobile indicating the establishment of the call.

4.6.1.4 Traffic Channel

After the mobile is successfully attached (to the base station), the traffic channels (forward and reverse) start communicating with the base station. This starts with the traffic channel initialization state where the mobile checks if it can receive the information from the base station (i.e. two consecutive good frames on the forward channel within 200 ms). After this, transmission begins on the reverse channel. For a mobile-originated call, the conversation will start while for a mobile-terminated call, the 'waiting stage' will start. For releasing a call, the mobile send the release message on the reverse traffic channel while for a base station to release the call, a similar message is sent on the forward channel. A simplified process for a mobile-originated call is shown in Figure 4.6.

4.6.2 Power Control

As mentioned before, CDMA is an interference-limited system and the total interference power at the base station is the summation of all the power received from all subscriber mobiles in the cell. Power and its control play a very important role in defining the interference and capacity of the CDMA system. Now as all subscribers use the same frequency band, power from one subscriber is treated as interference by another; hence the power transmitted by the subscribers needs to be controlled in an extremely efficiently matter. The power control is used to rectify the near–far effect. What is the near–far effect? Let us assume that there are two subscribers in the network, A and B, where B is a source of interference for A, as shown in Figure 4.7. If B is closer to the base station, in the absence of a power control feature, the received power would be higher for B than for A. This would result in a disparity between the SNRs of the two subscribers and B would be more 'clear' than the intended subscriber A. This jams the systems and blocks the entry of the subscriber C. This kind of problem is more prevalent in the DS systems than in FH systems. It is quite difficult to mitigate the far–near effect in a CDMA system; however, by using the fast power control mechanisms this problem can be overcome. The open loop power control mechanism is used for the initial coarse power control and for the fine power level tuning the closed loop power control is used. In the open loop power control, the mobile is able to set its transmitted output power (forward and reverse) to

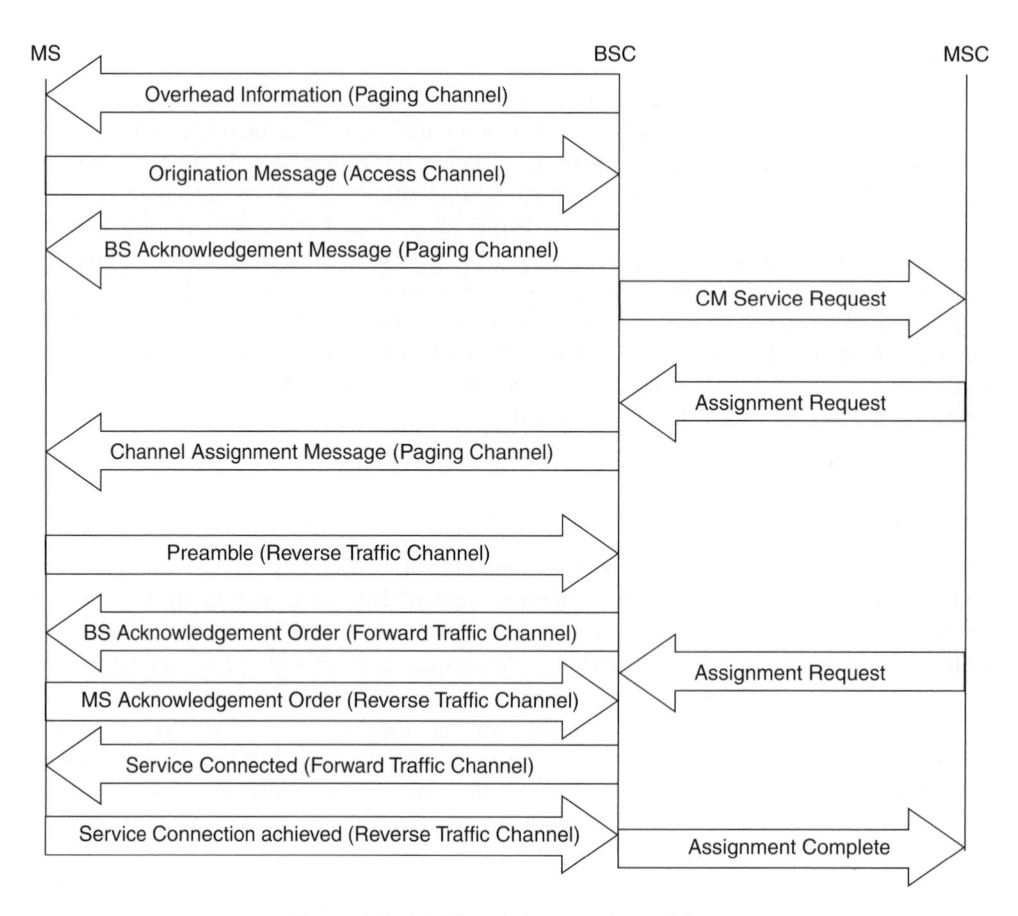

Figure 4.6 Mobile-originated voice-call flow.

a specific value when accessing the network. In order to keep the *S/I* ratio within limits, the mobile equipment receives the instruction in downlink to adjust its output power. This is called 'closed loop power control'. This is done in steps of 1, 2 and 3 dBs at a frequency of 1500 MHz. For maintaining the quality of communications (both in forward and reverse directions) the outer loop power control is used.

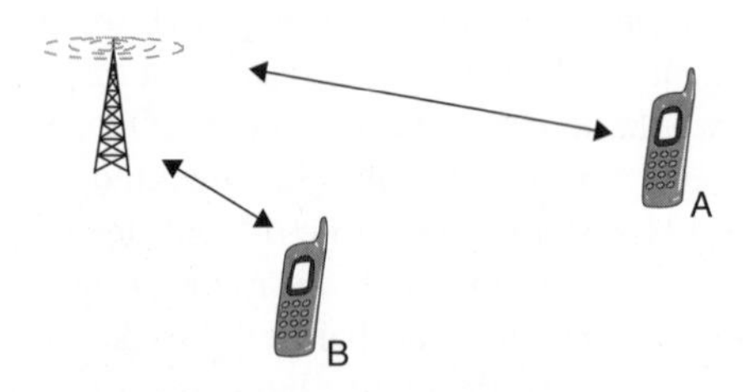

Figure 4.7 Near–far effect.

The power control phenomenon starts right at the time when the mobile is trying to get connected to the network. It can try to connect faster by transmitting a higher power and thereby increasing the interference or it can transmit a lower power to keep the interference at a lower level but at the same time decreasing its chances to get connected. However, in practice the mobiles transmit access probes wherein a small amount of power is transmitted towards the base station and this is increased progressively until the mobile is connected ('receives acknowledgement') to the base station. Since this is a mobile controlled operation, this is called 'open loop power control'. Even after the connection, the process of open loop power control goes on by the mobile in order to check the received power and adjusts the transmitted power accordingly and counters the slow fading effects. Fast fading or Rayleigh fading is frequency-dependent and cannot be controlled by the open loop power control. It is done by using the closed loop power control which unlike the open loop (which is controlled by the mobile only) is controlled by both the mobile and base station. The base station monitors the reverse link to measure the quality. When the quality degrades, on the information provided by the base station through the forward link, the mobile decreases its transmitted power. As the base station transmits all channels coherently in the same band, power control is not needed in the forward link. However, practically this is needed as sometimes the signal power coming from the mobile to the base station might suffer fading in such a way that the signal power equals that of the thermal noise. The forward link power control is not that stringent as the reverse link.

4.6.3 Handoff

This is also called 'handover' in UMTS networks and happens when the mobile station moves from the coverage area of one base station to another. The types of handovers in CDMA and WCDMA are similar. We have already understood various types of handovers in Chapter 3. The handover process works on the forward link (downlink) channel. The forward channel carries traffic and overhead information. The overhead information consists of pilot, synchronization and paging channels. The pilot channel is an un-modulated spread signal. When a mobile is switched on, it latches onto the nearest base station, that is the strongest pilot. When a mobile moves from one base station to another, the pilot level goes down and the mobile searches for a stronger pilot to latch onto, making sure that the information signal flow does not stop when the movement happens from one base station to another. There is a pilot database and the mobile maintains four sets of it: active, candidate, neighbour and the remaining set. The active set pilots are the ones that are currently being used to latch the mobile ones, the candidates are the ones that are very strong but not as strong as the active ones, the neighbours are the next in line as likely candidates for being used in handovers, while the rest all form part of the remaining set list.

4.7 Planning a CDMA Network

4.7.1 Capacity Planning

Although CDMA is an interference-limited system, the capacity of the system is dependent on various other factors such as power control accuracy, demodulation, etc. How do we calculate the total number of users in a given band?

The energy per bit user per noise power density is given as:

$$\frac{E_b}{N_o} = \frac{S}{RN_o}$$

where E_b can be understood as the average modulating signal power (S) allocated to each bit duration (T). The bit rate R is the inverse of the bit duration and hence E_b can be calculated as:

$$E_b = \frac{S}{R}$$

The noise power density N_o can be calculated as the total noise power (N) divided by the bandwidth (W):

$$N_o = \frac{N}{W}$$

Thus, E_b/N_o can be calculated as:

$$\frac{E_b}{N_o} = \frac{S}{N}\frac{W}{R}$$

Thus, there are two factors on which E_b/N_o depends: the signal-to-noise ratio (S/R) and the processing gain (W/R). The signal power transmitted by all mobiles is controlled in a way that the power received from all mobiles at the base station is equal. If T_u is the total number of users in the network, then the total interference power in the band is equal to the total power received from all the users. For one user, the signal-to-noise ratio is given as:

$$\frac{S}{N} = \frac{1}{T_u - 1}$$

Thus, E_b/N_o is given as:

$$\frac{E_b}{N_o} = \frac{1}{T_u - 1} \times \frac{W}{R}$$

Therefore, the capacity (total number of users in a cell) can be calculated as:

$$T_u \approx \frac{W/R}{E_b/N_o}$$

This is the capacity of one single cell; however the subscribers from other cells cause interference to this cell. Hence, we need to take a loading effect into consideration. The loading factor η is taking into consideration via the following equation:

$$\frac{E_b}{N_o} = \frac{1}{T_u - 1} \times \frac{W}{R} \times \frac{1}{\eta + 1}$$

The loading factor η is between 0 and 100 %. Usually a loading factor of 50 % is used in the network design. Some factors decrease interference, such as sectorization. A three-sector base station would experience less interference than the omni-directional cells. This is possible as the antennae in various sectors reject the subscribers that are not within its antenna pattern. A three-sectored base station would reduce the interference by a factor of three and a six-sector station would reduce the interference by a factor of six. This is called 'sectorization gain' (λ), making E_b/N_o:

$$\frac{E_b}{N_o} = \frac{1}{T_u - 1} \times \frac{W}{R} \times \frac{1}{\eta + 1} \times \lambda$$

The use of another factor impacts the interference – this is called the 'voice activity factor' (υ). The output of the vocoder is adjusted according to the speech pattern of the subscriber, that is it is higher during speech and lower during silence (i.e. when the subscriber is not speaking). By employing this technique, the total interference power is lowered in the system:

$$\frac{E_b}{N_o} = \frac{1}{T_u - 1} \times \frac{W}{R} \times \frac{1}{\eta + 1} \times \lambda \times \frac{1}{\nu}$$

As seen in the above equation, by varying the interference in the system, the total capacity can be varied.

4.7.2 Parameters in a CDMA Network

The optimization process in a CDMA network is similar to the optimization processes discussed in Chapters 2 and 3. So, we will not go into the process details here. However, let us discuss the few important parameters that would need to be tuned and optimized in CDMA networks. These parameters include the following:

- neighbour list entries;
- antenna configurations;
- hard handoff and soft handoff thresholds;
- active set and neighbour set search window sizes;
- access channel nominal and initial power settings;
- digital gain settings for pilot, page and synchronization channels;
- power settings/ BCR attenuation.

Neighbour lists should be complete and updated all the time. As we have seen before, they contain information related to the pilots and facilitate idle handoff pilot searching. The parameter used is the *Maximum Neighbour List Number sent to the Mobile* which is sent on the traffic channel. The neighbour list is constructed by using a neighbour list algorithm. This assumes importance in cases where there is a big neighbour list as the pilots are ranked during the construction of a combined neighbour list. Each entry in the neighbour list has an associated translation value called the *priority group* which supports in breaking ties during the construction of a combined neighbour list.

Antennae play a very important role in the link budget calculations which affect the coverage and capacity of the network. To improve the performance of the network, various techniques are used, for example to transmit diversity on the forward link. In this, the transmit signal from the base station is transmitted to the receiving (mobile) antenna via two antenna paths. Both the transmit antennae are spatially separated or 'polarization separated'. This improves the performance of the stationary/low mobility user. Two techniques are used: 'orthogonal transmit diversity' and 'space time spreading'. In the former, alternate coded and interleaved symbols are transmitted on alternate antennae while in the latter *all* coded and interleaved symbols are transmitted on alternate antennae. In both techniques, additional Walsh channels are not used. SMART antennae are used to enhance system performance. They use a multiple element system and combine the base station architecture and signal processing. This is capable in suppressing interference, thereby improving capacity and coverage. 'Antenna tilting' is another way to improve the performance as it is one way of controlling overshooting and equate forward/reverse coverage. Antenna azimuth adjustments also help in improving coverage and reduce overlap.

Intra-CDMA network and inter-CDMA network handoffs play an important role in deciding the network performance. The channel list message contains the list of all the appropriate carriers and the mobiles can *hash* across carriers before attempting to access the system. A feature like *hashing over paging carriers* allows the system to perform *hashing functions* which means that planning engineers would have the flexibility of planning paging features and at the same time reduce the load on the paging channels in areas with multiple carriers. Algorithms that would send information related to the strongest pilot sector in the active set are implemented. Features such as *triggered handoff escalation* would allow the inter-CDMA handoff and inter-generation (2G–3G) handoff as well. Other features include the distance-based handoffs, multiple pilots and inter-frequency handoffs. Soft handoff parameters include the ones related to adding the base station to the active set candidates, the threshold used to remove the base station from the active set candidates, related timers and related pilots. There has been software developments related to soft handoffs. IS95B brings dynamic threshold parameters to increase the efficiency of soft handoffs that includes features like soft swap (used to swap the outgoing pilot active set with a new one), early soft handoff (takes place as soon as the cell site call processing acquires forward and reverse traffic), six-way soft handoff (increases the number of soft handoff legs) and base station-assisted soft handoff (this minimizes the number of soft/softer handoffs – increases link capacity).

Some key parameters are related to the PN offset for pilot channel identities. The mobile maintains four sets of the pilot during its communication with the sector, namely active, candidate, neighbour and remaining set. For demodulation, the active set is used by the mobile and the remaining three set of pilots are handoff candidates which are scanned continuously for the strongest signal by the mobile. The pilots of the candidate and neighbour sets are defined by their offsets while that of the remaining set are defined by a parameter with an integer value less than 16. If the strength of the pilot is more than the threshold, then this is reported to the serving sector and this sector initiates the handoff procedure. In the case of interference from other base stations, the mobile combines the strongest of the three/four multipath signals by the use of 'rake receivers'. An average SIR of about 27 dB is used to prevent interference from the neighboring PN pilot sequence. This is quite high but is needed as the PN pilot sequence is time-aligned from the serving base station. Although it should not be time-aligned, in big networks this is not possible as there are only 512 PN offset indices and a unique PN offset index cannot be assigned in a way that interference can be controlled only by reduction in processing gain.

To gain an access channel, access probes are sent by mobiles on sector/carriers using W32-access channel public PN long code. To minimize the collision between the signals from subscribers generating a call or responding to a page, randomization is implemented. This optimization is necessary to minimize interference level and delay while increasing the access success rate. Some key parameters include the following: maximum access probe sequence request; maximum access probe sequence response; number of access probes (recommended value 5); access channel probe back-off range; initial power offset for access; nominal transmit power offset; transmit and receive antenna propagation delay, etc. In important places in a city, for example 'hotspots', there is always a possibility of overloading of the access channel. Another parameter takes care of such a situation and makes sure that access channel overload does not happen by automatically adjusting persistence parameters on a per-sector basis.

The qualities of the forward and reverse links are also monitored to maintain overall quality of the network. The forward link quality is needed for FER performance and rake receivers play an important role in reducing the fading effects. The link quality can degrade due to variations in the SIR required for the forward link FER, for example 1 %. Variations due to mobiles can be over 16 dB. In peak hours the forward links are under pressure due to lower speeds on the mobiles. Sometimes the forward overload control can degrade the transmit power of the mobile, thereby impacting its capability to reach the desired FER. The forward and reverse coverage should be adjusted similarly. The poor quality of the reverse link can be detected by measuring the RFER. Some reasons for this include loss of base station antenna, improper balance between the forward and reverse links, problems in the closed loop power control, etc.

Proper settings are required for the coverage and performance of the forward link in terms of *maximum power output* and *pilot power*. The pilot channels need to have adequate power (E_c) for overcoming the interference (I_o) around them. The transmit power of the cell can be adjusted by using baseband and RF unit (BCR) attenuation. BCR should be used when both the forward and reverse links are balanced.

4.8 CDMA2000

The key capability of IMT-2000 is to deliver a system with high spectrum efficiency and capacity (better than the second-generation systems). Five radio interfaces were approved by the ITU and three of them are based on CDMA (CDMA2000, TD-SCDMA and WCDMA). The world's first commercial 3G system was launched by SK Telecom (Korea) in October 2000 and was based on CDMA2000 1x technology (see Figure 4.8).

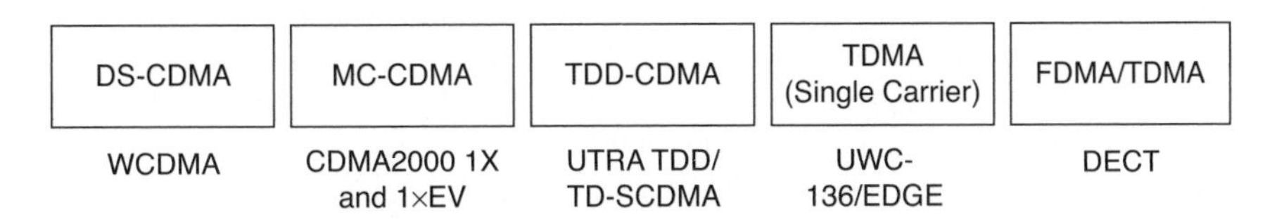

Figure 4.8 IMT-2000 terrestrial radio interfaces.

The CDMA2000 3G wireless system is based on the code-division multiple access (CDMA) system. The CDMA2000 system delivers high-bandwidth data and voice services to users of mobile equipment. CDMA 2000 comprises the following technologies:

- CDMA2000 1X.
- CDMA2000 1xEV-DO Technologies
  - CDMA2000 1xEV-DO Rel 0;
  - CDMA2000 1xEV-DO Rev A;
  - CDMA2000 1xEV-DO Rev B.

4.8.1 CDMA2000 1X

CDMA2000 1X was the first IMT 2000 technology, deployed in 2000. This is a spectral-efficient wide area network technology. Plus it supports packet data speeds more than 300 kbps in a single channel, 1.25 MHz channel. It also supports bi-directional peak data rates up to 153 kbps and an average of 60–100 kbps in a 1.25 MHz channel. It can also support up to 40 simultaneous voice calls per single 1.25 MHz FDD channel.

4.8.2 CDMA2000 1XEV-DO Technologies

CDMA2000 is a hybrid 2.5/3G technology – it considered 2.5G technology in 1xRTT and a 3G technology in EVDO and is standardized by 3GPP2. The CDMA2000 standards CDMA2000 1xRTT, CDMA2000 EV-DO and CDMA2000 EV-DV are approved radio interfaces for IMT-2000 standards and a direct successor to 2G CDMA, IS-95. CDMA2000 is loaded with advantages such as higher voice quality, increased data throughput capacity, differentiated VAS, spectrum efficiency, improved security, flexibility in architecture, etc. CDMA2000 is capable of providing voice capacity that is almost three times that of CDMAone (IS-95) by using selectable mode vocoders and various antenna diversity schemes. This technology delivers low data rate services at 64 kpbs to 144 kbps. For indoor environments, 2 Mbps would be possible while 144 kbps would be possible for all environments. The spectrum consists of a 1.25 MHz channel. Chip rates are multiples of the CDMAone system of 1.2288 Mcps with a carrier spacing of 1.25N MHz where N is 1, 3, 6, 9 and 12. There are 35 traffic channels per sector per radio frequency. The number of voice subscribers that can be supported by CDMA2000 is double than that of CDMAone (for $N = 1$) as it uses the QPSK modulation. Other features include lower code rates (1/4 rate), transmit diversity, faster power control, etc. This technology can be deployed in most of the cellular/PCS spectrums on frequency bands such as 450 MHz, 800 MHz, 1700 MHz, 1900 MHz and 2100 MHz. A comparison of the system parameters of CDMA2000 and WCDMA is shown in Table 4.3.

CDMA2000 1xRTT (one-time Radio Transmission Technology) uses the same frequency band as the IS-95, having a bandwidth of 1.25 MHz. The capacity of CDMA2000 is much higher than the IS-95; actually it is doubled with the addition of 64 more channels. The evolution of CDMA 2000 is shown in Figure 4.9.

CDMA2000 EV-DO or the CDMA2000 Evolution Data Optimized, as the name mentions is capable of carrying data on the air interface. It is suited for providing high speed data services to both the mobile and broadband subscribers. 1x voice operators prefer this technology as it

Table 4.3 Comparison of system parameters of CDMA2000 and WCDMA

Parameters	CDMA2000	WCDMA
Carrier Spacing	3.75 MHz	5 MHz
Chip rate	3.6864 MHz	4.096 MHz
Power Control Frequency	800 Hz	1500 MH
Frame Duration	20 ms	10 ms
Data Modulation	F-QPSK; R-BPSK	BPSK
Base Station Synchronized	Synchronous	Asynchronous

is not backward compatible as it uses an IP network rather than the SS7 and complex switches such as MSC. With the introduction of Release 0, multicast services can be offered that permit the same information to be transmitted to unlimited users. This saves a lot of work in terms of the same information being broadcasted multiple times. Applications include transmitting MP3 files, movies, etc. over the network.

CDMA2000 EV-DV or CDMA2000 evolution data/voice is able to carry data rates up to 3.1 Mbps on the forward link and 1.8 Mbps on the reverse link. Around 2004–2005, much discussion happened on the relative benefits of DV and DO. Traditional operators with an existing voice network preferred deploying DV, since it does not require an overlay. Other design engineers, and newer operators without a 1x voice network, preferred EV-DO because it did not have to be backward-compatible, and so could explore different pilot structures, reverse link silence periods, improved control channels, etc. The costs were lower, since EV-DO uses an IP network and does not require a SS7 network and complex network switches such as a mobile switching center (MSC). Also, equipment was not available for EV-DV in time to meet market demands whereas the EV-DO equipment and mobile application-specific integrated circuits (ASIC) were available and tested by the time the EV-DV standard was completed. As a result, the EV-DV standard was less attractive to operators, and has not

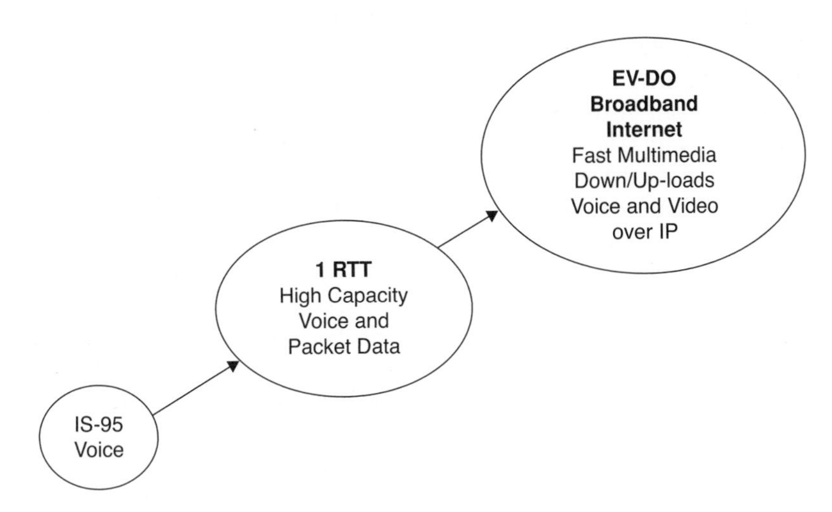

Figure 4.9 CDMA2000 evolution.

Table 4.4 Average data throughput rates of CDMA2000

CDMA2000 1X	60–100 kbps
CDMA2000 1xEV-DO	400–800 kbps

been implemented. Verizon Wireless (then Sprint Nextel in 2004) and smaller operators in 2005 announced their plans to deploy EV-DO. The key equipment for EV-DV technology was not available at this time and hence the development was suspended by Qualcomm (to focus on EV-DO development). The average throughput rates of CDMA 2000 are shown in Table 4.4.

There are revisions in this technology that are called Revision A (rev A) and Revision B (rev B). Rev A is the evolution of 1xEV–DO release 0. It takes in the OFDM technology, enabling multi-casting and also has increased peak rates on the forward and reverse links. It also supports symmetric, delay sensitive, real time and voice and data applications. It is also able to support the time-sensitive applications such as Voice over IP, etc. The further evolution of Release A is called Release B. It aggregates multiple EV-DO rev A channels, thus providing a higher performance, for example for Voice over IP based services, multimedia delivery, etc. The rev B standard was published by the Third-Generation Partnership Project 2 (3GPP2) under document number 3GPP2 C.S0024-B and by the Telecommunications Industry Association (TIA) and Electronics Industry Association as TIA/EIA/IS-856-B.

4.8.3 Channel Structure in CDMA2000

The forward traffic channels constitute three kinds of physical channels: fundamental channels (F-FCHs), supplemental channels (F-SCHs) and dedicated control channels (F-DCCHs). The Traffic Channel (TCH) in IS-95 is called the Fundamental channel in CDMA2000, supporting voice, data and signalling (750 bps to 14.4 kbps). A supplementary channel supports higher data rates while dedicated channels are used for signalling and 'bursty' data sessions. A mobile can support a maximum of two SCHs. Supplementary channels also improve modulations, coding and power control. These are transmitted by each sector of the base station if sufficient power and Walsh codes exist. The common channels – the pilot channels in CDMA2000 are also quite a few of these: the forward pilot channel (F-PICH), forward common area pilot channel (F-CAPICH) and forward dedicated area pilot channel (F-DAPICH). F-CAPICH is used by all mobiles within the coverage area of a sector while F-DAPICH is used for a dedicated mobile only. The synchronization channels in CDMA2000 are the same as that of IS-95 (F-SYNC, F-PCH and F-CCCH). The reverse link channels are similar to that of the forward link and include common channels such as the Reverse Access Channel (R-ACH) and Reverse Pilot Channel (R-PICH). R-PICH does not carry any data. Then there are reverse-dedicated channels, reverse-fundamental channels (R-FCHs) and reverse-supplementary channels (R-DCCHs). These channels are similar to that of the forward link except that the Walsh codes 2 and 4 are used. Services that need lower bit rates use one R-FCH while those needing higher bit rates use more dedicated channels.

4.8.4 Power Control

As shown in Table 4.3, the frame length in CDMA2000 is 20 ms which is divided into 16 equal power control groups. Other defined frame lengths include 5 ms, 40 ms and 80 ms. The 5 ms frame length supports the signaling burst, 40 ms supports interleaving and 80 ms the diversity gains for data services. Power control in terms of a fast closed loop is possible both in the forward and reverse directions (in IS-95, it was possible only in the reverse direction). Open loop power control is also possible in CDMA2000 and helps in reducing the impact of slow fading. In case there is a problem related to closed loop power control, open loop power control takes control of the situation by reducing the terminal output power and limiting the impact on the system. From a channel perspective, the base station is informed of the quality of the forward link by the power control bits that are multiplexed in the R-PICH.

4.8.5 Soft Handoff

The handoff process is similar in CDMA2000 as understood before. CDMA2000 provides a framework to the terminal in support of inter-frequency handover measurements consisting of the identity and system parameters to be measured. These measurements are done by the mobile itself. For single receiver terminals, channel interruption takes place when making measurements while with dual receivers measurements are done without any interruptions.

4.8.6 Transmit Diversity

This concept is similar to space diversity wherein the receiver has two antennae to receive the signals. In this, the signal is de-multiplexed and demodulated into two orthogonal signals – each is then transmitted on the same frequency and different antennae. The signals generated use Orthogonal Transmit Diversity or the Space Time Spreading. Plus when it reaches the receiving end, it is re-constructed by using space or frequency diversity mechanisms. The link performances improve by an order of 5 dB.

4.8.7 Security

The CDMA systems have better security than their TDMA and FDMA counterparts. Primarily this is due to the spread spectrum technique that is used in these networks. The codes spread through the full channel bandwidth of 1.25 MHz. The access to the network is only provided to the mobile stations/subscribers. The signals are not only difficult to intercept (due to the uplink and downlink using different encodings making demodulation difficult for the noise like signals) but the process is quite complicated and expensive. One reason is also the code mask used that provides security to the physical layer.

4.8.8 CDMA2000 Network Architecture

The CDMA2000 3G wireless system is based on the code-division multiple access (CDMA) system. CDMA2000 delivers high-bandwidth data and voice services to users of mobile equipment. Figure 4.10 shows the infrastructure of a CDMA2000 wireless network. The

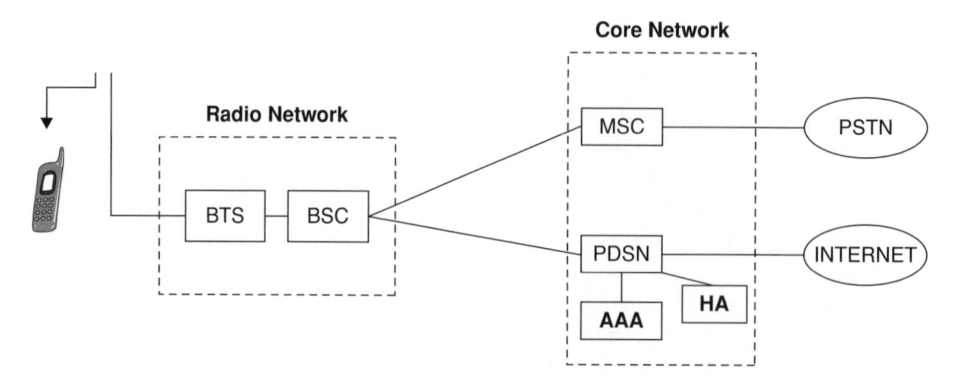

Figure 4.10 CDMA2000 (1XEV-DO Rel 0) network architecture.

CDMA2000 network comprises three major parts: the core network (CN), the radio access network (RAN) and the mobile station (MS). The core network is further decomposed into two parts, one interfacing to external networks such as the Public Switched Telephone Network (PSTN) and the other interfacing to the IP-based network such as the Internet. The mobile station terminates the radio path on the user side of the network and enables subscribers to access network services over the Um interface.

The CDMA2000 access network may perform mobility management functions for registering: authorizing, authenticating and paging IP based terminals, independent of circuit-based terminals. The access network may perform handoffs within an access network and between access networks of the same technology and may support handoffs between access networks of differing technologies. The main elements that are a part of the RAN are BTS, packet control function (PCF) and BSC.

The core network consists of an MSC, HLR, VLR and AC. In addition, message centres (MCs) and short message centres (SMEs) are a part of the core network.

4.8.9 Key Network Elements (CDMA2000)

4.8.9.1 Mobile Station

The mobile station, as in the GSM or UMTS networks, interacts with the base station to exchange the packet data. When a mobile station registers with the HLR, it is considered to be capable of making voice and data calls. The data calls are made using the point-to-point protocol between the mobile and the packet data serving node. Mobile stations that are based on IS-95 standards can access calls via circuit-switch data while the ones that are based on IS-2000 are capable of making both the circuit- and packet-switch data calls. Circuit-switch calls have a maximum rate of 19.2 kbps while the packet-switch data services have a maximum rate of 144 kbps.

4.8.9.2 Base Transceiver Station (BTS)

This is an interface between the mobile and BSC/RNC and performs functions such as transmit power control, sector separation, managing back haul (to minimize delays), frequency assignments, etc.

4.8.9.3 Base Station Controller (BSC)

This is the interface between the base station and the core network. It is responsible for mobility management functions such as handoffs. For EV-DO, security functionality is also managed by the BSC (or RNC).

4.8.9.4 Packet Control Function (PCF)

This is responsible for the routing of the IP packet traffic between the mobile and the core network (PDSN). It buffer packets and relays them by supplying the supplementary channels as needed between the mobile and PDSN.

4.8.9.5 Packer Data Serving Node (PDSN)/Foreign Agent

This is the gateway from the access network to the core network. As mentioned before, the PDSN terminates the PPP for a mobile station (i.e. session initiated by the subscriber). It also performs functions such as authorization, accounting, authentication, managing subscribers, etc. For mobile IP subscribers, a 'foreign agent' handles packet data routing and encryption.

4.8.9.6 AAA/Home Agent

These are used for the authentication, authorization and accounting services. The 'home agent' allows the 'roaming' of its own subscribers in other networks as it provides them with 'anchor IP addresses' for the mobile station while forwarding its traffic for delivery to the appropriate network.

4.8.10 Interfaces of the CDMA2000 Network

As mentioned above, the interfaces are also called 'reference points' in the CDMA networks.

4.8.10.1 Um Reference Point

This is the reference point between the user equipment and the BTS. It performs functions such as the mobility management, resource management and connection management.

4.8.10.2 Rm Reference Point

This is the interface between the terminal equipment and mobile terminal (both are within the mobile station). It is responsible for PPP (point-to-point protocol), IP and the upper layer.

4.8.10.3 Abis Reference Point (A7/A3)

This is the interface between the BTS and BSC that is used for data exchange. An A7 interface is used for signalling while A3 is used for establishing/removing the traffic connections.

4.8.10.4 A8/A9 Reference Point

This is the interface between the BSC and PCF. A8 carries user traffic while the A9 carries the signalling traffic.

4.8.10.5 A1/A2/A5 Reference Points

A1, A2 and A5 are the interfaces between the BSC and MSC. A1 is used for signalling, A2 for voice while A5 is used for the CS data traffic.

4.8.11 Call Set Up Processes

Let us now understand the call set-up in CDMA2000 networks.

4.8.11.1 1xRTT Network

As shown in Figure 4.11, before the mobile is allowed to operate, it collects sets of configuration messages from the BSC. This is done with the guidance of the base stations. As the mobile

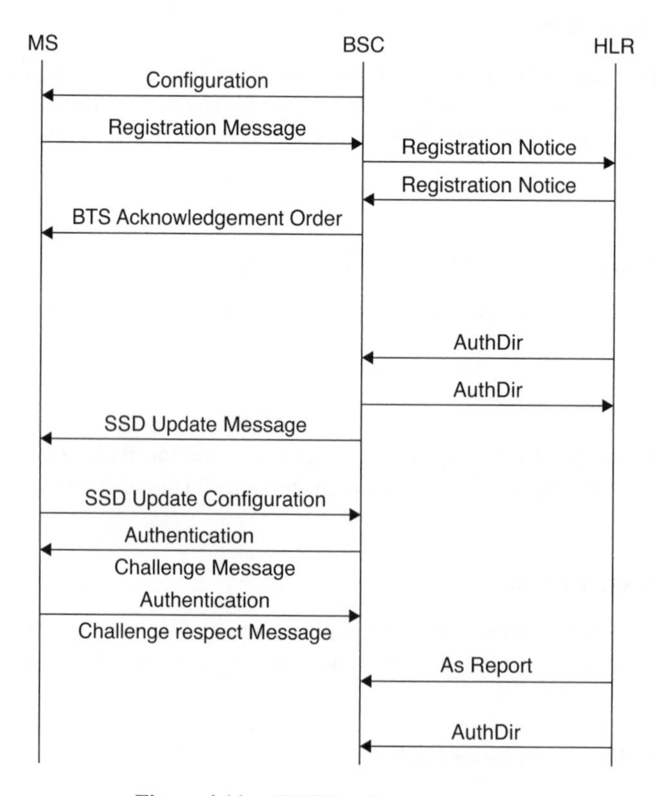

Figure 4.11 1XRTT call set up process.

needs to do this, a registration message is sent by it to the BSC. In response, the registration notification request goes to the MSC/VLR and is responded to by the HLR along with the MS service profile. On receiving this, the BSC sends the successful registration acknowledgement to the mobile through the base station. At the same time as receiving the authentication request, the AC initiates the authentication algorithms to start the SSD update process. This process generates the SSD sub-key. This is used for authentication and session key derivation. This SSD update process happens in parallel with the registration process. The authentication directive is then sent to the BSC. This is then acknowledged by the system and the SSD update message is sent to the mobile. This generates the confirmation message of the SSD update from the mobile. The mobile also extracts the message from the authentication directive and returns it to the system. The results are sent to the authentication centre and are verified by the ASreport which if correct is sent to the VLR. In the case of fraud, this is sent to the 'Fraud Information Gathering System'.

4.8.11.2 EV-DO Network

A 'unicast' access terminal identifier request (UATI) is sent by the mobile to the RNC. This is assigned by the RNC to the mobile. The mobile sends the message when assignment is complete. Following this session establishment, PPP/Link Control protocol negotiations are complete between the mobile and RNC. A challenge handshake authentication protocol (CHAP) is sent to the mobile by the RNC. The mobile responds with an A12 CHAP key to the RNC. This also includes the network access identifier. This is now forwarded by the RNC to the local AAA server, which are forwarded to the home AAA server for validation. If valid, the home AAA sends the access accept message to the visitor AAA from where is it routed to the RNC. The RNC informs the mobile of the results. The process is shown in Figure 4.12.

4.9 TD-SCDMA

'Time Division Synchronous Code Division Multiple Access' (TD-SCDMA) or UTRA/UMTS-TDD, 1.28 Mcps Low Chip Rate (LCR), is an air-interface found in UMTS mobile telecommunications networks in China as an alternative to W-CDMA. Together with TD-CDMA, it is also known as UMTS-TDD or IMT 2000 Time-Division (IMT-TD). The term TD-SCDMA is a bit misleading. While it suggests covering only a channel access method based on CDMA, it is actually the common name for the whole air-interface specification.

TD-SCDMA uses the S-CDMA channel access method across multiple time slots. The TD-SCDMA combines the advanced TDD/TDMA system and the adaptive CDMA component working in a synchronous mode.

ITU accepted the Chinese standards TD-SCDMA or Time Division-Synchronous Code Division Multiple Access as one of its international standards in November 1999 and this was adopted in May 2000. In March 2001, it was included in 3GPP Release 4. It is also an active participant of the Rel4 and Rel5 specifications. According to the 3GPP, a six-phase harmonization of the TD-SCDMA standards with WLAN (seamless handoff of the CS domain) is on its way.

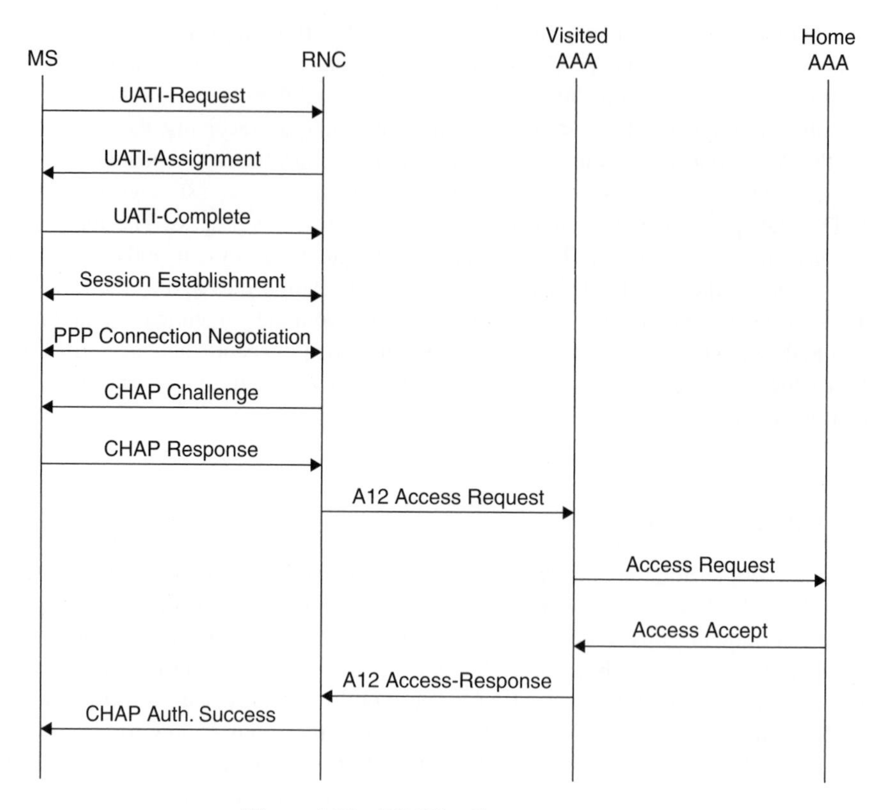

Figure 4.12 EV-DO call set up process.

The key features of TD-SCDMA include the following:

- Access to a new UMTS spectrum for enhanced network capacity.
- The spectrum is 3 to 5 times higher than the GSM (the TD-SCDMA spectrum is shown in Figure 4.13).
- It has a bandwidth of 1.6 MHz, thereby handling more traffic with less base stations.
- It supports all types of radio network resources (wide area, local area, hot spots, etc.).
- Flexibility for asymmetric traffic data rates.
- High-speed data (including packet and multimedia).

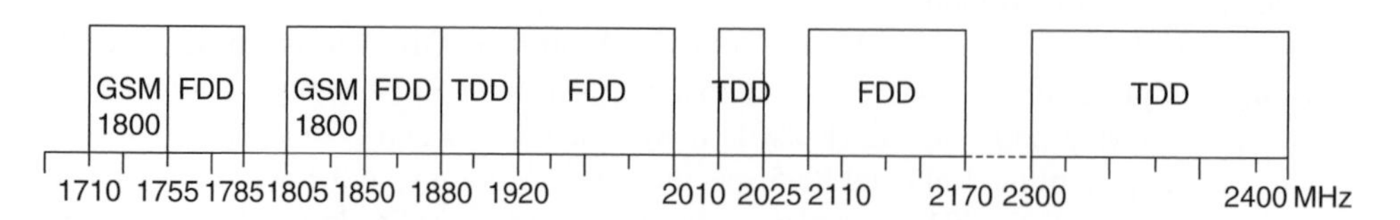

Figure 4.13 Frequency spectrum for a TD-SCDMA network.

The technical specifications of TD-SCDMA networks are as follows:

- Frequency band: 2010 MHz–2025 MHz in China (WLL 1900 MHz–1920 MHz).
- Minimum frequency band required: 1.6 MHz.
- Frame length: 5 ms.
- Multi-carrier option.
- Handovers: hard.
- 'Smart' antennae.
- Frequency re-use: 1 (or 3).
- Chip rate: 1.28 Mcps.
- Frame length: 10 ms.
- Number of slots: 7.
- Modulation: QPSK or 8-PSK.
- Voice data rate: 8 kbit/s.
- Receiver: joint detection (mobile, rake).
- Power control period: 200 Hz.
- Number of slots/frames: 7.
- 'Baton handover'.
- Uplink synchronization.
- Circuit-switched services: 12.2 kbits/s, 64 kbits/s, 144 kbits/s, 384 kbits/s, 2048 kbits/s.
- Packet data: 9.6 kbits/s, 64 kbits/s, 144 kbits/s, 384 kbits/s, 2048 kbits/s.

The main technical advantages of the TD-SCDMA include the following:

- Easy upgrade of GSM to the 3G network.
- Greater network efficiency.
- Increased capacity.
- Reduced migration risks, etc.

The 3G spectrum is allocated between the FDD and TDD, while the TD-SCDMA uses the TDD spectrum. The bandwidth of the TD-SCDMA systems is 1.6 Mbps and the chip rate per carrier is 1.28 Mcps. The TDMA uses a 5 ms frame which is sub-divided into seven timeslots that are assigned either to single or several users to be used flexibly. For symmetric services, for example the telephony timeslots are divided equally between the downlink and the uplink. For applications such as the Internet where the data transfer is asymmetric, more time slots are used for the downlink. Due to this capability, the system is more efficient. The number of voice channels per time slot is 16 as compared to 8 for the UTRA TDD. Spectral efficiency is higher in these systems as compared to UTRA systems.

The TD-SCDMA system employs smart antennae that help in reducing inter-cell interference. The SIR improves by about 8 dB. These antennae also assist in optimizing the link budget and thereby increase the capacity of the air-interface. The soft handover function does not exist in this system as unlike in WCDMA; the cell breathing concept does not exist here. Another feature that improves the performance of the systems in terms of coverage is the joint detection and terminal synchronization. The former eliminates the multiple access interference while the latter improves the uplink signal quality.

4.9.1 Services in TD-SCDMA

There are symmetric and asymmetric services in TD-SCDMA networks. The former ones are used when the same amount of data is transferred in both directions, for example telephone and video calls, while in the latter ones unequal transmission takes place – more data are transferred from the base station to the mobile. In the symmetric service, the time slots are split equally between the uplink and downlink, while in the asymmetric service, more time slots are used in the downlink direction. Here the TDD and TDMA support in assigning the radio network resources to move the traffic in the uplink and downlink directions. Combined with the CDMA, the capacity of the radio interface increases significantly. In WCDMA, when using a chip rate of 1.6 Mcps this will permit a carrier bandwidth of 1.6 MHz while in the TD-SCDMA, multiple 1.6 MHz carriers can be deployed. Also, the FDD used in CDMA network needs a pair of frequency bands for uplink and downlink (this leaves a portion of the spectrum that is not used for data transfer) while in TD-SCDMA unpaired frequency bands are used. This optimizes the capacity of the air-interface making it more 'spectrum efficient'.

4.9.1.1 Dynamic Channel Allocation

As mentioned before, TD-SCDMA utilizes TDMA, TDD, CDMA and SDMA modes. Thus depending on the interference scenario, the resources are allocated, thereby minimizing the interference. There are a few methodologies to implement the DCA, as follows:

- Time domain allocation (TDMA).
- Frequency domain allocation (FDMA).
- Space domain allocation (SDMA).
- Code domain allocation (CDMA).

In TDMA, allocation of traffic is allocated to the 'least-interfered timeslots' while in FDMA it is allocated to the 'least-interfered carrier'. The 'best direction de-coupling' on a per user basis is done in the SDMA network while the 'least-interfered code' is preferred in the CDMA scheme.

4.9.1.2 Mobile Synchronization

Synchronization in TD-SCDMA systems is much more accurate as the mobile sends an advance time offset between the reception and transmission, thus making sure that the signal arriving at the base stations are 'frame-synchronous'. This improves the multi-user joint detection. Not only does the mobile traceability improves but it also makes possible the link-quality measurements of the neighbouring base stations during the non-activity state, thus making the handovers a faster process. Hence, conventional handovers are used instead of the soft handovers in W/CDMA networks.

4.9.1.3 Joint Detection

In TD-SCDMA, joint detection receivers are implemented in the receivers – both base stations and the mobile. Due to this, the multiple access interference (MAI) is minimized, allowing a higher CDMA loading factor. MAI happens as the received codes in CDMA are not orthogonal to each other, making the co-relation process difficult. The received signal is not quite that different from the interfering signal. Thus, joint detection receivers are used to reduce this impact. This technique uses algorithms to extract all the CDMA codes in parallel and removes the interference codes by MAI. The total number of 16 codes per time slot per radio carrier can be processed and detected. Also, due to joint detection, the system is more robust against the fast signal fluctuations, thereby reducing the complexity of the power control systems.

4.9.1.4 Smart Antennae

These are used to make the system more robust against interference. Unlink omni-antenna systems, these antennae transmit the signals to the specific mobiles. This leads to an increased received power by the mobiles, increased sensitivity of the base station receivers and improvements in interference (inter- and intra-cell). In TD-SCDMA, advanced beam forming bi-directional adaptive antennae are used. Due to this, an improvement of about 8 dB in *C/I* can be achieved, thereby reducing the power transmitted by the mobile. This has a positive impact on the total budget of the project as well as the number of base stations can be reduced.

A comparison of the features of two CDMA TDDs, that is UTRA TDD and TD-SCDMA, is shown in Table 4.5.

A comparison between WCDMA and TD-SCDMA is outlined in the following:

- TD-SCDMA uses TDD, while WCDMA uses the FDD scheme. TD-SCDMA can accommodate more asymmetric traffic as compared to the WCDMA due to dynamic allocation of time slots in the downlink and uplink directions.
- WCDMA is more suited towards the symmetric traffic and 'macro-cells' while TD-SCDMA is used for the 'micropico' cell environment.
- Spectrum allocation flexibility is more as the TD-SCDMA does not require the paired spectrum for uplinks and downlinks.
- The usage of the same carrier frequency for the uplink and downlink directions allows the base station to deduce the downlink channel information from the uplink estimates. This is useful to the application of beam forming techniques.

Table 4.5 Comparison of key features between UTRA UDD and TD-SCDMA

Feature	UTRA TDD	TD-SCDMA
Bandwidth	5 MHz	1.6 MHz
Chip rate/Carrier	3.84 Mcps	1.28 Mcps
Number of time slots	15	7
Capacity: No. of voice channels per times slot	8	16
No. of voice channels per carrier	$7 \times 8 = 56$	$3 \times 16 = 48$
Spectrum efficiency	10Erl./MHz (possible increase 50% by voice stimulation)	25Erl./MHz
Capacity: Total transmission rate per time slot	220.8 kbps	281.6 kbps
Total transmission rate provided by each carrier	3.31 Mbps	1.971 Mbps
Spectrum efficiency	0.662 Mbps/MHz	1.232 Mbps

- The uplink signals are synchronized at the base station receivers. This is done by continuous timing adjustments. This reduces interference between users of the same time slot by improving the orthogonality between the codes. This improves system capacity and hence performance. As uplink synchronization is needed, the system hardware costs rises to achieve this function.

4.9.2 Network Planning and Optimization

The fundamental processes related to network planning and network optimization remain the same (as discussed above in Chapters 2 and 3). However, let us try to understand some aspects related to the TD-SCDMA network. For various kinds of users/areas/demands, macro (rural/remote areas) and micro (urban/hotspots) sites are used. For dense urban areas, macro-base stations are used along with micro-base stations that cover the blind areas and hotspots. The objectives of coverage include for deep indoor and indoor first wall (95 %) while for in-car coverage the objective can be 75 %.

Optimization in TD-SCDMA networks has few aspects that are different from the WCDMA network optimization. The handoff in TD-SCDMA is called 'baton handoff' unlike in other networks where it is called 'soft handoff'. In a *normal* handoff the exact location of the mobile is not required while in the baton handoff the exact location of the mobile is needed. This means that the measurements of all the cells are not needed and the process can work with few neighbouring cells. Thus, proper and accurate measurements are needed for handoffs in the TD-SCDMA network. During the baton handoff process communication links with the target cell is made at the same time with the communication link with the former base station. Unlike, as we have seen before in this chapter, the pilot cell with the strongest power is not always the service cell in this system. Channel allocation is a more complex procedure in TD-SCDMA. In WCDMA, the same carrier frequency channel needs to be allocated properly while in TD-SCDMA a special algorithm called DCA is used to allocate the channels. This algorithm works and delivers results based on the real time interference measurements of the mobile and the network. This helps in allocation of the channel resources flexibly and at the same time increases the frequency band utilization ratio. In TD-SCDMA, 16 codes are used for each time slot and each carrier. Smart antennae and DCA minimize the inter-cell interference while the intra-cell is removed by the joint detection. Unlike in WCDMA where the 'cell-breathing' phenomenon makes its impact felt on the coverage when the load increases, in TD-SCDMA, the cell breathing does not have an effect. The traffic load can be increased without decreasing the coverage. Also, no soft handover is needed in TD-SCDMA due to the joint-detection, synchronization and DCA. Just like in GSM, conventional handover techniques are used, thereby reducing the complexity of the radio planning process.

For indoor coverage, there are a few issues in TD-SCDMA networks. In these networks, handoffs work on knowing the accurate position of the mobile. The gains of the indoor antennae are used which are different to that of the smart antennae. Thus if the subscriber is moving under different antenna overages the distance of the user to the base station cannot be judged all the time. If the subscriber is not moving, then also the signal received by the base station comes via different antennae, so making it difficult to judge the distance and for synchronization the base station needs to compensate for the delay which is not possible.

5

HSPA and LTE

5.1 HSPA (High Speed Packet Access)

5.1.1 Introduction to HSPA

HSPA is also called 3.5G technology (beyond 3G) that increases the capability of the UMTS system and the performance of the network becomes better in terms of end user experience such as file download/upload streaming services, VoIP, etc. The peak data rates increase to 14 Mbps (downlink) and 5.8 Mbps (uplink), thus providing five time more capacity in downlink and double capacity in the uplink direction. HSPA is the 3G release 5 solutions for expanding the throughput of the system (release 99 provides data rates up to 384 kbps).The increase in data rate is done through an additional downlink transport channel called the High Speed Downlink Shared Channel (HS-DSCH). This improves latency and increase the throughput.

5.1.2 Standardization of HSPA

HSPA constitutes two major advancements, in the downlink and uplink directions. In the downlink direction, it is called HSDPA or High Speed Downlink Packet Access while in the uplink it is called HSUPA or High Speed Uplink Packet Access. HSDPA was a part of release 5 and HSUPA was a part of release 6 specifications. The HSPA is deployed on the WCDMA network and the network elements are the same as in the WCDMA network. With some hardware upgrades in the base station and RNC and software upgrades, a WCDMA network is ready to function as a HSPA network. The data rates in the HSDPA networks range from 1.8 Mbps to 10 Mbps (and 20 Mbps for MIMO systems) while in the HSUPA network it is up to 4 Mbps. HSDPA is mainly for NRT (Non-Real Time) traffic and offers lower cost per bit due to the extended capacity in the downlink. The channelization codes define the number of mobile subscribers in the HSDPA network for associated DPCHs.

5.2 HSDPA Technology

As mentioned above, HSDPA is a apart of release 5 of the 3GPP. Though the network elements are similar to that of the WCDMA network, new user equipment is used in the HSDPA network.

Cellular Technologies for Emerging Markets: 2G, 3G and Beyond Ajay R. Mishra

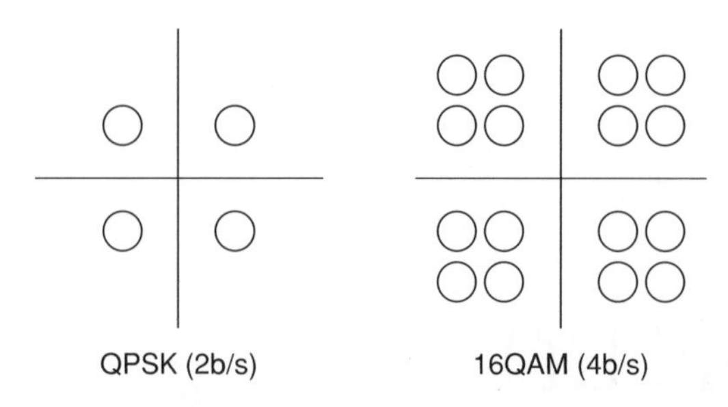

Figure 5.1 Modulations in HSPA.

The new UE is able to handle both the QPSK and 16QAM modulation schemes. Apart from this, the UE should also have a capability to handle multi-code processing, fast automatic requests, etc.

The main features on the basis of which HSDPA achieves its performance are:

- Higher order modulations: adaptive modulation and coding schemes.
- Short transmission time intervals (TTIs).
- Hybrid Automatic Repeat reQuest (HARQ) re-transmission protocol.
- High speed channels shared both in the code and time domains.
- Fast-packet scheduling controlled by the Medium Access Control.
- Fast link adaptation.

Quadrature Phase Shift Keying (QPSK) is used as the modulation scheme in release 5. HSDPA also uses 16QAM as the modulation scheme, thereby making more efficient use of the bandwidth which results in a higher data rate as 16QAM has a peak rate capability that is higher than the QPSK. 16QAM uses four bits per symbol against QPSK that uses two bits per symbol (shown in Figure 5.1).

The transmission time interval in the HSDPA network is reduced to 2 ms as compared to the release 99 where it is 10 ms, 20 ms or 40 ms. This reduces the round trip time, thereby increasing the rate of tracking channel variations – thus supporting link adaptation and channel dependent scheduling. The channels from the shared resources are allocated dynamically 500 times per second (i.e. 2 ms).

We have seen re-transmission in EDGE networks. Similarly in HSDPA also, a soft combining technique is used in which the mobile terminal is able to request re-transmission of the data if the received data are not correct before it attempts to decode the received data. This technique improves the system performance. This function in WCDMA was handled by the RNC whereas in HSDPA it is handled by the base stations thereby making it closer to the mobile station.

In shared transmission, some channels codes and transmission power are shared among the various users – both in terms of codes and time. HSDPA is based on this technique. This is better than the usage of dedicated channels. As mentioned above, data rate increase is possible due to HS-DSCH. This is mapped on shared code resources that consist of up to 15 codes, though the actual number depends upon the system itself.

Instead of round-robin allocation of resources, channel-dependent allocation of resources is done. This is done by using fast packet scheduling feature. Using this, the capacity of the downlink channel can be increased toward users that have better channel conditions. The downlink conditions are measured by the scheduler by looking into the quality reports that each mobile sends to the base station. This helps in determining the HSDPA network performance. The scheduler decides to which mobile the HS-DSCH should be transmitted. By using the link adaptation technique, modulation and the codes are also determined. In the case where there are many users that are in a queue, that is high load on the cell, the probability of the scheduling users with good channel quality is high. In such cases, the traffic priorities take over, that is the conversation class takes a priority over the interactive class. This means that the HS-DSCH does not use the soft handover technique that is used by the release 99. It is not possible to implement soft handover in the network as this technique (fast scheduling) is done by one base station.

In the WCDMA network, fast power control is used to compensate the variations in power control, but in HSDPA bit rate adjustments are used. Various signals experience different conditions in the downlink direction and variations are both in terms of position and time. Based on the quality reports, fast link adaptation adjusts various transmission parameters and in some cases it even works on modulation aspects as well. Power control features do compensate for differences in signal levels; however, the output power available is not fully utilized. Thus, power control is not the most efficient way to handle this. In the bit rate adjustment, the data rate is adjusted keeping the transmission power constant. At the same time, QAM16 modulation can be used when possible.

5.2.1 WCDMA to HSDPA

Let us now look at the summary of changes done in the WCDMA network to realize a HSDPA network:

- A new bearer HS-DSCH which allows several users to be multiplexed and hence effective utilization of resources.
- A fixed spreading factor of 16 allowing a maximum of 15 parallel codes – both for user and signalling traffic.
- A transmission time interval of 2 ms, increasing the rate of tracking the variations and allowing the system to adapt faster. Adaptive modulation and coding (AMC) is used to compensate for the variations in radio transmission conditions.
- To handle new features, a new MAC-hs entity is added.
- HARQ (Hybrid Automatic Repeat Request) is used in the re-transmission mechanism and is associated directly with the base station rather than the RNC. The same is the case with the scheduler mechanism.

5.2.2 HSDPA Protocol Structure

Adding new elements to the UMTS network would need some changes in the protocol structure of the HSDPA network (shown in Figure 5.2). The MAC layer is now split into two: MAC-d and MAC-hs. The former remains in the RNC while the latter is placed in the base station.

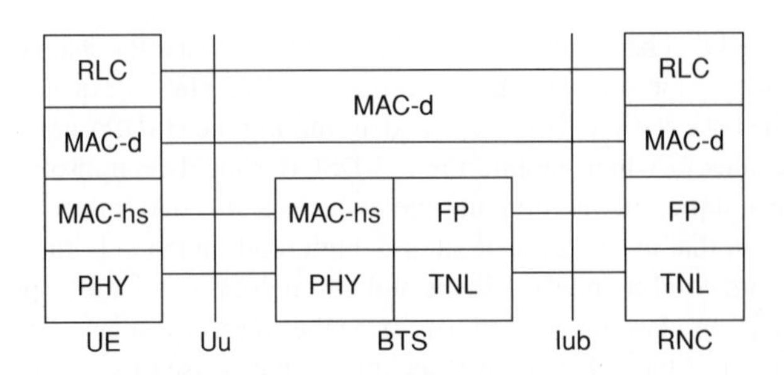

Figure 5.2 HSDPA user-plane protocol.

5.2.2.1 MAC-d

This is located at the RNC and is for dedicated channels. It is used for mapping between the transport and logical channels. Other functions include multiplexing/de-multiplexing upper layer PDUs, ciphering for the transport mode RLC and identifying user equipment in common transport channels.

5.2.2.2 MAC-hs

The fast scheduling is done at the base station and is responsible for features such as packet scheduling, link adaptation and error correction (layer 1).

5.2.3 User Equipment

The characteristics of the HSDPA handsets are more complex than the 3G ones due to several new features, such as:

- equalizer and diversity;
- 16 QAM modulation;
- hybrid ARQ;
- increased and faster buffer memory;
- faster turbo decoder.

After MAC-d flow is established, the capability of the UE is sent to the base station via S-RNC. They include the maximum number of HS-DSCH codes that can be received simultaneously, maximum number of bits that the US can receive in one TTI, minimum inter TTI arrival, total buffer size (without RLC AM buffer) and maximum number of soft handover bits over all the HARQ process. Also parameters like the UE's capability to support 16QAM are also transferred to the UE. Categories 1–10 support the 16QAM while the last two support the QPSK modulations. The maximum capability is provided by the category 10 terminal that provides a capacity of 14.4 Mbps data rate as shown in Table 5.1.

Table 5.1 HSPDA UE capabilities

Category	Supported Modulations	Max # of HS-PDSCH Codes	Inter-TTI Interval	Max TB size (bits)	# of soft bits (Memory size)	Max Data Rate (Mbit/s)
1	16QAM and QPSK	5	3	7298	19200	1.2
2	16QAM and QPSK	5	3	7298	28800	1.2
3	16QAM and QPSK	5	2	7298	28800	1.8
4	16QAM and QPSK	5	2	7298	38400	1.8
5	16QAM and QPSK	5	1	7298	57600	3.6
6	16QAM and QPSK	5	1	7298	67200	3.6
7	16QAM and QPSK	10	1	14411	115200	7.2
8	16QAM and QPSK	10	1	14411	134400	7.2
9	16QAM and QPSK	15	1	20251	172800	10.8
10	16QAM and QPSK	15	1	27952	172800	14.4
11	QPSK	5	2	3630	14400	0.9
12	QPSK	5	1	3630		1.8

5.3 HSDPA Channels

The packet data are transferred in the downlink direction through the DCH (data control channel), FACH (forward access channel) and DSCH (downlink shared channel) in release 99. In HSDPA, the DSCH is replaced by a high speed DSCH. In HSDPA networks, when no activity takes place in UE movement, the FACH is used for signalling. However, the DCH is always used in the HSDPA network. For CS data, the services run on DCH while for PS data, the signalling radio bearer runs on the DCH. Thus, the DCH can be used for any data (CS/PS) and has a fixed spreading factor. There are some more channels that have been added to the HSDPA, namely HS-DSCH, HS-SCCH and HS-DPCCH.

5.3.1 HS-DSCH (High Speed Downlink Shared Channel)

HSDPA data are carried over the HS-DSCH. The HS-DSCH is the transport channel that is mapped onto the HS-PDSCH (High Speed Physical Downlink Shared Channel). It can use multi codes (maximum feasible, 15) with a spreading factor of 16. The higher order modulation scheme of 16QAM is used, thereby allowing more bits to be carried per symbol (over 2 ms TTI slots). The signal quality should be better when using 16QAM than the QPSK and both the amplitude and phase need to be estimated correctly. The synchronization of transmission and scheduling is complex and hence soft handover is not supported. Also, power control is not supported.

5.3.2 HS-SCCH (High Speed Shared Control Channel)

HS-SCCH uses QPSK modulation and a fixed spreading factor of 128 which allows carrying of 40 bits per slot. This carries information about decoding the next HS-PDSCH frame. One

HS-SCCH only is configured when HSDPA is operated on the time multiplexing while more than one is needed when it is being operated in the code multiplexing format though only four are supported by a single UE. Data rates available for the user are dependent upon the type of terminal, power allocation and environment.

5.3.3 HS-DPCCH (High Speed Dedicated Physical Control Channel)

HS-DPCCH carries the uplink feedback information. It carries the Ack/Nack (repetition) and CQI (Channel Quality Information) which informs the base station about the data rates that the UE expects to receive at a given time. This information is used for link adaptation and physical layer re-transmission purposes. The HARQ channel provides the information about the accuracy of the decoded information. Using a modulation of BPSK and a spreading factor of 256, the channel has a three lost (2 ms) structure. The first is used for HARQ while the remaining two slots are used for CQI information use. In release 6, there are some enhancements that would improve the cell edge operations.

5.4 Dimensioning in HSDPA

The most important part of dimensioning is the link budget. The output of dimensioning in an average throughput depends upon the power allocated for HSDPA. The power allocated should be able to fulfill two targets: achieve throughput and reserve power for R99 DCH traffic. There are quite a few variables that should be considered in HSDPA dimensioning as the downlink data are carried on the shared channel. The throughput in HSDPA comprises, on average, HSDPA cell throughput for single users and minimum throughput at the cell edge for single user and average HSDPA users throughput. The average throughput at a certain point can be estimated using the average SINR value. However, practically, this throughput changes all the time due to fast fading of the radio channels (base stations use the link adaption feature). For uplink calculations, one should remember that coverage is uplink limited. The uplink data are carried on the associated DPCH (this is the normal NRT bearer). Additional margin is needed in the UL link budget to meet the requirements for HS-DPCCH as well. An example of a HSDPA uplink DPCH link budget is shown in Table 5.2.

For the downlink budget, an average SINR allows creation of the DL link budget for HSDPA.

- It does not depend on bit rate or number codes for HSDPA.
- Can be calculated when the HSDPA power, BS total TX power, orthogonality and G-factor are known.

$$SINR = SF_{HSDPA} \times \frac{P_{HSDPA}}{P_{TOT\_TX}\left(1 - \alpha + \frac{1}{G}\right)}$$

where:

- SF_{HSDPA} is the spreading factor;
- P_{HSDPA} is the allocated transmission power of HS-PDSCH;

Table 5.2 Examples of HSDPA uplink DPCH and HS-PDSCH link budgets

Channel	DPCH	DPCH	DPCH		Channel	HS-PDSCH
Service	PS Data	PS Data	PS Data		Service	PS Data
Service Rate	64	128	384	kbps	Service Rate	384
Transmitter – UE					Transmitter – Node B	
Max Tx Power	21	21	21	dBm	Max Tx Power	38
HS-DPCCH Overhead	1.3	0.6	0.4	dB	Tx Antenna Gain	18
Tx Antenna Gain	2	2	2	dBi	MHA Insertion Loss	0.5
Body Loss	0	0	0	dB	Feeder Loss	2
EIRP	21.7	22.4	22.6	dBm	EIRP	53.5
Receiver – Node B					Receiver – Handset	
Node B Noises Figure		3.3		dB	Handset Noise Figure	8
Thermal Noise		−108		dBm	Thermal Noise	−108
Uplink Load		50		%	Downlink Load	80
Interference Margin		3.0		dB	Interference Margin	7.0
Interference Floor		−101.7		dBm	Interference Floor	−93.0
Service Eb/No	5	2.8	2.5	dB	SINR Requirement	4.5
Service PG	17.8	14.8	10.0	dB	Spreading Gain	12.0
Receiver Sensitivity	−114.5	−113.7	−109.2	dBm	Receiver Sensitivity	−100.5
Rx Antenna Gain	18.0	18.0	18.0	dB	Rx Antenna Gain	2
Feeder Loss	0	0	0	dB	Body Loss	1
Fast Fade Margin	1.8	1.8	1.8	dB	Fast Fade Margin	0
Soft Handover Gain	0	0	0	dB	Soft Handover Gain	0
Macro Diversity Gain	0	0	0	dB	Macro Diversity Gain	0
Slow Fade Margin	8	8	8	dB	Slow Fade Margin	8
Building Penetration Loss	0	0	0	dB	Building Penetration Loss	0
Max. Allowed Path Loss	144.4	144.2	140.0	dB	Max. Allowed Path Loss	143.0

- P_{TOT_TX} is the total BS power (including HS-PDSCH and HSSCCH powers);
- α is the DL orthogonality in certain locations in the cell;
- G is the geometry factor.

5.5 Radio Resource Management in HSDPA

5.5.1 Physical Layer Operations

The physical layer operation in a HSDPA network from a data perspective is synchronous from the UE side while it is asynchronous from the network side.

Each data user is evaluated by the base station scheduler every 2 seconds. After a data UE is identified, HS-DSCH parameters are determined by the base station. HS-SCCH (from a set of four channels) starts getting transmitted by the base station. UE decodes the HS-SCCH, first part 1 and then the rest. After decoding part 2 of HS-SCCH, the UE determines the 'belonging' of data to the ARQ process and subsequently combines the data in the soft buffer. After decoding, Ack/NAck is sent in uplink for THE needed L1 procedures. The call ends with the sending of the Ack/NAck field.

5.5.2 Adaptive Modulation and Coding Scheme

Another important feature in HSDPA is link adaptation (LA) which is done by changing the modulations and number of codes. With the absence of a power control feature in HS-PDSCH, the impact of LA is so high that it leads to cell capacity being lost and fast scheduling not working. The UE uses the CQI (that is signalled through HS-DPCCH) to tell the network of the highest data rate it can accept. Based on this information, the scheduler chooses a coding and modulation format for the next TTI that would work better in poor conditions. This is based on an inner loop algorithm. The major input parameters are the CQI and DPCH power measurement reports. If the performance of the inner loop algorithm gets biased due to reasons such as improved receiver architecture, an outer loop algorithm exists, based on the BLER target obtained from RLC acknowledge information and the CQI, to correct it.

5.5.3 Power Control

The power allocated to HS-SCCH should be sufficient as HS-DSCH can be decoded if HS-SCCH is correctly received, though more power would mean more interference. Thus, HS-SCCH should be power controlled at each transmission time interval (TTI). As is usually the case with power control phenomena, a higher power is needed at the cell edge and less near the base station. HS-SCCH power control mechanisms use inputs such as CQI reports and the downlink DPCCH power.

5.5.4 H-ARQ (Hybrid Automatic Repeat reQuest)

H-ARQ adds tremendous robustness to the HSDPA system. There are two techniques for this: incremental redundancy (IR) and chase combining (CC). IR is also called non-identical re-transmission as it uses different rate matching between re-transmissions, that is the relative number of parity bits to systematic bits varies between re-transmissions. CC is also called identical re-transmission as the same rate matching is used between different re-transmissions. After these bits have been re-transmitted, they are combined (weighted by SNR) by a decoder. In IR, the additional information is incrementally transmitted by using various puncturing schemes if the decoding fails the first time. A 1/3 punctured turbo code is used in IR for re-transmissions. New RLC packets are given priority by the MAC-hs layer. In the H-ARQ multiple stop-and-wait mode, that is when one H-ARQ process is waiting for Ack, then another can use the channel for sending data instead of just stop-and-wait as it would reduce the efficiency of the system.

- When the base station receives an Ack from a UE, then everything is fine.
- When the base station receives a Nack from a UE, then the packet was received but not detected properly. This re-transmission (IR) should take place from the base station.
- In cases where the base station does receive any Ack or Nack, but rather a DTX, it should re-transmit using another self-decodable rate matching scheme.

Table 5.3 Layer-1 transmission–re-transmission link budget (3GPP)

Delay Event	Delay
HS-DSCH transmission	2.0 ms (3 TS)
Before the UE sends Ack/Nack on the uplink HS-DPCCH	5.0 ms (7.5 TS)
For Ack/Nack transmission	0.67 ms (1 TS)
From reception of Ack/Nack until HS-SSCH transmission	3.0 ms (4.5 TS)
For HS-SCCH trans before starting the HS-DSCH transmission	1.33 ms (2 TS)
Total	**12 ms**

Table 5.3 shows the layer 1 transmission/re-transmission link budget.

5.5.5 Fast Packet Scheduling

To achieve the optimized cell capacity, along with required service experience, this is assisted by another feature: fast packet scheduling. The fast packet scheduling algorithms define how to share the available resources to the available UEs. However, the capability of the UE also affects the scheduling process. As can be seen below in Table 5.5, categories 1 and 2 UEs are capable of receiving data in every 3rd TTI while categories 3, 4 and 11 UEs can receive data in every 2nd TTI; other category UEs can receive data in every TTI. These algorithms are not specified in 3GPP and are left to the equipment manufacturers to define on how to design them. One key quality of the scheduler is flexibility in including the release 5 parameters, including complying with the power targets specified by the RNC. There are a few different types of algorithms that are present: round robin, proportional fair, minimum bit rate scheduling, maximum *C/I* and maximum delay scheduling. Round robin scheduling or 'best effort' is the most popular scheme and the least complex as it allocates the resources with equal probability without taking in quality conditions (as shown in Table 5.4). Diagrammatically opposite is

Table 5.4 Packet scheduling techniques

Packet scheduler	Scheduling rate	Fairness/Performance
Fair throughput (FT)	Fast (2–20 ms basis). In Rel99 architecture this is slow	Same average user throughput over activity time
Round-robin	Fast (2–20 ms basis). In Rel99 architecture this is slow	Same average physical resources but uneven user data rates
Maximum *C/I* (M-*C/I*) or maximum throughput (M-TP)	Fast (2–20 ms basis)	Optimum cell capacity but very uneven user data rates and limited coverage
Proportional fair (P-FR)	Fast (2–20 ms basis)	Approximately same average physical resources and uneven user data rates (fairer than FR PS) with very high capacity

the proportional fair scheduling which is more complex (though it gives a gain of 20–60 %) and uses the quality information before allocating the resources by the user. The maximum *C/I* scheduler gives the resources to a very small number of users. In this scheme, the users at the end of the cell might never get scheduled. A minimum bit rate scheduler would provide a more advanced quality of service differentiation.

5.5.6 Code Multiplexing

When several UEs are present in the same TTI, then data are sent to them by using the different set of codes. This is called code multiplexing. It is also used to optimize the performance of the data users. If there are 10, then 10 codes are used in the HS-PDSCH, 5 codes could be used to send data for UE1 and the remaining 5 codes could be used for UE2 in the same TTI. In the case of time multiplexing all 10 codes would be scheduled for one UE in a TTI. A typical example that requires code multiplexing is VoIP over HSDPA. The costs of the systems implementing code multiplexing also increase as there are overheads increases. Also, implementation of both the time and code multiplexing is expensive.

5.5.7 Handover

As stated before, in a HSDPA network there is no soft handover. This is because HS-DSCH and HS-SCCH transmission take place in a single cell. The serving HS-DSCH cell is determined by the RNC. There is a synchronized change in the serving cell which is based on measurement procedures that are based in CPICH measurement reports from the UE. The algorithms can be based on the concepts that all user candidate sets are taken into account. Both the downlink and uplink base station measurements are taken into account for the serving cell measurements. When the handover procedure between the two base stations takes place, it means it takes place between two HS-DSCHs. Assume that one UE is moving from one base station to another. The signalling radio bearer is mapped to both the HS-DSCH and E-DCH (Enhanced Dedicated Channel). A measurement report is sent onto the SRB based on which the RNC allocates another base station (to which handover will take place). When resources (from the second base station) are ready, reconfiguration messages are sent to the UE after which it starts receiving HS-SCCH from the second base station. Based on the channel quality, CQI reports are sent. Packet data are requested from the second base station via the RNC and subsequently the data are transmitted on HS-DSCH to the UE. Sometimes, duplicate messages are sent as well by the RNC via both the base stations. Handover is complete when the RNC receives the reconfiguration complete message from the UE. Handovers, such as those between the sectors of the same base station and between HSDPA and non-HSDPA, are also possible.

5.5.8 Resource Allocation

The protocols used in HSDPA are similar to the ones used in the UMTS network. The RNC and base station talk to each other by using the Node B Application Protocol (NBAP). The resource allocation by the RNC takes place by sending the NBAP message from the RNC to the base station. Channelization codes are however allocated for transmission though the signalling between the base station and RNC. In order to increase the spectral efficiency many HS-PDSCH codes are allocated to the base station. This may result in blocking as the channelization codes that are reserved for HS-PDSCH cannot be used for transmission of release 99; however RNC can release some allocated HS-PDSCH channels to prevent this blocking. The RRM algorithms try to keep the total power below a target level. In HSDPA networks, reports of the average power of non-HSDPA carriers form the basis of such measurements that lead to admission control and scheduling decisions of the RNC. Though there are two ways to control power in HSDPA networks – one through the base station (it can adjust any unused power in the cell) and the other through the RNC – the final control of power is done by the RNC.

5.5.9 Admission Control

As mentioned in Chapter 3 (UMTS), the admission control principle remains the same. The RNC decides, based on various measurements, whether or not the user can be granted access to the network. For CS services, for example speech, the algorithm uses DCH while for PS services it uses QoS parameters. Some key parameters that are used in AC algorithms are the average carrier transmit power, non-HSDPA transmit power, HS-DSCH power needed to serve all HSDPA users, HS-DSCH signal quality and the QoS parameters. Some of these QoS parameters in HSDPA in I_{ub} are guaranteed bit rate (GBR), discard timer (DT) and scheduling priority indicator (SPI).

5.6 High Speed Uplink Packet Access (HSUPA)

HSUPA or High Speed Uplink Packet Access is a complementary technology in the uplink direction to the downlink HSDPA. In the initial phase of development in 3GPP, it was called the Enhanced Uplink Dedicated Channel (E-DCH). The specifications for HSUPA are included in release 6 that include investigations related to shorter TTI for faster link adaptation, H-ARQ for making re-transmissions more effective, faster scheduling, faster DCH set up and higher order modulations.

5.6.1 HSUPA Technology

HSUPA technology works in tandem with release 99. So, many of the features of release 99 are used in HSUPA networks. Fundamental features such as that of radio resource management, random access, cell selection, etc. remain the same. The uplink transport channel E-DCH is similar to that of HS-DSCH for downlink though this is a dedicated channel and not a shared one. Some new signalling channels are required in HSUPA. Higher order modulations are not used in HSUPA as they need more energy per bit (unlike in HSDPA where complex modulations are supported).

5.6.2 HSUPA Protocol Structure

Like in HSDPA, there are some new aspects in the protocol structure in HSUPA. A new MAC layer called MAC-e is added in the base station, RNC and UE (as shown in Figure 5.3). This

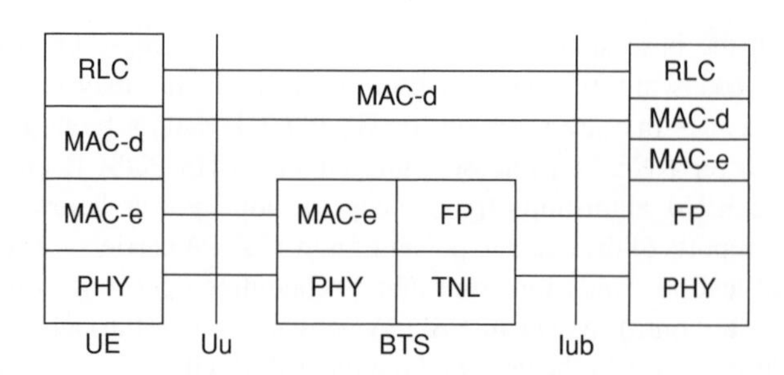

Figure 5.3 HSUPA user-plane protocol.

is because scheduling is controlled by the base station and the combining functionality is also at the base station. This new layer's main task is to make sure that packets are sent to the upper layers in a sequence that was sent to the base station. After all, the data that are received by the base station would be in various orders and this re-ordering is performed by the MAC-e layer. The RLC layer performs the task of re-transmission when the physical layer transmits them incorrectly.

5.6.3 HSUPA User Terminal

There are six categories in the HSUPA user terminal, as shown in Table 5.5. The key differentiation comes from the number of parallel codes supported (multi code capability) and support to 2 ms TTI although all categories support 10 ms TTI.

5.7 HSUPA Channels

There is a new channel in the HSUPA called E-DCH or Enhanced Dedicated Channel. The main features of this channel include support to faster base station scheduling, shorter TTI and faster HARQ (with incremental redundancy). This channel is similar to the HSDPA's

Table 5.5 HSUPA UE categories

HSUPA Category	Codes Spreading	TTI	Transport Block Size	Data Rate
1	1 × SF4	10	7110	0.71 Mbps
2	2 × SF4	10	14484	0.71 Mbps
2	2 × SF4	2	2798	1.40 Mbps
3	2 × SF4	10	14484	1.45 Mbps
4	2 × SF2	10	20000	2 Mbps
4	2 × SF2	2	5772	2.89 Mbps
5	2 × SF2	10	20000	2 Mbps
6	2 × SF2 + 2 × SF4	10	20000	2 Mbps
6	2 × SF2 + 2 × SF4	2	11484	5.74 Mbps

HS-DSCH but is different from the perspective of soft handover (it is possible in E-DCH), no adaptive modulation (no higher order modulation is supported), faster power control and a variable spreading factor. Data are carried on the E-DPDCH (E-DCH Dedicated Physical Data Channel) and control information is sent on the D-DPCCH (E-DCH Data Physical Control Channel). For scheduling purposes, the channels used are E-AGCH (E-DCH Absolute Grant Channel) and E-RGCH (E-DCH Relative Grant Channel). Re-transmission indication is done with the help of E-HICH (E-DCH HARQ Indicator Channel).

5.7.1 E-DPDCH

The E-DCH Dedicated Physical Data Channel or E-DPDCH transmits the data bits from the mobile to the base station using the BPSK modulation and faster power control. To adjust the number of bits to the actual data transmitted, it supports the orthogonal variable spreading factor (OSVF). Due to its support of a spreading factor of 2, double channel bits per code are delivered. As mentioned above, fast HARQ and fast base station scheduling are supported by E-DPDCH. A 2 ms TTI is supported by E-DPDCH.

5.7.2 E-DPCCH

The E-DCH Dedicated Physical Control Channel or E-DPCCH exists to support the E-DPDCH by providing the information needed to decode the data channel transmission along with the information related to channel estimation and power control. It has one slot format of SF 256 delivering 30 channel bits in a 2 ms sub-frame with 10 bits of information transmitted per TTI. This contains three segments: the E-DCH Transport Format Combination Indicator (E-TFCI), indicating the format of transportation of E-DPDCH, the re-transmission sequence number, which informs the HARQ sequence number, and a single bit called the ‘happy bit’ which indicates if the mobile is content with the data rate or not.

5.7.3 E-AGCH

The E-DCH Absolute Grant Channel or E-AGCH is the downlink physical channel. It tells the mobile about the maximum transmission power that it can use for data transmission. The 5-bit information tells the power level that E-DPDCH can use. One bit is used for indicating the absolute grant scope through which the scheduler can allow/not-allow the transmission of the HARQ process.

5.7.4 E-RGCH

The E-DCH Relative Grant Channel or E-RGCH is the downlink physical channel. It is used for adjusting the uplink data rate by scheduling the relative transmission power that the mobile is allowed to use. It is BPSK-modulated and transmitting cell-dependent.

5.7.5 E-HICH

The E-DCH HARQ Indicator Channel is a downlink channel. It is used for acknowledgments (positive/negative) for uplink packet transmission. Positive acknowledgment is used when the base station receives TTI correctly and negative when TTI is received incorrectly. As in the case of E-RGCH, this is BPSK-modulated and transmitting cell-dependent. The structure is similar to that of E-AGCH.

5.8 HSUPA Radio Resource Management

5.8.1 HARQ

Hybrid Automatic Repeat reQuest or HARQ in HSUPA is similar to that of HSDPA except that in the case of HSUPA it is fully synchronous and with IR (Incremental Redundancy) making re-transmissions more effective. Also, the buffer is maintained by the base station rather than by the mobile. IR and CC are present in the HSUPA as in the HSDPA (explained before in this chapter). The process is synchronous in the HSUPA. All the timing sequences are defined with respect to 2 ms and 10 ms (4 HARQ processes) and these tell which HARQ process is being used. The number of these processes is not required to be configured.

5.8.2 Scheduling

As in HSDPA, the scheduling is moved to the base station in HSUPA as well. But unlike in HSDPA where all of the power can be re-directed to a single user, in HSUPA resources are distributed evenly across all users, that is the approach is a dedicated one unlike in HSDPA where a shared approach is used. The scheduler in the HSUPA system does the task of allocating resources of existing users when admitting new users into the network. It has information about the interference in the uplink direction and is faster in controlling interference as compared to the scheduler in the RNC. The scheduling in HSUPA has better spectral efficiency (L1 HARQ) and is faster (base station-based scheduling). Three physical channels are used for this purpose: E-DPCCH in uplink with E-AGCH and E-RGCH in the downlink direction. There are two scheduling methods, long-term grants and short-term grants. Long-term grants are issued to many mobiles that can send data simultaneously, thereby the grants are increasing or decreasing according to the load of the cell and is done in the code domain. Short-term grants allow multiplexing mobiles in the time domain. Scrambling and channelization codes are not shared between various mobiles in order to allow the multiplexing uplink transmissions of many mobiles.

5.8.3 Soft Handover

In release 99, scheduling and HARQ are not required to operate in many base stations (to reduce the near–far problem) while in HSUPA, as such a problem does not exist, it is needed in a maximum of 4 cells. This number may increase in a mixed network, that is some base stations are not HSUPA-active. Scheduling is impacted by the uplink soft handover function. All the base stations are impacted by noise as all of them receive the signals from the mobiles – thereby an impact on transmission takes place. The handover control in RNC decides on the cells that are in 'active set' and the cell that is a serving HSUPA cell (this decides the cell that is in control of the HSUPA subscriber).

5.9 HSPA Network Dimensioning

Most of the factors remain the same for the HSUPA as in WCDMA networks. However, as there is change in link efficiency, the bit energy to noise density ratio does changes (E_b/N_t). EIRP can be calculated as:

$$\text{EIRP} = (\text{UE Maximum Transmit Power}) - (\text{UE cable, connector, combiner losses}) \\ + (\text{UE Transmit Antenna gain}) - (\text{Maximum Power Reduction})$$

The class of the mobile determines the UE maximum transmit power, for example a class 3 UE has a maximum transmit power of 24 dBm while a class 4 UE has a maximum transmit power of 21 dBm. The cable, connector losses are 0 dBi as the form factor is quite low. Similarly for internal mobile antennae, the antenna gain is taken to be 0 dBi. MPR is used in HSUPA to lower the operating point of the power amplifier. Multi-code operations of HSUPA causes significant time variations of power, which results in a large peak-to-average ratio (PAR) which puts extra constraints on the power amplifier of the UE.

$$\text{Receiver Sensitivity (dBm)} = (\text{Thermal Noise Density}) + (\text{Noise Figure}) + (\text{ROT}) \\ + (E_b/N_t) + (\text{Information Rate (in dB)})$$

Thermal Noise Density = kT (where k is the Boltzmann's constant and the temperature (T) is 290 K).

The receiver noise figure is taken to be 5.0 dB. The ROT or Rise of Thermal measures the interference level at the cell receiver antenna before the considered radio link is added into the system. ROT is the ratio between the total received power and thermal noise power. The summation of the signal, thermal and interference noise power is called the total received power. Interference is the sum total of background thermal noise and all the interference generated by users from the same and other cells. The minimum signal to noise ratio for the system to maintain the desired quality is E_b/N_t. This depends on factors such as outer loop power control block error rate (target 1 % to 10 %), the number of HARQ transmissions, mobility channels, traffic to pilot ratio and transport block sizes. Factors such as macro diversity, penetration loss, antenna gains and power amplifier back-off factor will as well be taken into account. The maximum distance for a mobile from the base station can be calculated from the maximum path loss.

Also, the propagation models applied are similar to those used in UMTS networks, that is the Walfish–Ikegami model, the Hata model, etc. (described earlier in this book). The HSUPA is operated with three different channel configurations, for example E-DCH, E-DPDCH and HS-DPCCH. The power consumption of all these three is taken into consideration for link budget analysis. Examples of link budgets are shown in Table 5.6. As mentioned above, the

Table 5.6 Examples of uplink and downlink HSUPA link budgets

Cell Edge Throughput		384	kbps	Cell Edge Throughput	64	kbps
RNC Databuild				Target BLER	10	%
PtxMax HSDPA		37.8	dBm	Propagation Channel	Pedest. A 3 km/hr	
Channel	HS-PDSCH	HS-SCCH		Channel	HSUPA	
Service	PS Data	Control		Service	PS Data	
Service Rate	384	–	kbps	Service Rate	64	kbps
Transmitter – Node B				**Transmitter – UE**		
Max Tx Power	37.4	27	dBm	Max Tx Power	24	dBm
Cable Loss	2	2	dBi	HS-DPCCH Overhead	2.5	dB
MHA Insertion Loss	0.5	0.5	dB	Tx Antenna Gain	2	dBi
Tx Antenna Gain	18.5	18.5	dB	Body Loss	0	dB
EIRP	53.4	43.0	dBm	EIRP	23.5	dBm
Receiver – Handset				**Receiver – Node B**		
Handset Noise Figure	8	8	dB	Node B Noise Figure	2	dB
Thermal Noise	−108	−108	dBm	Thermal Noise	−108	dBm
Downlink Load	80	80	%	Uplink Load	50	%
Interference Margin	7.0	7.0	dB	Interference Margin	3.0	dB
				Own Connection Interf.	0.06	
Interference Floor	−93.0	−93.0	dBm	Interference Floor	−103.1	dBm
SINR Requirement	4.5	1.5	dB	Service Eb/No	−0.5	dB
Spreading Gain	12.0	21.0	dB	Service PG	17.78	dB
Receiver Sensitivity	−100.5	−112.5	dBm	Receiver Sensitivity	−121.30	dBm
Rx Antenna Gain	2	2	dB	Rx Antenna Gain	18	dBi
Body Loss	0	0	dB	Cable Loss	0.5	dB
DL Fast Fade Margin	0	0	dB	Benefit of using MHA	0	dB
DL Soft Handover Gain	0	0	dB	UL Fast Fade Margin	1.8	dB
MDC Gain	0	0	dB	UL Handover Gain	1	dB
Building Penetration Loss	12	12	dB	Building Penetration Loss	12	dB
Indoor Location Prob.	90	90	dB	Indoor Location Prob.	90	%
Indoor Standard Dev.	10	10	dB	Indoor Standard Dev.	10	dB
Slow Fade Margin	7.8	7.8	dB	Slow Fade Margin	7.8	dB
Isotropic Power Required	−82.8	−94.7	dB	Isotropic Power Required	−118.2	dB
Allowed Prop. Loss	136.2	137.7	dB	Allowed Prop. Loss	141.7	dB
				Cell range		
				Frequency	2100	MHz
				Antenna height BS	30	m
				Antenna height MS	1.5	m
				Correction factor	0	dB
				Cell radius	1.23	km
				Site coverage area (clover	2.97	km^2

majority of uplink link budgets are similar to that of a R99 DCH HSUPA uplink link budget which makes use of E_b/N_o figures rather than SINR figures.

5.10 LTE (Long Term Evolution)

5.10.1 Introduction to LTE

Long Term Evolution or LTE is a project within the 3GPP standards which focuses on the improvement of the UMTS network in terms of spectral efficiency, improving services, lowering costs, etc. 3GPP is one of the four major wireless standards (HSPA+, 3GPP EDGE Evolution, 3GPP Ultra Mobile Broadband, Mobile WiMax-based on 802.16) that is referred to as 3.9G. This is not a standard but results in a new evolution of release 8 of the 3GPP specification. The work group focusing on various areas of LTE completed a study in 2006 resulting in the creation of a 'work item' for 3G LTE. Some of the specification areas that were focused on are described in the following.

5.10.1.1 Peak Data Rate

An instantaneous DL peak data rate of 100 Mb/s within a 20 MHz downlink spectrum allocation (5 bps/Hz) and an instantaneous UL peak data rate of 50 Mb/s (2.5 bps/Hz) within a 20 MHz uplink spectrum allocation.

5.10.1.2 User Throughput

The average user throughput in the uplink should be 2–3 times higher than release 6 (enhanced uplink), while the average throughput in the downlink should be 3–4 times higher than release 6 of the HSDPA.

5.10.1.3 Spectrum Efficiency

In a fully loaded network, the target for spectrum efficiency in the UL is 2–3 times higher than release 6 (enhanced uplink) while in the Dl, it is 3–4 times that of release 6 (HSDPA).

5.10.1.4 Mobility

The network across the network is to be maintained from 120 km/h to 350 km/h. Higher performance should support higher speed. For lower speeds, the E-UTRAN needs to be optimized for low mobile speeds.

5.10.1.5 Coverage

Targets related to mobility, throughout and spectrum efficiency to be met for cells with a range of 5 kms. For increased cell ranges up to 30 kms, similar targets with a slight degradation can be used.

5.10.1.6 User Plane Latency

In unloaded conditions and small IP packets, less than 5 ms.

5.10.1.7 Control Plane Capacity and Latency

In a 5 MHz bandwidth, at least 200 active state users should be supported. Transition times of less than 100 ms from the idle to the active state, and 50 ms between the dormant and active states should be present.

5.10.1.8 Spectrum

The operating frequencies include 1.25 MHz, 16 MHz, 2.5 MHz, 5 MHz, 10 MHz, 15 MHz and 20 MHz in UL and DL. The system should support content delivery over aggregated resources.

5.10.1.9 Multi Broadcast Multicast Services (MBMSs)

Both the voice and MBMS service delivery should be possible while reducing terminal complexity in terms of modulation, coding, UE bandwidth, etc.

5.10.1.10 Architecture

There will be a single E-UTRAN packet-based architecture. The support for E2E QoS will be present including the optimized back-haul communications protocols.

5.10.1.11 Inter-Operability

The system should co-exist with GERAN/ UTRAN-including terminals. Features like inter-system handovers should be possible with interruption times of less than 300 ms.

LTE is the next step towards the so called '4G' systems. It uses the existing work done in the fields of GSM, EGPRS, WCDMA, HSPA, etc. to reach up to higher user data rates with lower latency. The LTE networks are totally packet-switched, thereby allowing a wider service range that is supported by using the TCP/IP based standards. SAE or System Architecture Evolution of an All-IP network simplifies the network architecture (as discussed later and in the following chapter).

The specifications, although not complete, were approved in December 2007. Some of the key aspects of these specifications are:

- Increased spectrum flexibility: 1.25–20 MHz.
- Peak data rates: 100 Mbps DL/ 50 Mbps UL within a 20 MHz bandwidth.
- At least 200 active users in every 5 MHz cell. (i.e. 200 active data clients).
- Large cell: cell size of 5 km (optimal), 30 km sizes with reasonable performance and up to 100 km cell sizes supported with acceptable performance.
- High bandwidth/low latency to enable new services and revenue streams (sub-5 ms latency for small IP packets).

5.11 LTE Technology

5.11.1 Access Technology

As the bandwidth increased to 20 MHz, new access schemes are used in this technology: OFDM (Orthogonal Frequency Division Multiplexing) for the downlink and SC-FDMA for the uplink directions. OFDMA was chosen because of several characteristics: high potential for throughput due to link adaptation and frequency domain scheduling, it provides high performance in frequency selective fading channels, the base band receiver is less complex, less interference due to interference rejection combining and higher spectral efficiency. The frequency bandwidths that are used are 1.4 MHz, 3 MHz, 5 MHz, 10 MHz, 15 MHz and 20 MHz. E-UTRA is designed to operate in the frequency bands listed in Table 5.7. The sub-carrier spacing is 15 KHz in all of these bandwidths with clock 2N (∗x) multiples of

Table 5.7 Frequency bands in which E-UTRA can operate. © 2008. 3GPP™ TSs and TRs are the property of ARIB, ATIS, CCSA, ETSI, TTA AND TTC who jointly own the copyright in them. They are subject to further modifications and are therefore provided to you "as is" for information purposes only. Further use is strictly prohibited

	Uplink (UL) eNode B receive UE transmit			Downlink (DL) eNode B transmit UE receive			UL-DL Band separation	
E-UTRA Band	F_{UL_low}	–	F_{UL_high}	F_{DL_low}	–	F_{DL_high}	$F_{DL\_low} - F_{UL\_high}$	Duplex Mode
1	1920 MHz	–	1980 MHz	2110 MHz	–	2170 MHz	130 MHz	FDD
2	1850 MHz	–	1910 MHz	1930 MHz	–	1990 MHz	20 MHz	FDD
3	1710 MHz	–	1785 MHz	1805 MHz	–	1880 MHz	20 MHz	FDD
4	1710 MHz	–	1755 MHz	2110 MHz	–	2155 MHz	355 MHz	FDD
5	824 MHz	–	849 MHz	869 MHz	–	894 MHz	20 MHz	FDD
6	830 MHz	–	840 MHz	875 MHz	–	885 MHz	35 MHz	FDD
7	2500 MHz	–	2570 MHz	2620 MHz	–	2690 MHz	50 MHz	FDD
8	880 MHz	–	915 MHz	925 MHz	–	960 MHz	10 MHz	FDD
9	1749.9 MHz	–	1784.9 MHz	1844.9 MHz	–	1879.9 MHz	60 MHz	FDD
10	1710 MHz	–	1770 MHz	2110 MHz	–	2170 MHz	340 MHz	FDD
11	1427.9 MHz	–	1452.9 MHz	1475.9 MHz	–	1500.9 MHz	23 MHz	FDD
12	[TBD]	–	[TBD]	[TBD]	–	[TBD]	[TBD]	FDD
13	777 MHz	–	787 MHz	746 MHz	–	756 MHz	21	FDD
14	788 MHz	–	798 MHz	758 MHz	–	768 MHz	20	FDD
...								
33	1900 MHz	–	1920 MHz	1900 MHz	–	1920 MHz	N/A	TDD
34	2010 MHz	–	2025 MHz	2010 MHz	–	2025 MHz	N/A	TDD
35	1850 MHz	–	1910 MHz	1850 MHz	–	1910 MHz	N/A	TDD
36	1930 MHz	–	1990 MHz	1930 MHz	–	1990 MHz	N/A	TDD
37	1910 MHz	–	1930 MHz	1910 MHz	–	1930 MHz	N/A	TDD
38	2570 MHz	–	2620 MHz	2570 MHz	–	2620 MHz	N/A	TDD
39	1880 MHz	–	1920 MHz	1880 MHz	–	1920 MHz	N/A	TDD
40	2300 MHz	–	2400 MHz	2300 MHz	–	2400 MHz	N/A	TDD

3.84 MHz. OFDM is a system where the allocated spectrum is divided into thousands of carriers, each of different frequency, carrying a part of the signal. The OFDMA offers certain other advantages due to which it is chosen for LTE. Some of these advantages include: flat architecture, that is capacity increases up to 50 % over CDMA due to frequency domain scheduling and it is possible to have this along with the base station scheduling; OFDMA can be used in conjunction with the SSC-FDMA in the uplink direction; the MIMO (Multiple Inputs Multiple Output) technology is easily compatible with the OFDMA that allows LTE to achieve data rates; it is easy to provide different bandwidths with OFDMA. However, OFDMA cannot be used in the uplink direction as it has high peak-to-average ratios that require a linear transmitter. As these amplifiers have low efficiency, OFDMA remains unsuitable for uplink directions. Hence, SC-FDMA, which has a better power amplifier efficiency, is used for the uplink direction.

5.11.1.1 Downlink

Modulations that are supported in the downlink direction are QPSK, 16QAM and 64QAM. Though the number of OFDM sub-carriers is dependent upon the used bandwidth, a maximum of 1200 sub-carriers are available. The sub-carriers do not occupy the whole of the bandwidth as both in the uplink and downlink directions CPs (cyclic pre-fixes) are present. The duration of one radio frame is 10 ms while the transmitting timeslots are 0.5 ms long while the sub-frames are 1.0 ms long. The downlink signal generation is shown in Figure 5.4. The scrambling coded bits transmitted on the physical channel are modulated to generate complex symbols. These are them mapped onto one or many transmission layers where these are pre-coded and mapped for each antenna port, resulting in the generation of a complex value time domain OFDM signal. Layer mapping and pre-coding are MIMO-related processing functions. MIMO systems can be defined as a general wireless communication link where a link for the transmission and receiver is equipped with multiple antenna elements. MIMO systems are chosen ahead of the single antenna systems because of their ability to turn multi-path propagation, a pitfall of wireless communication, into a benefit for the user. They take advantage of random fading and multi-path delay spread for multiplying transfer rates.

5.11.1.2 Uplink

As mentioned above, SC-FDMA multiplexing is used in the uplink direction. As in the downlink direction, QPSK, 16QAM and 64QAM are used as modulation techniques. The data rate in the uplink direction can be increased by usage of MIMO/Spatial Division Multiple Access (SDMA). This increase in data rate is also dependent upon the number of antennae

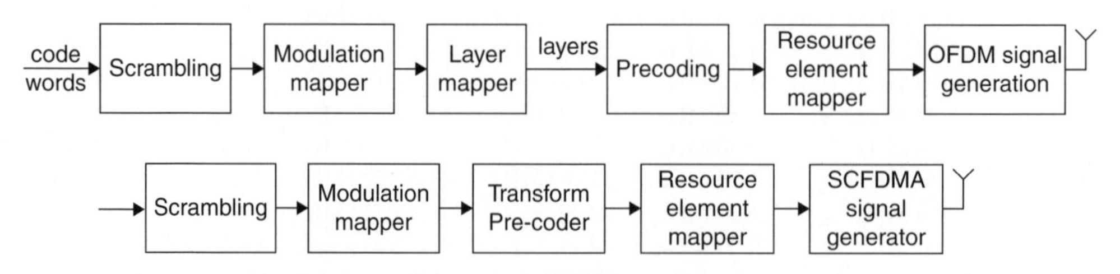

Figure 5.4 OFDM and SC-FDMA signal generation.

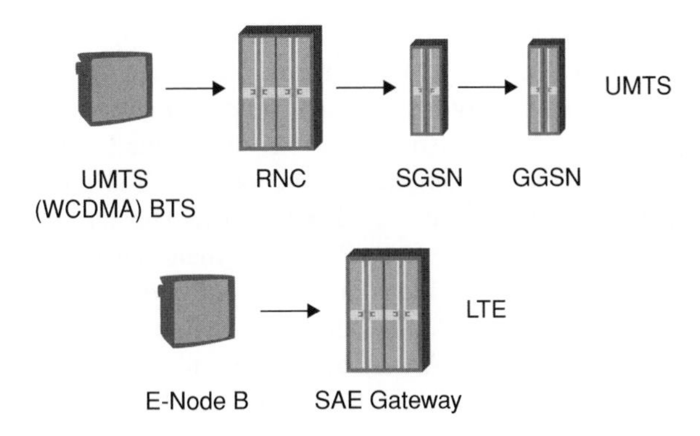

Figure 5.5 UTMS and LTE.

used at the E-Node Bs. The uplink signal generation process (as shown in Figure 5.4) starts with scrambling followed by modulation of the scrambled signal. This is then pre-coded to generate a complex value signal which is mapped onto resource elements, leading to the generation of a complex-valued time domain SC-FDMA signal.

5.11.2 LTE Network Architecture

The LTE network architecture is an overall flat architecture. As seen in Figure 5.5, the UMTS packet core network consists of the base station, RNC, SGSG and GGSN. The network architecture in LTE is much more flat. It consists of an e-Node B and SAE gateway. This network is based on a TCP/IP protocol with higher service levels like voice, video, messaging, etc. built on it. Based on this, feasibility studies related to All IP Networks (AIPNs) were started in 2004 by the 3GPP. The air-interface consisting of OFDMA and SC-FDMA together is called E-UTRA or Enhanced UTRA. The prefix E- is to be used in evolved networks beyond the former UMTS components. In the air-interface, a MIMO (Multiple Input and Multiple Outputs) system is used that employs up to four antennae per base station while turbo channel coding is used. The overall architecture is described in technical specification 36.300 and 36.401. The E-Node Bs are connected to each other by X2 interfaces and are also connected to the core network that is an evolved packet core (EPC) network via S1 interfaces.

5.11.2.1 E-Node B

The functions of E-Node B in LTE are similar to those of the base station in the GSM/UMTS (shown in Figure 5.5). They include RRM functions such as admission control, mobility control, radio bearer control, dynamic allocation of resources to UEs, etc. Other functions include routing of the user data to the serving gateway, scheduling and transmission of the paging and broadcast messages, IP header compression and encryption, measurement and reporting (e.g. for mobility and scheduling-related issues, etc.).

5.11.2.2 SAE or System Architecture Evolution

SAE is also called an Evolved Packet Core (EPC) or Evolved Packet Services (EPSs). According to 3GPP release 8, EPS is a framework to migrate the 3GPP systems to a low latency and higher data rate system. It is a packet switched system that supports data, including voice. There are two nodes: MME (Mobility Management Entity) and GW (Gateway). Mobility Management Entity, as the name suggests, is responsible for managing the mobility and parameters related to security. MME is also responsible for the idle UE tracking, roaming and authentication-like features. Bearer-management functions are also handled by the MME. The gateway is of two types: Serving Gateway and PDN Gateway. The Serving Gateway is the one that terminates the interface towards the E-UTRAN. The main functionalities include packet routing and forwarding, mobility anchoring for inter 3GPP mobility (also called the 3GPP anchor function), idle mode packet buffering, local mobility anchor point for e-Node B handovers, LI (Lawful Interception), etc. At any given time, for each UE, there is one single Serving Gateway. The other gateway is the P-GW or the one that terminates the interface towards the PDN (Packet Data Network). Its main functions include anchoring mobility between 3GPP and non-3GPP networks (also called the SAE anchor function), charging support, LI, packet screening, policy enforcements, IP address allocation to user equipment, etc.

5.11.3 Channel Structure

The physical channels used in LTE are listed in the following.

5.11.3.1 Downlink Channels

PBCH: Physical Broadcast Channel.

PHICH: Physical Hybrid ARQ Indicator Channel.

PDSCH: Physical Downlink Shared Channel.

PMCH: Physical Multicast Channel.

PCFICH: Physical Control Format Indicator Channel.

PDCCH: Physical Downlink Control Channel.

5.11.3.2 Uplink Channels

PUCCH: Physical Uplink Control Channel.

PUSCH: Physical Uplink Shared Channel.

PRACH: Physical Random Access Channel.

Information transfer to the MAC and higher layers are offered through the physical layers. The channels that are used for this purpose are listed in the following.

5.11.3.3 Downlink

BCH: Broadcast Channel.

DL-SCH: Downlink Shared Channel.

PCH: Paging Channel.

MCH: Multicast Channel.

5.11.3.4 Uplink

UL-SCH: Uplink Shared Channel.

RACH: Random Access Channel.

The MAC layer offers the following logical channels.

5.11.3.5 Control Channels

BCCH: Broadcast Control Channel.

PCCH: Paging Control Channel.

CCCH: Common Control Channel.

MCCH: Multicast Control Channel.

DCCH: Dedicated Control Channel.

5.11.3.6 Traffic Channels

DTCH: Dedicated Traffic Channel.

MTCH: Multicast Traffic Channel.

5.11.4 LTE Protocol Structure

As mentioned above, the E-Node Bs are connected to each other via X2 while they are connected to the EPC via S1. As shown in Figure 5.6, the PDPC, RLC, MAC sub-layers that terminate on the network side perform function such as ciphering, header compression, ARQ, HARQ, etc. In the control plane protocol as shown in Figure 5.7, the PDPC sub-layer performs functions such as ciphering and integrity protection. The functions performed by the RLC and MAC are similar to that of the user plane protocol, for example compressions, ARQ, etc. The Radio Resource Controller performs functions such as radio bearer control, connection/mobility management, UE measurement reporting and control, paging, etc.

The NAS protocol is used for communications between the UE and MME only.

Figure 5.8 shows a functional split between the E-UTRAN and EPC (both connected via a S1 link). There is a logical separation between the data and signalling. Both the E-UTRAN and EPC are separated from transport functions and are not tied to its (transport functions) addressing schemes as well. As shown in the protocol stack, the E-Node B fully controls the mobilities of the RRC connections (in UMTS, it was done by RNC). Security issues

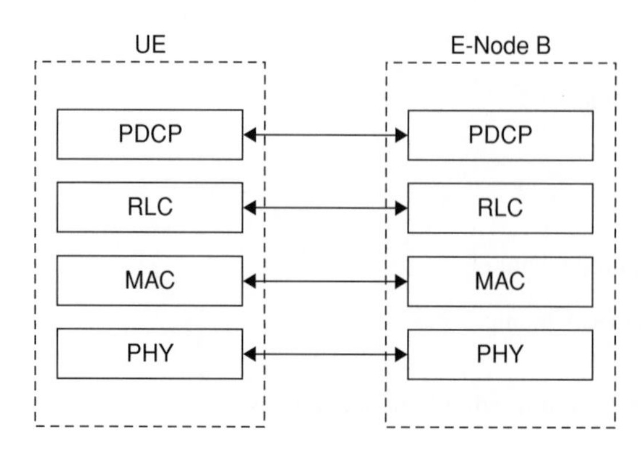

Figure 5.6 User-plane protocol stack.

are handled by the MME. The anchoring function is handled by the gateways, including the address allocation for the UEs.

5.11.4.1 Layer 1

Layer 1 can adapt to various frequency spectra. As mentioned before, the radio frame is of 10 ms duration. It has 20 slots with each slot of 0.5 ms. Two adjacent slots, that is 1 ms, form one sub-frame. This is called Type 1 and is used for the FDD mode. In Type 2, which is used for the TDD mode, two half-frames of 5 ms are used and have eight slots of 0.5 ms.

5.11.4.2 Layer 2

It contains the three sub-layers: MAC, RLC and PDPC. As mentioned above, they perform functions such as RRC connection/mobility management, UE measurement reporting and control, paging, radio bearer management, etc.

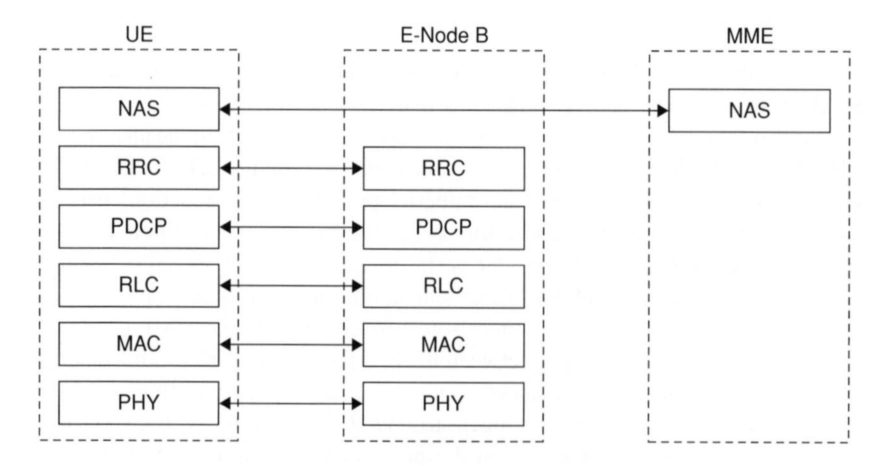

Figure 5.7 Control-plane protocol stack.

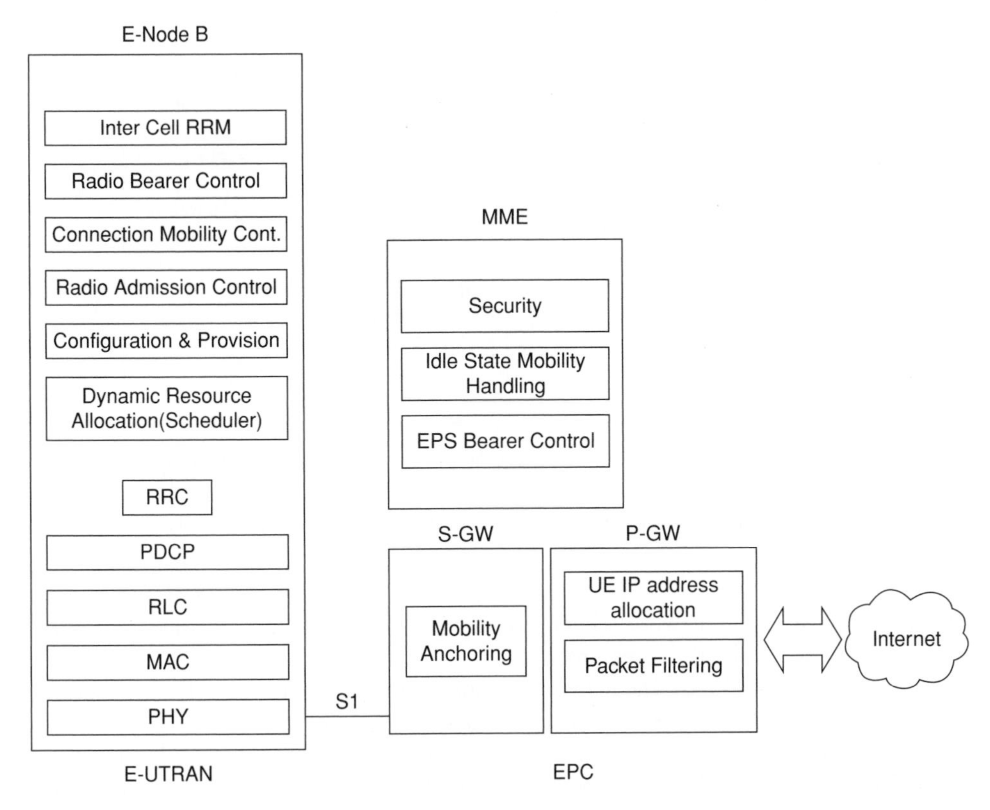

Figure 5.8 Protocol structure in LTE.

5.12 Radio Resource Management

The key aspects of radio resource management include functions such as radio admission control, radio bearer control, mobility control, dynamic resource allocation, load balancing, dis-continuous reception, quality of services, multi-media broadcast and multi-cast services (MBMSs), security, etc. Many of these functions have been explained in this or previous chapters. However, let us have some insight into MBMS and security. In LTE, it is possible to multicast/broadcast over a single frequency network (MBSFN). In this, a time-synchronized waveform is transmitted from multiple cells for a given duration. The whole process works in a way that multi-cell combining happens over the air in a way that it appears (to UE) as its transmission is taking place from a large cell. The CP is used as a filler to cover the difference in propagation delays. There are two layers: the access stratum (AS) and NAS. The AS security layer is used for RRC security and user plane protection while the NAS is used for the evolved packet layer.

5.13 Security in LTE

There are five security feature groups which are defined in the 3GPP recommendation TR 33.102, as described in the following (see Figure 5.9).

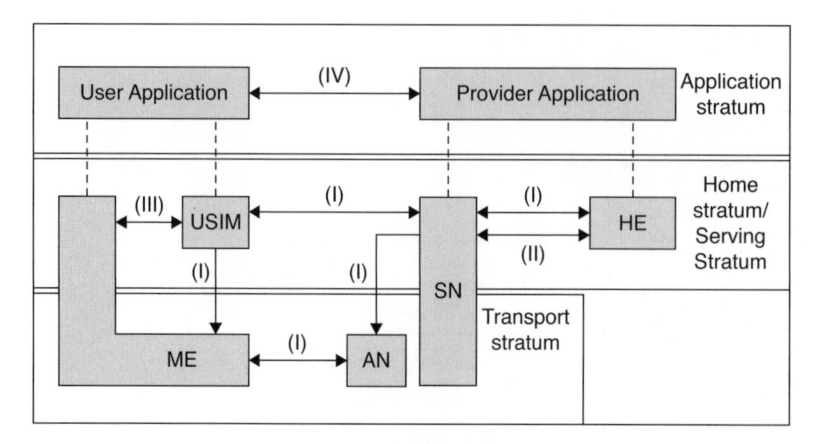

Figure 5.9 Security architecture in LTE. © 2002. 3GPP™ TSs and TRs are the property of ARIB, ATIS, CCSA, ETSI, TTA AND TTC who jointly own the copyright in them. They are subject to further modifications and are therefore provided to you "as is" for information purposes only. Further use is strictly prohibited.

Security to Users. The network access provides users with secure access to services and protects against attacks on the access interfaces.

Security to Network Nodes. The network domain enables nodes to securely exchange signalling data and user data and protects against attacks on the wire line network.

Security to Mobile Stations. The user domain provides secure access to mobile stations.

Security to Application Domain. The application domain security enables applications in the user and provider domains to securely exchange messages.

Security Related to Visibility and Configurability. The visibility and configurability of security allow the user to learn whether a security feature is in operation or not and whether the use and provision of services should depend on the security feature.

5.13.1 Network Access Security

There are a few features that are related to network access related security. These include the ones related to user – identity confidentiality, location confidentiality and un-traceability. The identity, location and the different services delivered cannot be 'eavesdropping' on the radio access link. Entity authentication includes both the user and network authentication. These take place at each connection set up between the user and the network. Confidentiality of the data on the network access link is very much necessary from a security perspective. This confidentiality is in terms of both user data and signalling. The following confidentiality is the data integrity wherein the receiving entity is able to verify that the signalling data has not been modified in an unauthorized way. The mobile equipment identification is done as well by the serving network.

5.13.1.1 Network Domain Security

This will involve the features being developed that will enable the information exchange between 3GMS providers.

5.13.1.2 User Domain Security

The access to the USIM is restricted until the USIM has authenticated the user. This makes sure that access to the USIM is restricted to authorized user(s). This is done securely by enabling a secret code that is known only to the user and stored in the USIM. Thus, when the code is inserted, access to the USIM is given to the user. Another feature is the enablement of the USIM-terminal link. This is again done by the secret code that is stored in the USIM and the terminal.

5.13.1.3 Application Security

There is a possibility to provide the capability for operators/third parties to create applications that are resident on the USIM. There exists the need to secure messages which are transferred over the network to applications on the USIM. These security mechanisms are developed and described in 3GPP TS 23.048.

5.13.1.4 Secure Visibility and Configurability

Usually, the security features are 'transparent' to the user. However, in certain cases, these features are visible to the user and should be provided to them. Some of these include information given to the user on the confidential data being protected or the level of security that is provided when the user moves from a higher to a lower level security, for example UMTS-GSM. Apart from this, the user is able to configure the use or provision service depending on the availability of the security feature. Some of the features include enabling/disabling of USIM authentication, accepting/rejecting incoming non-ciphered calls, setting up/not setting up non-ciphered calls and accepting/rejecting the use of certain ciphering algorithms.

6

OFDM and All-IP

6.1 Introduction to OFDM

OFDM stands for Orthogonal Frequency Division Multiplexing. A special case of Frequency Division Multiplexing or FDM, OFDM is both a modulation and multiplexing technique. The concept of OFDM has existed since around 1966; however, it has really found its importance beyond the research laboratories only recently in cellular communications. OFDM is used as a modulation technique in many systems, such as DAB (Digital Audio Broadcasting), DVB (Digital Video Broadcasting), WLAN (Wireless Local Area Network), PCS (Personal Communication System), etc. Some major developmental milestones include the first OFDM schemes, including the orthogonal QAM proposed in 1996. Although the first patents were issued in 1970 it was not until 1987 that OFDM was employed for digital broadcasting. OFDM was used for ADSL, HDSL, DAB, DVB, etc. in 1991–1997, along with the 802.11a/g standards for WLAN (1999–2002). It is expected that OFDM will become a key technology for 4G systems.

In OFDM, the signal is split into several narrow band channels at different frequencies, modulated (e.g. PSK modulation) by data and then re-multiplexed by the OFDM carrier to create a sub-carrier. There have been many ways to explain this but a 'truck-load' analogy might suit the best. Consider a truck that is full of a load, for example 400 kg, that needs to be transported from point A to point B. A large truck can carry the load of 400 kg or four smaller trucks can carry 100 kg each, as shown in Figure 6.1. The total data transferred is still the same. However, if an accident occurs, only a quarter of the load gets damaged. These four small trucks are OFDM sub-carriers. However, they still need to be orthogonal and multiplexed, for example using Frequency Division Multiplexing. PSK is quite similar to FDM, but modulation makes a difference between the two with more focus on the interference reduction. Data that are carried by the number of orthogonal sub-carriers are divided into several parallel data channels – one for each sub-carrier. These are then modulated at a lower rate thereby maintaining the data rates to a level similar to a conventional single carrier. OFDM is a special form of the multi-carrier multiplexing (MCM – single data stream multiplexed over sub-carriers that have lower data) scheme.

Cellular Technologies for Emerging Markets: 2G, 3G and Beyond Ajay R. Mishra

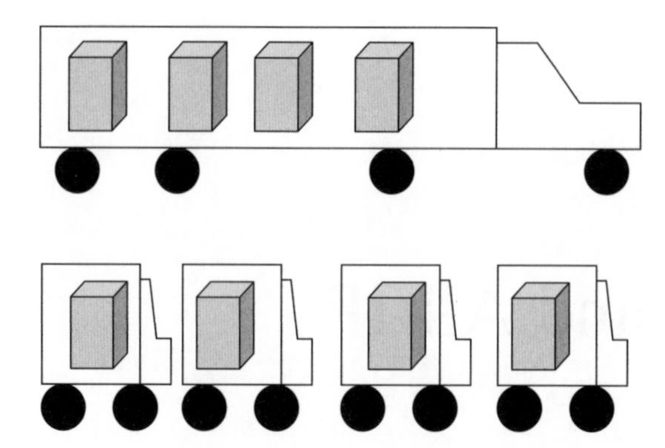

Figure 6.1 One big truck carrying all cargoes versus four small trucks carrying equally split cargoes.

Another important aspect to understand here is that no channel equalizer is required and these multiple narrow band signals can be implemented by using Fast Fourier Transformation (FFT) instead of modems. Later on, Discrete Fourier Transformation (DFT) was used (in the 1970s). Multi-carrier schemes are able to cope with bad channel conditions such as interference, attenuation due to multi-path, etc. better than single carrier signals.

OFDM is a system where the signal is a superimposition of many modulated sub-channel signals. This means that the resultant signal may show a peak value against the average signal level resulting in harmonic distortion of power. This induces adjacent channel interference. Some ways and means have been identified to counter this problem such as creation of headroom by not driving the power amplifier to saturation. This fundamentally would lead to reducing the peak-to-mean power ratio. Another issue is the synchronization between the transmitter and receiver. The higher the time and frequency synchronization, the higher is the link quality. There is also a possibility of inter-carrier interference, that is between the sub-carriers. These are due to a frequency mismatch between the transmitter and receiver oscillators. This can be a serious issue if multi-path is added to it, resulting in problems related to speedily moving vehicles. There are again ways and means developed to correct any problems related to synchronization, for example algorithms related to pilot bits and timing tracking, etc.

OFDM offers certain advantages such as:

- High spectral efficiency.
- Single frequency network (several transmitters transmit the same signal over the same frequency channel).
- Efficient implementation using FFT and DFT.
- Capable to adapt itself to severe conditions and robust against Interference (ISI, Co-channel), fading, etc.
- Low latency (latency is defined as the time it takes for the network to respond to a user command).
- Complex channel equalizers not required.

OFDM also offers some areas of concern such as:

- Problems resulting from the instantaneous signal peaks due to superimposition of sub-carrier signals.
- Performance related problems due to synchronization.

6.2 OFDM Principles

6.2.1 Frequency Division Multiplexing

FDM or the frequency division multiplexing technique is one in which many signals are combined on single communication channels wherein each of these sub-channels has a different frequency within the main channel. Due to this multiplexing, the result is as expected: a high bandwidth channel that is complex in nature (due to all the sub-channels being a part of it). At the receiver, a de-multiplexer is used to separate the channels. The sub-channel signal transmission is done at maximum speed. Let us understand this with an example. Assume that there are four channels, each having a centre frequency of 49 Hz, 50 Hz, 51 Hz and 52 Hz, as shown in Figure 6.2. This is to be transmitted over a single channel. This channel then should have a bandwidth that can carry these four frequencies. Both at the transmitter and receiver ends, a multiplexer and de-multiplexer are used. Also, band-pass filters are used at the receiver end so that the correct frequency is then received.

The FDM system is not sensitive to delays due to propagation; hence the channel equalization techniques used in TDM are not as complex. On the contrary, bandpass filters which are more complex and costly are needed to be used, thus making TDM more attractive in this aspect. Linear amplifiers that are more complex and expensive are used in these systems compared to the 'non-liners' (less complex and less expensive) which are used in TDM systems.

6.2.2 Orthogonality

As mentioned above, the sub-carriers are orthogonal to each other. As the carrier is a sine/cosine wave, the total areas under a sine/cosine wave is zero. And thus, the area under n sine waves

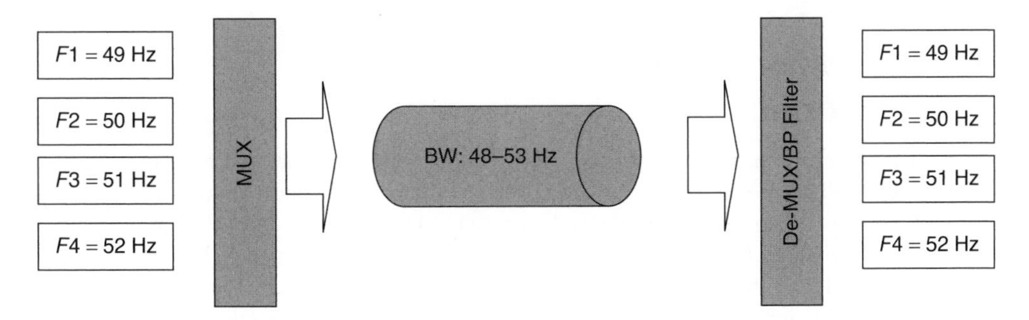

Figure 6.2 Frequency division multiplexing.

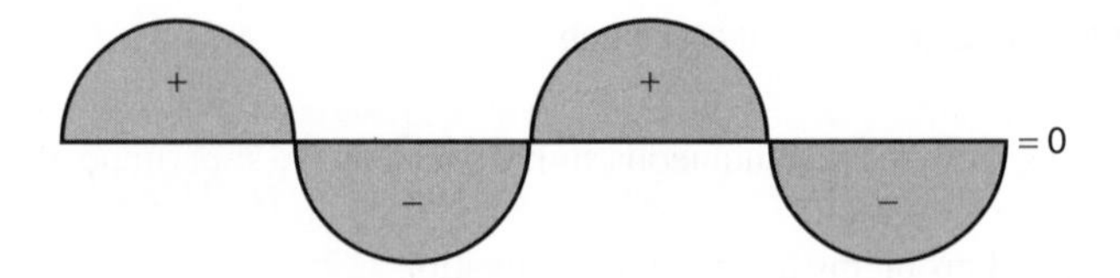

Figure 6.3 Area under a sine/cosine wave.

is also zero. Mathematically speaking, if all the sinusoids of frequency n are multiplied by a sinusoid of frequency m/n, then the result is zero. The area under the product of two waves is also zero as shown mathematically below (see Figure 6.3).

$$\begin{aligned} p(t) &= \sin mwt \times \sin nwt \\ &= \frac{1}{2}\cos(m-n) - \frac{1}{2}\cos(m+n) \end{aligned}$$

As they are sinusoids, they hence can be written as:

$$= \int_0^{2\pi} \frac{1}{2}\cos(m-n)wt - \int_0^{2\pi} \frac{1}{2}\cos(m+n)wt = 0 - 0$$

All sine and cosine waves of the m and n integers called harmonics are orthogonal to each other.

Even in tight spectral conditions, due to orthogonality, many sub-carriers can be transmitted without interference from each other. When the orthogonality concept is applied to FDM, the transmitter and receiver design gets simplified as no filter is needed (shown in Figure 6.2). The term orthogonality signifies the fact that the sub-carriers are 90 degrees to each other. This is possible be assigning the sub-carrier frequencies to these channels. The application of orthogonality not only results in reduction of cross-talk between these carriers but also negates the presence of a guard band. Also, due to orthogonality, higher spectrum efficiency is utilized. If we assume that there are N sub-carriers, while the difference between them is Δf, then the total bandwidth requirement is $N\Delta f$.

6.2.3 Modulation in OFDM

6.2.3.1 QAM-OFDM

As explained before, in OFDM there are sub-channels that are separated by a guard band. For efficient utilization of the spectrum, the overlap of the sub-channels is done. In the QAM method, the input data stream is encoded using QAM symbols. A QAM symbol block is inputted to the parallel-to-serial converter. This result is an in-phase signal. These signals are then transmitted over the radio channel and de-multiplexed at the receiver (shown in Figure 6.4). These QAM signals are filtered and recovered at the receiver using the parallel-to-serial converter and QAM de-modulator. This system requires a narrower bandwidth, that is better spectral efficiency than most of the other systems. However, this is not the most spectrally

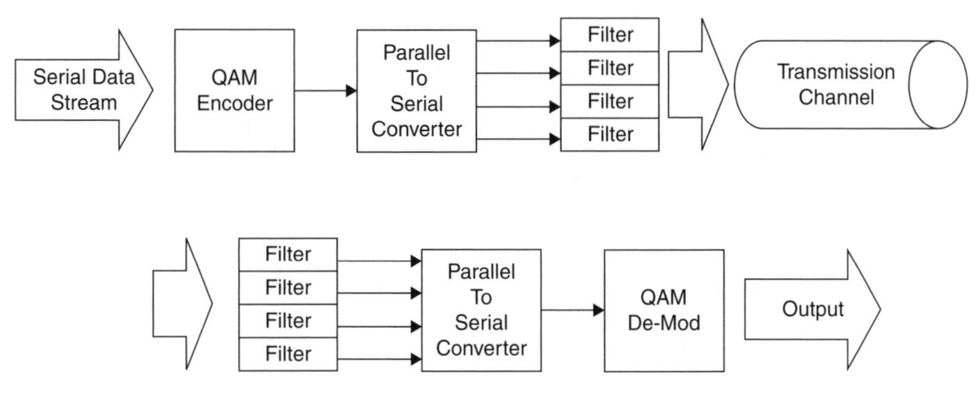

Figure 6.4 QAM-OFDM system.

efficient system as there is frequency spillage due to adjacent frequency sub-bands (in sub-channels). This leads to requirements of some amount of a guard band and increased spacing between the sub-bands – leading to lower spectral efficiency.

6.2.3.2 DFT-OFDM

What is the role of FFT? The output of FFT is a frequency domain signal. Fast Fourier Transform (FFT) takes a random signal, multiplies it with complex exponentials over the range of frequencies. After this, it sums each product and plots the results that are coefficient of frequencies. These coefficients are called the spectrum and signify the amount of frequency in that signal. Inverse FFT as the name suggests converts the signal back into the time domain. Thus, FFT and IFFT are a matched linear pair. However, one needs to understand that IFFT is only a mathematical concept and that it really doesn't matter what is the input and output of it, that is both the FFT and IFFT give the same outputs for the same inputs. As discussed above (QAM-OFDM), the fundamental transmission technique remains the same as before. The only difference is the addition of FFT. An inverse FFT is applied to the output of the QAM encoder, resulting in complex time domain samples. To obtain a final transmission signal a series of conversions happens, such as samples being quadrature mixed; the result of which is conversion to an analogue signal. These analogue signals at their carrier frequency are modulated sine and cosine waves and this is the resultant signal that is used for transmission. At the receiver end, a similar procedure happens: quadrature mixing–sine/cosine modulation–low pass filtration–analogue to digital conversion–FFT to convert back into the frequency domain–N parallel streams–parallel to serial converter resulting in the original data stream. This helps the signal to achieve a higher level of protection against the fading effect. The analogue concept of FFT can be applied to the digital domain by using the Discrete Fourier Transform (DFT). The IFFT counterpart is the IDFT or Inverse Digital Fourier Transform. However, practically the OFDM systems are implemented using the FFT/IFFT that is mathematically equivalent to DFT/IDFT. Frequency domain and time domain signals are shown in Figure 6.5.

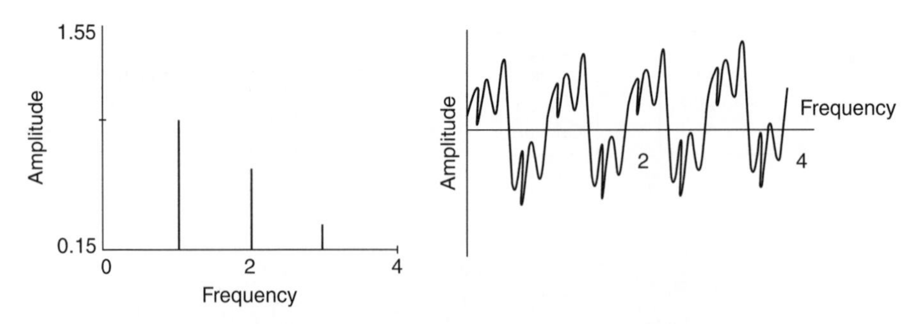

Figure 6.5 Frequency domain and time-domain signal.

6.2.4 Inter-Symbol and Inter-Carrier Interference

Let us now understand inter-symbol interference (ISI) and inter-carrier Interference (ICI) before we look into the importance of cyclic prefix. Most of the wireless systems with multipath face the problem of inter-symbol interference. When the signal is transmitted from a transmitting antenna, it travels on different paths to reach the receiving antenna. As the signal travels through different path lengths, each signal arrives at the receiving antenna with a delay that is different from other signals reaching the receiver. This leads to distortions in the received signals. When the signal distortions take place due to previous signals, this results in inter-symbol interference. There are other symbols as well involved in this system; hence ISI is not only due to previous symbols but also due to other symbols. There is interference between symbols of the own sub-carriers resulting in inter-carrier interference.

6.2.5 Cyclic Prefix

Two things happen when the transmission of the signal happens over a dispersive channel: one is the destruction of channel orthogonality (due to channel dispersion) and the other is transmission of multiple OFDM symbols. The former causes inter-carrier interference (ICI) and the latter causes inter-symbol interference (ISI). The problem loss of orthogonality can be resolved by adding a cyclic prefix. Due to addition of the cyclic prefix, the ICI can be prevented while introduction of a silent guard band prevents ISI. In principle, cyclic prefix, as the name suggests, is added at the beginning and is similar to the end of the symbol – thus making it look cyclic (block diagram shown in Figure 6.6). This is used with modulation to counter the effect of multipath by making it settle before the arrival of the main data. The signal is sinusoidal of infinite duration. As the real signals are time limited, by adding a prefix of the end of the signal, it is made to appear as circular. The length of the prefix matches that

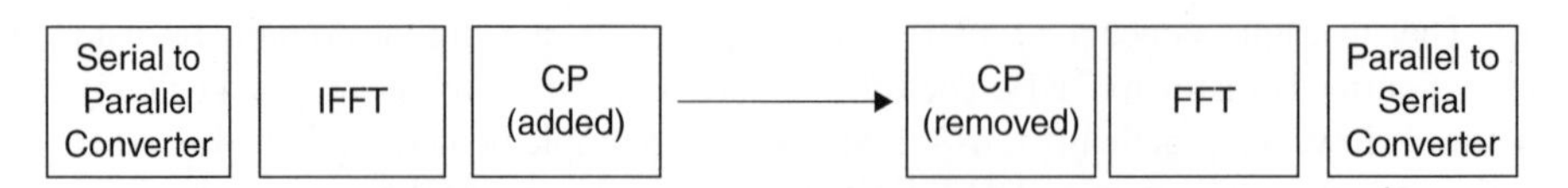

Figure 6.6 Block diagram of CP in the OFDM signal flow.

of the guard interval thus making sure that the signals become orthogonal to each other before actually it is 'used'; hence the system becomes more robust towards multipath.

6.2.6 Coded OFDM (C-OFDM)

This is a combination of error coding and OFDM modulation results in a coded OFDM system. During the transmission process of the sub-carriers, one or more may be impaired or even lost. This results in a stream of bit errors. Though this might be difficult to correct, interleaving at source would result in bit errors being far apart and easy to correct by the decoder. Interleaving and coding the bits prior to the IFFT are the parts of the process in C-OFDM resulting in the adjacent bits being taken in the source data and spreading across the sub-carriers. In the C-OFDM system, channel information from the equalizer is used to find the reliability of the received bits. This information is used by the decoding blocks to assign proper weights (e.g. marking bits as 'low confidence' in the case of 'frequency low') when making decoding decisions.

6.3 MIMO Technology

For countering the effects of multipath fading and increasing the performance – quality and capacity of the wireless systems – smart antennae are used. There are four kind of smart antennae: beam forming, spatial diversity and space time spreading, SDMA (Space Division Multiple Access) and Multiple Input and Multiple Output (MIMO). Beam-forming smart antennae are half wavelengths that are used for creating spatially selective transmitters or receiver beams helping to reduce the impact of co-channel interference and increase the number of users in the system. When multiple antennae are placed quite far from each other (typically 10λ), the individual received signals are the ones that have experienced independent fading and at the same time, achieved maximum diversity gain. This smart antenna system increases system robustness. The SDMA system distinguishes between the individual subscribers utilizing subscriber-specific spatial signatures, thus using the same frequency band to support many subscribers. And finally, multiple antenna systems like MIMO (Multiple Input and Multiple Output) increase the throughput of the system in terms of number of bits that can be transmitted by a subscriber for a given bandwidth. The improvement of the system performance is multi-fold and hence is found to be utilized in high rate WLAN, 3G and next generation 4G systems.

6.3.1 MIMO System

The multiple input and multiple output system is shown in Figure 6.7. In these systems, time is complemented with spatial dimension inherent in the use of multiple spatially distributed antennae. There are multiple antennae on the transmitter side and multiple antennae on the receiver side. Thus for each of the x transmitting antennae, there are y receiving antennae. Thus there are xy propagation paths. The combination takes place in such a way that BER and data rate are improved for the user. As shown in Figure 6.7, the transmitting block is fed with the compressed binary stream signal. After processes such as error coding and modulation

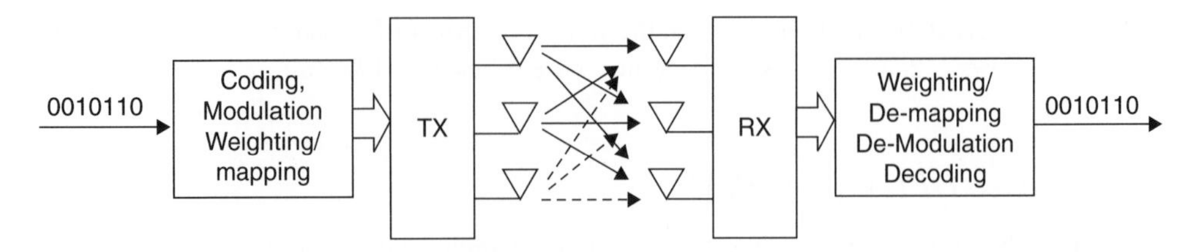

Figure 6.7 MIMO system.

mapping, the separate symbol streams are mapped to one of the antennae. The signal is then frequency modified, filtered, amplified and finally transmitted. At the receiving end, processes such as demodulation and de-mapping take place to receive the original data. The main aspect of the whole process is the weighting algorithm which is the reason for intelligence of the system. The response of each antenna element to a given signal, along with the interference signals, can be optimally combined to weights selected as a function of each element response. Another concept used is that of the beam forming which helps in increasing the focus of the energy, thereby increasing the SNR.

In MIMO systems the improvement happens due to the ability of the system to turn around the multipath propagation for the benefit of the subscriber by taking advantage of the random fading and multipath delay spread for multiplying transfer rates. The other advantage of the MIMO system is the ability to increase the link capacity. Due to this, many benefits happen, such as increase in capacity and throughput, high data rates, improvement in signal capacity, low power consumption and increase in spectral efficiency.

6.3.2 MIMO Mode of Operation

The MIMO system operation takes place in two modes: the diversity mode and the spatial multiplexing mode. In the diversity mode, multiple antennae are used, as shown in Figure 6.8.

In the diversity mode, the signal quality is higher due to implementation of multiple antennae at the receiver or transmitter or both of the ends. We have seen the concept of diversity in earlier chapters as well (e.g. Chapter 2). There are many diversity schemes that include simpler ones having multiple antennae at the receiver end. In this the receiver system automatically attaches itself to the antennae having better receiver power. Other schemes include the combining of all the received signals over multiple antennae. This is called maximum radio combining (MRC) and results in maximizing of the signal-to-noise ratio. Transmit diversity is also possible, but is quite complicated to implement. The simplest way to implement this is to use the same antenna

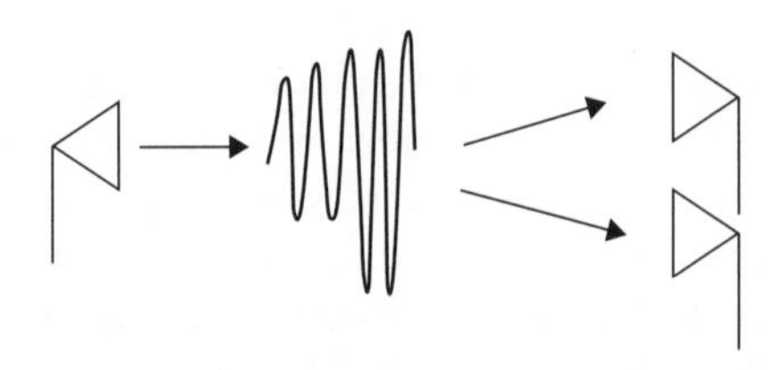

Figure 6.8 Diversity in MIMO systems.

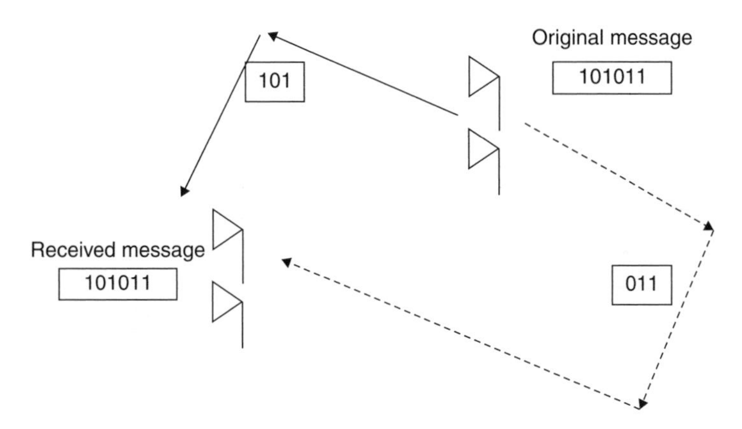

Figure 6.9 Spatial multiplexing.

for transmission from which the signal comes from the target receiver before. The information is transformed to RF to avoid interference. Application of diversity schemes enables an increase in the coverage areas and throughput of the system. In spatial multiplexing systems (shown in Figure 6.9), different parts of the data are sent on multiple paths in parallel. At the receiver end, a de-multiplexing algorithm is used to recover the data. This method is used for the multipath environments with multiple correlated signal paths. This method increases the capacity of the system. However, this does not work very well with the low SNR environments as interference plays a critical role in helping non-recognition of the uncorrelated signals.

6.4 OFDM System

After having understood the basic principles of the OFDM technology, let us understand the OFDM system. The OFDM uses differential modulation schemes or adaptive modulation and coding schemes. In this, the modulation and coding scheme of each channel is changed according to the variation in channel conditions. This change is based on the inputs from the receiver. The subscribers close to the cell site are assigned higher order modulations (e.g. QAM) while subscribers at the cell edge are assigned lower order modulation schemes (e.g. QPSK). Channel coding/decoding and interleaving techniques are used as they improve BER. After the process of IFFT and parallel-to-serial conversion, the signal is transferred over MIMO systems. At the receiving end, the process to retrieve the signal back starts that includes FFT, de-mapping/de-modulation, de-puncturing, de-interleaving and channel decoding (as shown in Figure 6.10).

6.4.1 OFDM Variants

There are a few variants of OFDM on the market. A fast-hopped form of OFDM that uses multiple tones and fast hopping to spread the signals is called Flash OFDM (developed by Flarion). Vector OFDM or VOFDM is OFDM plus MIMO technology. This has been developed by Cisco. Wideband OFDM creates a large space between the channels and so the

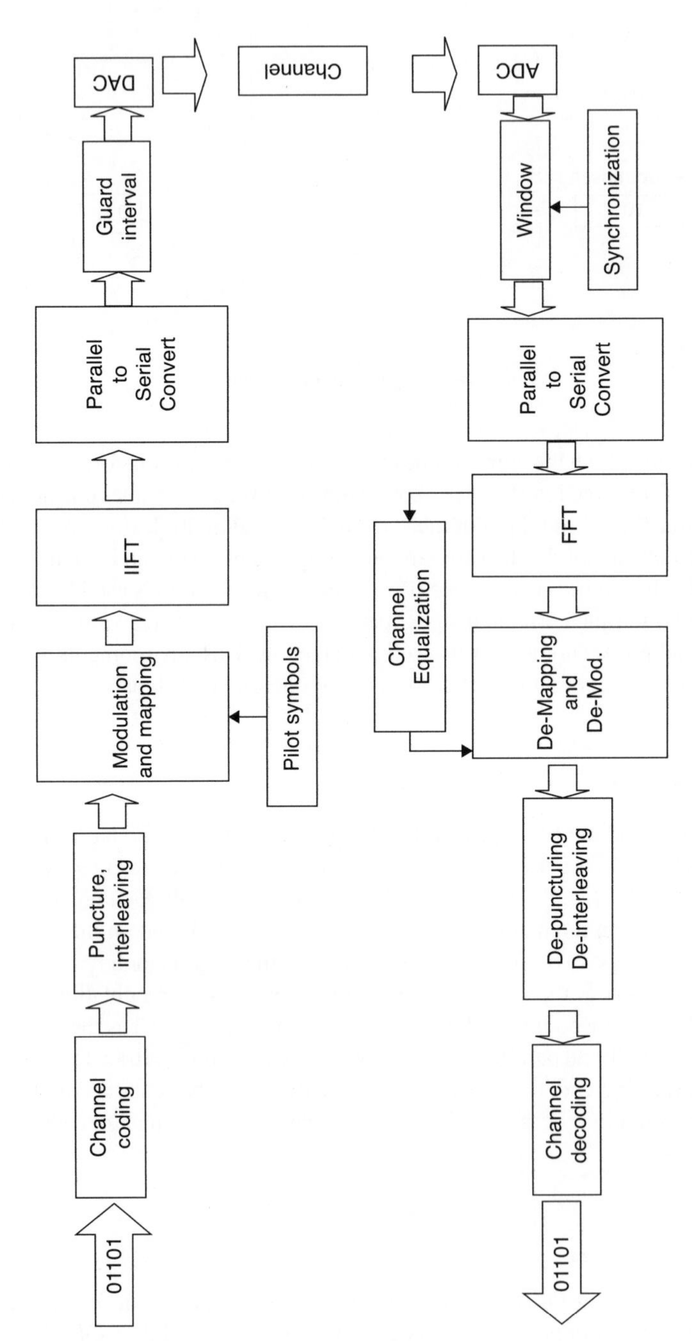

Figure 6.10 OFDM system architecture.

performance is not affected by the frequency errors between transmitter/receivers (applicable to WiFi Systems).

6.5 Design of OFDM Channel

There are a few parameters that are important for the design of OFDM systems. These include the system bandwidth, sub-carrier bandwidth, number of sub-carriers and length of the cyclic prefix (CP).

- Frequency selectivity: maximum multiple path delay $\tau_{\max}$.
- Time variance: maximum Doppler frequency $f_{D,\max}$.
- OFDM symbol duration is T_s.
- Choose the CP that is a small fraction of symbol length (this minimizes SNR; though it should be larger than the channel impulse response).
- Length of sub-carriers to be equivalent to the number of sub-carriers (it should not be too long or ICI caused by the Doppler frequency becomes the limiting factor).
- Calculate the lower and higher limits of the OFDM signal using the guard interval and Doppler frequency, respectively.
- The signalling overhead involved in the various multiple access schemes basically depends on the following factors:
  - OFDM symbols per frame;
  - size of modulation block;
  - number of users;
  - used sub-carriers N.

6.6 Multi-User OFDM Environment

All that we have seen before in this chapter is a single-user environment. The same concept of single user is now extended for a multi-user environment, as shown in Figure 6.11. The multi-user technique is used for achieving high downlink capabilities for future systems. The total capacity of the system is maximized for each sub-channel that is assigned to the user with the highest channel-to-noise ratio (for that sub-channel). If we look closely, the overview block diagram remains the same except for the fact on which technique needs to used in block '?', for example should it be TDMA, FDMA or CDMA.

Multi-user OFDM provides a quite flexible radio interface that allows users to be allocated, using TDM, FDM or a hybrid of FDM/TDM. The principle behind the OFDM-TDMA is that every user is allocated all the sub-carriers in a certain number of time slots. In OFDM-FDMA, every user transmits on some number of sub-carriers on all the time slots. In the absence of channel information, a frequency-hopping scheme is used. In OFDM-CDMA every user transmits on all sub-carriers using an orthogonal code. In the multi-carrier system it is possible to monitor the radio channels in a fast and easy way. This feature allows optimizing the system performance by allowing sub-carriers to minimize the effect of the fading (frequency selective) and SIR. Also, it is possible to dynamically allocate or change the bandwidth based on link quality and/or dynamically change the sub-carrier to matching environmental conditions.

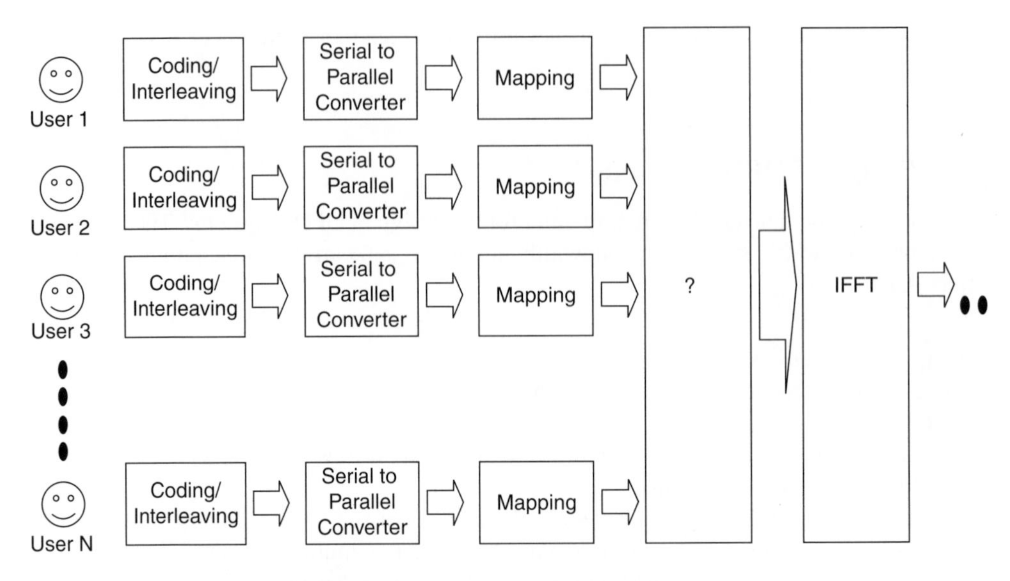

Figure 6.11 Multi-user OFDM environment.

Though most OFDM systems use a fixed modulation scheme over all sub-carriers for simplicity purposes, adaptive modulation is used for maximizing the data throughput of the sub-carriers that are allocated to the users. This is based on the SNR measurements of each sub-carrier followed by the application of the modulation scheme that maximizes the spectral efficiency while at the same time maintaining an acceptable BER. Though this technique has been used in ADSL networks, it has not found extensive application in the wireless domains. This is due to the difficulties faced in effective tracking of radio channels. However, the adaptive modulation approach has some limitations. There is always a trade-off between power control and adaptive modulation. A good channel path would mean that a high modulation scheme can be used. A reduced power would mean a reduced modulation scheme. Overhead information is also needed in adaptive modulation and mobility increases the required overheads.

6.7 All-IP Networks

Internet Protocol (IP) has played a great role in providing network layer protocol for the wired telecommunications world and is now being considered (and used) for a similar role in the wireless telecommunications world. Its characteristics of providing an open infrastructure for creating and providing services and applications has made it an attractive contender for the role. Using IP to a level that the network can be called an all-IP network is going to make the networks (both wired and wireless) more robust, scalable and cost effective. In All-IP networks, the current networks that are circuit switched-based are moved to the packet based using the Internet protocols and technology. Both the wireless and wireline networks are moving towards an All-IP network scenario (Figures 6.12 and 6.13).

There have been quite a few activities going on for standardization at the 3GPP level. Under 3GPP1, there has been global specification work for GSM/MAP evolution to the 3G and

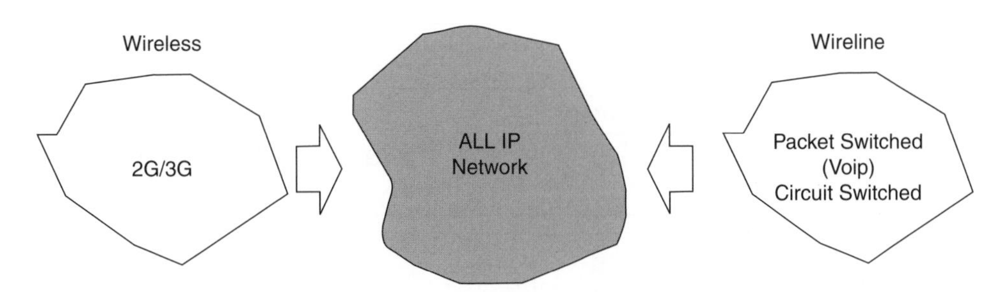

Figure 6.12 Flows of the wireless and wireline networks towards an All-IP network.

UTRA RTT and these standards are quite mature now. Under 3GPP2, global specifications for ANSI/TIA/EIA networks are under development.

The current work for IP networks under 3GPP2 contains development of specifications for IP based transport networks for BTS to BSC and access network re-designing to support IP based networks. For CDMA2000, the packet data specification is based on mobile IPv4 and is currently working on IPv6. Another group is working on the specification development in areas such as working on current mobiles over the IP based core network.

From a wireless network perspective, IP has been more prominent in the core side. Let us understand the core network evolution from the 3GPP release perspective.

6.7.1 Core/IP Network Evolution in Cellular Networks

6.7.1.1 3GPP Release 99

The first specification released by the **3G Partnership Project** for UMTS was Release 99 (3GPP R99). This specification set had a relatively strong 'GSM presence'. It introduces a system having a wideband radio access and a core network evolved from GSM. The original GSM platform with GPRS extensions for packet data services is used effectively. This is an obvious approach since it saves cost by utilizing the existing platform. This was also done so

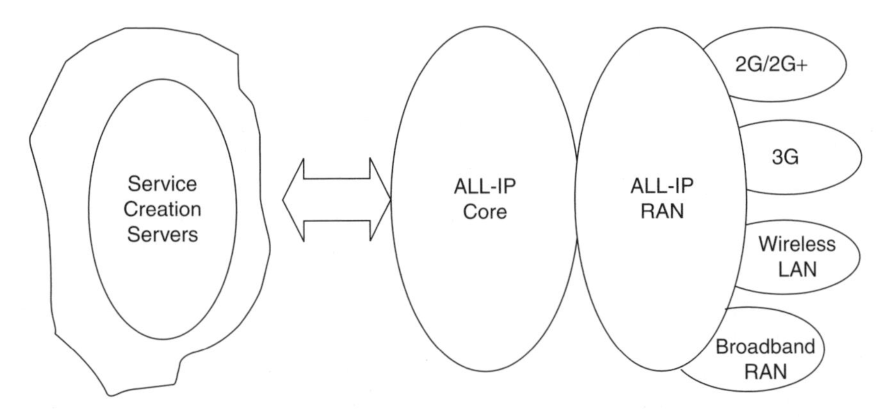

Figure 6.13 All-IP network overview.

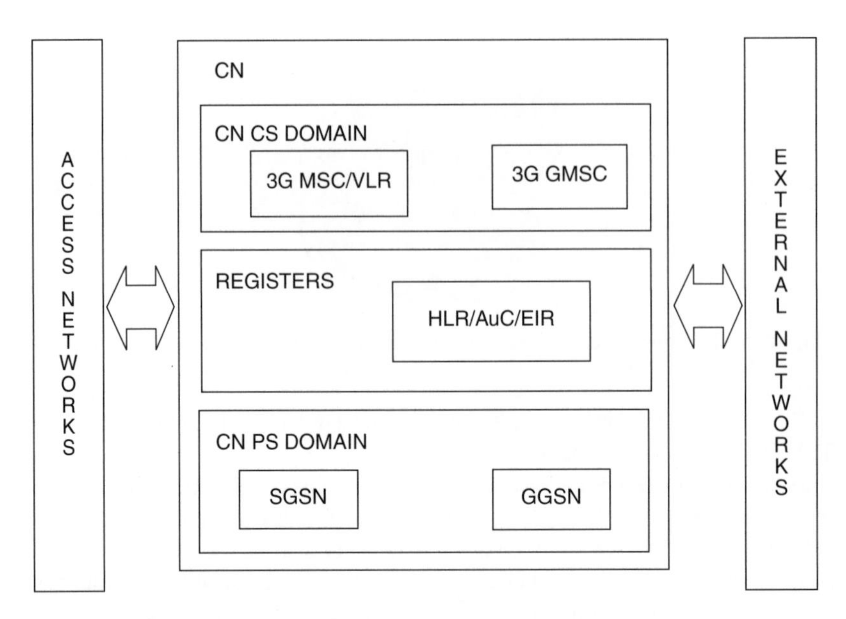

Figure 6.14 3GPP R99 core network architecture.

that firstly the UMTS network remains 'backward-compatible' with the existing GSM network and secondly, GSM and UMTS should be able to inter-operate together. Since the network subsystem of the GSM/GPRS is capable of providing the basic communication services for both circuit and packet switched traffic together with a rich set of supplementary and value-added services, it became an obvious choice for the basis of the UMTS Core Network. Also in the 3GPP R99 Core Network, the traffic will be either circuit switched or packet switched in nature, both of these traffic types require some specific arrangements to be done and this is why the Core Network is functionally further divided into two domains, the Circuit Switched Domain (CS Domain) and the Packet Switched Domain (PS Domain). The Core Network architecture is shown in Figure 6.14.

6.7.1.2 3GPP Release 4

The next release was earlier known by the name 3GPP R00 but because of the multiplicity of the changes proposed, the specification activities were scheduled into two specification releases, 3GPP R4 and 3GPP R5. The 3GPP R4 defines major changes in the UMTS core network circuit switched side and these are related to separation of user data flows and their control mechanisms. 3GPP R4 also introduces mechanisms and arrangements for multimedia.

The MSC/VLR evolved to MSC Server and Media Gateway (MGW). MSS takes care of Connection and Mobility Management functionality. It also contains VLR. The Media Gateway contains the facilities to perform actual switching and network inter-working functions. For example, transcoders, echo cancelling equipment and modems are located at the MGW. Depending on the network configuration the MGW may contain plenty of other functionality like for instance functionality to perform circuit-packet conversions in the case of VoIP calls.

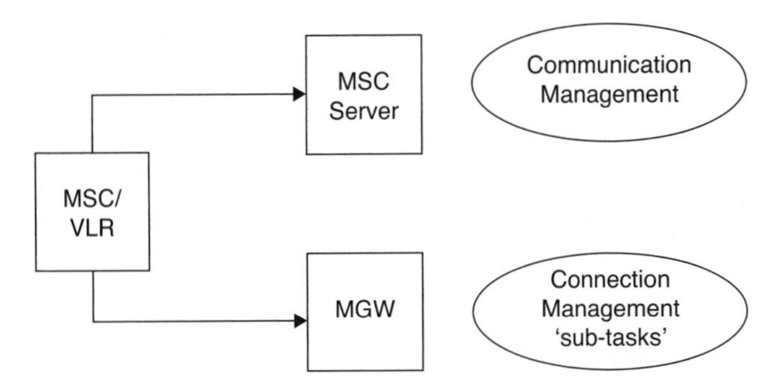

Figure 6.15 MSC server and media gateway.

The division between the MSC server and MGW is not one-to-one; one MSC server may control numerous MGWs which bring scalability into the system (shown in Figure 6.15). If the MSC server amount is under-dimensioned, the network may easily suffer from outages and relatively large amounts of subscribers will not gain a circuit switched service.

There are several alternatives to implement the connections between the user equipment and MSC server and between the MSC server and MGW. But one of the most attractive options is the IP based Session Initiation Protocol (SIP). SIP is also considered to be used in other connections with 3GPP R4/5. Details of the SIP protocol can be found in the IETF RFC 2543.

In future, IP is expected to be in use in as many occasions as possible and will eventually replace SS7 for transport signalling. When the traffic transport is IP based, it is natural that traffic control should also be IP based; this is why SIP is seen as a very interesting alternative to implement the call control protocol between the MSC server and MGW.

With user traffic and services the change is not necessarily so dramatic; there will be both packet- and circuit-switched traffic. The amount of circuit-switched traffic is expected to decrease but the pace of this development is very hard to predict. The operators like to utilize the existing investments made to circuit-switched technology first and when it is economically reasonable, the technology is changed.

From a transmission point of view, the circuit-switched voice call wastes transmission resources. In a normal voice call, the voice activity factor is, on average, 50 % that is one party is talking while the other one is listening. The system, however, reserves 100 % resources. By implementing VoIP and using IP as the transport technology, the transmission resources used for the call can be matched.

The 3GPP R4 contains an additional element called the IP Multimedia Subsystem (IMS). The IMS, when fully specified, will contain a uniform way to maintain VoIP calls, thus offering to the operators the way to deliver VoIP calls between UMTS networks. In addition to 'universal VoIP' the IMS offers a platform to other real-time and non-real-time IP services, like for instance, multimedia services.

The IMS consists of a Media Gateway Control Function (MGCF), Call State Control Function (CSCF) and MRF (Media Resource Function). These three functionalities form an extended model of connection management when compared to 3GPP R99. Basically the MGCF controls the MGW used in the connections; whether to use conversions, echo canceling, etc. The CSCF and MRF together form the logic for how the transaction using IMS is treated.

The MGCF–CSCF–MRF model brings a big and new aspect to the system which is related to the services, their creation and treatment. The 3GPP R4 contains a Service Platform and this is taken into use through the MGCF–CSCF–MRF chain. The basic concept exists in 3GPP R99 and actually the Service Capabilities are presented there, but in 3GPP R4 the Core Network structure supports the effective use of Service Capabilities.

3GPP R4 contains all the possibilities of traffic treatment. If the transaction coming from the access network is packet switched, it may be relayed to the external network either in a packet-switched or circuit-switched manner. Also, if the transaction coming from the access network is circuit switched, it may go to the external network either in a circuit- or packet-switched manner. In 3GPP R99 the nature of the connection (circuit/packet) remains the same through the network.

6.7.1.3 3GPP Release 5

The 3GPP R5 aims to introduce a UMTS network where the transport network utilizes IP networking as much as possible. Therefore this goal has been given the name 'All IP' network. IP will be used in both network control and user data flows. The mobile network implemented according to the 3GPP R5 specifications will be an end-to-end packet-switched cellular network using IP as the transport protocol instead of CCS7 which holds a major position in existing circuit-switched networks. IP-based networks will still support circuit-switched services. The 3GPP R4/R5 will also start to utilize the possibility for new radio access techniques. In 3GPP R99 the basis for the UMTS Terrestrial Access Network (UTRAN) is WCDMA radio access. In 3GPP R4/5 another radio access technology derived from GSM with Enhanced Data for GSM Evolution (EDGE) will be specified to create the GSM/EDGE Radio Access Network (GERAN) as an alternative to build a UMTS mobile network.

In 3GPP R5 the access network will experience more changes and the changes done in the Core Network are minor in nature. In 3GPP R5 the traffic is always packet switched. It may be real-time or non-real-time. In the development of R5 the focus has shifted to the PS Domain, which has been extended with the IMS functionality. The services, their accessibility and creation are emphasized in 3GPP R4/R5 core network implementation. The Service Capability Layer has already been introduced in 3GPP R99 but in further implementations its role will be increased by Open Service Architecture (OSA) based-solutions. The OSA provides mechanisms for universal service creation and management.

The changes performed between R4 and R5 should not be visible to the end-users. The UTRAN radio path still works the same as now; also the terminals being used are still working as such. Within the access network the transport technology could be IP instead of ATM.

Besides UTRAN the evolved GSM BSS named GERAN can be connected to the Core Network with an Iu Interface. Thus the traffic coming from GERAN gets the same treatment as the traffic coming from UTRAN in the sense of interfaces. If the operator has IMS in use, the Core Network CS Domain is not basically needed any more; the main step (besides the new radio access alternatives) between 3GPP R4 and R5 is whether to quit the CN CS Domain or not. This will depend significantly on the direction and maturing of the VoIP development.

6.7.1.4 3GPP Release 6

Release 6 is an enhancement of earlier 3GPP releases, aiming to bring mobile users a complete 3G experience. 3GPP R6 includes numerous new features, among them being High Speed

Uplink Packet Access (HSUPA), the second phase of IP Multimedia Subsystem (IMS), inter-working with Wireless Local Area Networks (WLANs), Multimedia Broadcast Multicast Service (MBMS) and enablers for Push-to-Talk (PoC). Release 6 brings many notable new services to mobile device users. Symmetrical data services, enabling two-way high-speed data communications for services such as video conferencing and mobile e-mail are specified. High Speed Uplink Packet Access (HSUPA) will complement the High Speed Downlink Packet Access (HSDPA) for a rich, two-way, interactive wireless broadband user experience. HSUPA and HSDPA together will enable symmetrical data communications at a high speed, supporting multimedia, Voice over IP, etc.

The second phase of IMS comprises all of the core network elements for offering multimedia services. IMS makes it possible for operators to offer mobile users multimedia services using Session Initiation Protocol (SIP). SIP enables mobile users to use services based on Internet applications. In Release 6, IMS is developed to support inter-working with circuit-switched networks, non-IMS networks and 3GPP2-based CDMA systems.

Release 6 also defines inter-working with Wireless Local Area Networks (WLANs). The inter-working is defined in a very flexible way, enabling different multi-radio scenarios. Multimedia Broadcast Multicast Service (MBMS) makes it possible to efficiently distribute multimedia content to multiple recipients. Such content could be, for example, video or music clips. Conversational services such as Push-to-Talk (PoC) are also specified in Release 6, together with Open Mobile Alliance (OMA).

6.7.2 Advantages of All-IP Network

Implementation of All-IP networks is a necessity as the data grow multi-fold since they present a lot of advantages. It is possible to deploy the voice services seamlessly across the packet network. New services development and deployment is quite easy. But most importantly, voice, data and video can be unified to be carried over a single network.

6.8 Architecture of All-IP Networks

Why do we want to move towards an All-IP network? As shown in Figure 6.16, the networks are moving towards a stage where the user would like to experience a feeling of ‘anytime and

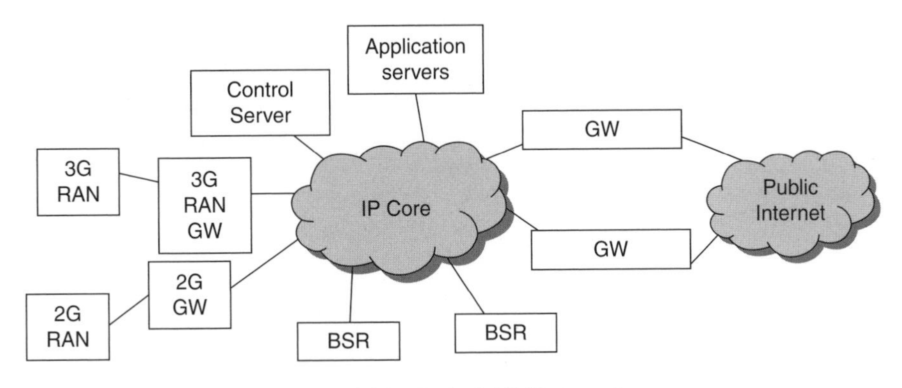

Figure 6.16 Typical All-IP network.

anywhere'. So, basically the user wants to have a flexible network – a network where the user equipment is able to choose between the best available networks. The user wants to move not only between various networks seamlessly, but also wants the highest possible QoS.

The All-IP network architecture is based on the principle that migration from IPv4 to IPv6 is not only feasible but inter-operability is also possible.

A typical All-IP network is shown in Figure 6.16. In such a network, the base stations themselves act as routers. These base stations are connected to the service provider while at the same time the GSM and UMTS networks can also be connected to the All-IP networks through gateways.

The key components of All-IP networks include:

- radio network;
- core network;
- control network;
- gateways;
- service architecture.

As seen in Figure 6.17, the radio access will contain the GSM, UMTS, etc. networks. Also, in future, additions such as WiMax would be natural. Though the existence of the all these networks on a single RAN transport network might not be there, in all probability, many of these networks would co-exist on the evolved access networks.

The requirements of all these access networks would define the transport network characteristics. The applications would be playing an important role in defining the bandwidth requirements and the quality of services of the network. The main aspect of such a network would be control and resource management. This will enable different types of networks to

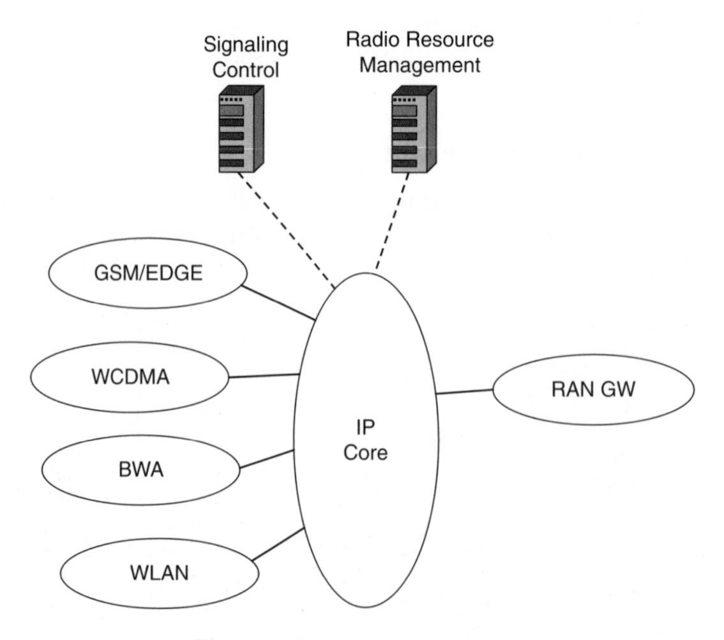

Figure 6.17 All-IP RAN network.

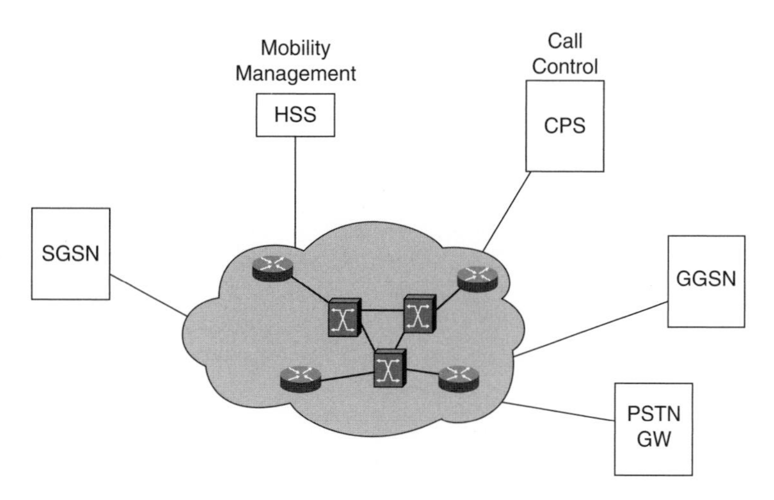

Figure 6.18 IPv6 core network (All-IP).

be connected to the core network. This core network is common to all various types of access networks. The calls are processed through the call processing servers. As Figure 6.17 shows, the backbone would be IP based (it will be based on IPv6). It is absolutely necessary to manage the multiple access radio technologies by a common server as it will lead to several inefficiencies. Also, due to common management systems, problems related to coverage, capacity and quality can be handled much well. The enablement would lead to proper load balancing, congestion control, Quality of Services, and optimized traffic distribution within the network. From an element perspective, the IP-RAN would consist of multi-radio base stations/routers (both integrated and stand alone), radio network controllers and control servers.

As mentioned above a core network is common to all the access networks. The IPv6 will be deployed for backbone network (shown in Figure 6.18). The home location register is now evolved to a combined HLR/HSS (Home Subscriber Server) which is capable of holding the subscriber data of all the access networks.

The fundamental difference between the conventional cellular networks and the All-IP network is that the functionality of the BTS and RNC is now divided into the IP-BTS and the servers. IP-BTS can be considered as a 'mini-RNC/BSC' (having macro-diversity combining) that is connected through Iu interfaces towards the gateways and the Iu interface towards other IP base stations. As mentioned in UMTS, there is a serving base station and drift base station. The base station functions include controlling the common channels access for group of cells. The gateways are used as access points to RAN from the core network and other access networks while the CS gateway is used to handle the CS traffic to the core networks. The servers are used to perform all those functions of RNC/BSC that cannot be handled by the IP-BTS. They handle functions such as paging, signalling/bearer, radio resource management (e.g. handovers), logical operations and maintenance, Quality of Service management, etc.

The Mobile Terminal consists of an enhanced IPv6 stack, able to perform marking according to the user-subscribed services. The Networking Control Panel performs AAA registration/de-registration. The Mobile Terminal Networking Manager takes the decision to execute the handover and attach procedures, based on information received from the user and information received from the networking devices (WCDMA, WLAN, Ethernet drivers).

7

Broadband Wireless Access: WLAN, Wi-Fi and WiMAX

7.1 Wireless Technology Differentiation

The wireless networks, under three categories, are shown in Figure 7.1: Mobile networks, Wireless LAN and Broadband Wireless networks. While cellular networks are about both voice and data, wireless LAN networks are only data. BWA networks are mainly data. Both WLAN and BWA fall predominantly under the fixed category. In this chapter we will take a closer look at the WLAN and BWA (Wi-Fi and WiMAX) networks.

7.1.1 Broadband Wireless Access

Broadband Wireless Access or BWA is a radio access technology that is used to deliver broadband services to the users' premises. Broadband refers to having an instantaneous bandwidth greater than 1 MHz while supporting data rates of/more than 1.5 Mbps (802.16-2004 standard). The services are delivered via radio and user-to-user connections are less. If there is a user-to-user connection, it is made via the core network. The available bandwidth is shared between the users that are covered under one coverage area. As the user requirements are many, BWA networks deliver both voice and other kinds of data services. Under deregulated markets, for operators without existing wired infrastructure, wireless broadband networks offer an exciting opportunity as they are quick and easy to deploy resulting in less time to market.

Generally, broadband describes high speed, high capacity data communication making use of DSL, a modem, fixed wireless access, optical fibres, VSAT, WLAN, etc. As per Broadband Policy 2004, broadband in India is defined as – 'Always-On' data connection that is able to support various interactive services including Internet access having the capacity of a minimum download speed of 256 kbps to an individual subscriber from the 'Point of Presence' of the service provider. Now in order to provide an alternative to the twisted pair local loop, technologies developed in the wireless domain came to be known as wireless local loop or fixed wireless access. In 1999, the IEEE 802 committee set up the 802.16 working group that worked towards developing the broadband wireless standards – thereby standardizing the

Cellular Technologies for Emerging Markets: 2G, 3G and Beyond Ajay R. Mishra

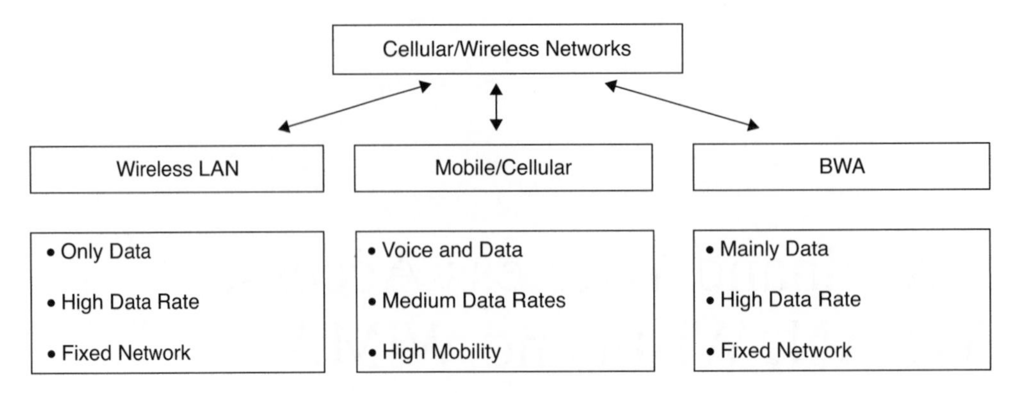

Figure 7.1 Wireless technology differentiation.

air-interface and related functions of the wireless local loop. There were three sub-groups formed:

- IEEE 802.16.1 – Air-interface for 10 to 66 GHz.
- IEEE 802.16.2 – Co-existence of broadband wireless access systems.
- IEEE 802.16.3 – Air-interface for licensed frequencies, 2 to 11 GHz.

7.1.2 IEEE 802.16

The 802.16 Standard defines how the traffic will move between the core network (e.g. the Internet) and a subscriber on the air-interface. The standard is based on a 3-layered structure, namely a physical layer, MAC (Media Access Control) layer and Service layer, as shown in Figure 7.2. The physical layer is the lowest layer that specifies the modulation scheme, synchronization, multiplexing structure, frequency bands, etc. The MAC layer holds the functions that are responsible for providing services to the users. This layer allocates the channel capacity to fulfill user requirements. The top or the service layer is responsible for providing various services, for example Internet access, digital multicast (audio/video), etc to the users. Utilizing the services of these three layers, the 802.16 Standard defines the traffic movement.

The transmission of the data from the user to the base station is done using the DAMA–TDMA (Demand Assignment Multiple Access–Time Division Multiple Access)

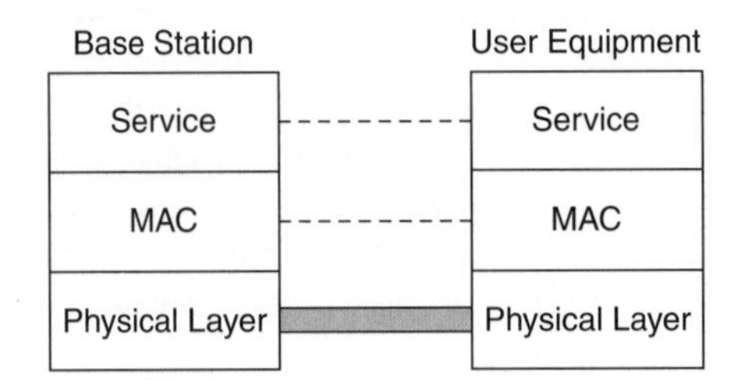

Figure 7.2 Protocol structure of 802.16.

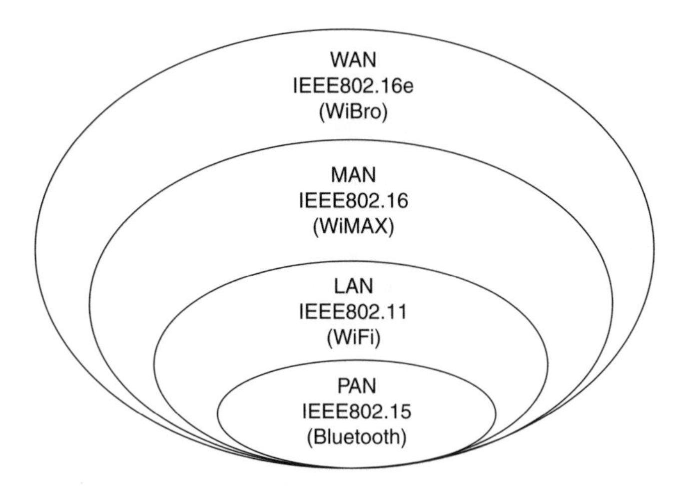

Figure 7.3 Wireless broadband access networks.

system, thereby making sure that the system adapts itself to the changes in demand for capacity by the user. This is done by dynamic variation of the timeslot assignment. There are two data streams from base station to subscriber – one is continuous transmission (for audio/video) and the other is 'burst' transmission (e.g. IP based traffic). There are several advantages of the systems that are based on 802.16 technology – fast service delivery, no high installation costs, no requirement of wired networks and they can reach areas where wired networks cannot do so.

7.1.3 BWA Technologies

There are various types of wireless networks (as shown in Figure 7.3). Let us have a look at them briefly.

7.1.3.1 Wide Area Networks

Wide Area Network or WAN is a data network that has a coverage over a very large geographical area. Such networks are interconnected to form a WAN. These networks have many such switching nodes. Typical networks are WiMAX, GSM, UMTS, etc. The user can access the network for services such as the Internet using an access card and a computer (or a similar device, e.g. PDA). They have higher data rates and speeds as compared to the networks accessed by mobile phones.

7.1.3.2 Metropolitan Area Network

A Metropolitan Area Network or MAN is spread over a few kilometres. Several Local Area Networks (LANs) can be interconnected to form a MAN. Several MANs can together form a WAN. A small township can have a MAN that interconnects several LANs. Few such townships when interconnected can form a WAN.

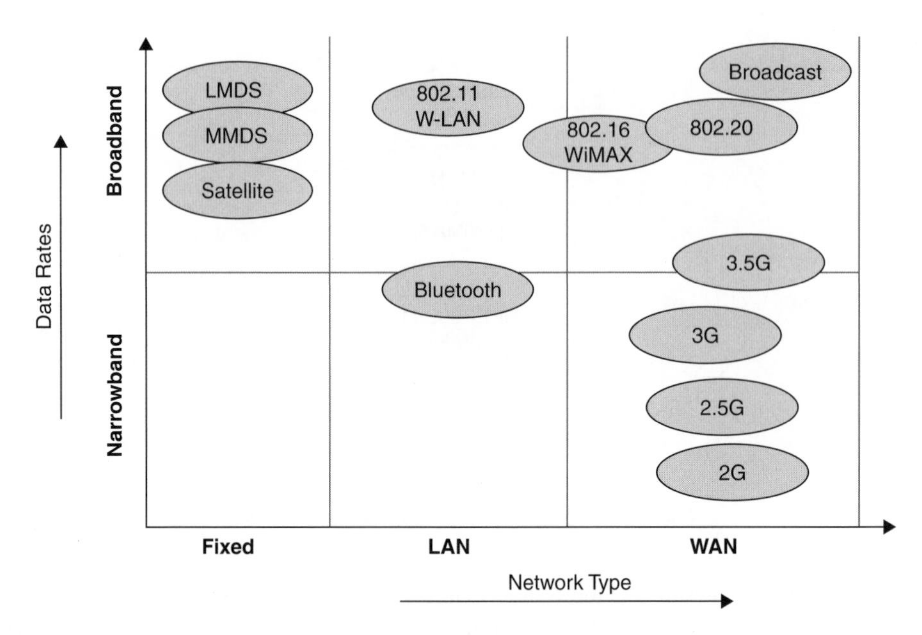

Figure 7.4 BWA functionalities.

7.1.3.3 Local Area Network

This is designed for a small area, for example hundred of metres, so that devices can be inter-connected. A typical example is a LAN in a company. Many computers and other communicating devices are connected to each other and they communicate with each other. A typical example is the Wi-Fi network.

7.1.3.4 Personal Area Network

With ranges of around tens of metres, the communication happens between two devices, for example between two laptops. A higher data speed and lower range are features of a PAN. A typical example of PAN technology is 'Bluetooth'.

As seen in Figure 7.4, the BWA covers almost all of the known wireless technologies. However, in this chapter, we will only cover technologies such as Wireless LAN, Wi-Fi and WiMAX.

7.2 Wireless LAN

7.2.1 IEEE 802.11

IEEE 802.11 is the original standard that allowed a bandwidth of 1–2 Mbps. Further development of this standard was done in order to optimize bandwidth requirements, improve security and compatibility. This standard characterizes the Wireless Local Area Network. Various 802.11 standards (variations) are shown in Table 7.1.

Table 7.1 Some IEEE 802.11

Name of standard	Network	Description
802.11a	Wifi5	This allows higher bandwidth (54 Mbps maximum through put, 30 Mbps in practice). It provides 8 radio channels in the 5 GHz frequency band
802.11b	WiFi	Most widely used one. It offers a maximum thorough put of 11 Mbps (6 Mbps in practice) and a reach of up to 300 meters in an open environment. It uses the 2.4 GHz frequency range, with 3 radio channels available
802.11c	Bridging 802.11 and 802.1d	It is only an amended version of the 802.1d standard that lets 802.1d bridge with 802.11-compatible devices (on the data link level)
802.11d	Internationalisation	It is a supplement to the 802.11 standard which is meant to allow international use of local 802.11 networks. It lets different devices trade information on frequency ranges depending on what is permitted in the country where the device is from
802.11e	Improving service quality	It is for improvement in the quality of service at the level of the data link layer. It defines the requirements of different packets in terms of bandwidth and transmission delay so as to allow better transmission of voice and video
802.11f	Roaming	It is a recommendation for access point vendors that allows products to be more compatible. It uses the Inter-Access Point Roaming Protocol, allowing roaming for the user
802.11g		It offers high bandwidth (54 Mbps maximum throughput, 30 Mbps in practice) on the 2.4 GHz frequency range
802.11h		This was developed to bring together the 802.11 standard and the European standard (HiperLAN 2. hence the h in 802.11h)
802.11i		This is developed to improve the security of data transfers
802.11		Not any more used; (was used for infra-red)
802.11j		Japanese Standard

The Wireless LAN is based on the IEEE 802.11b Standard on the direct sequence spread spectrum specification. We have seen the details of this spread spectrum technique in Chapter 4 on CDMA. The 802.11b is an enhancement of the 802.11 specification in terms of capacity – the capacity handled is now higher, namely 11 Mbps, up from 2 Mbps.

WLAN systems use the spread spectrum technique. We have already discussed the spread spectrum technique in earlier chapters. As a recollection, the technique involves spreading the original signal through modulation so that the resulting signal is of a wider bandwidth. This will result in avoiding a higher power density by spreading the signal over a wider frequency band; hence the signal becomes resistant to the interference. In the WLAN system, the users are allocated using the TDMA frames – thus sharing the capacity between multiple users. IEEE 802.11 specifies a spreading ratio of 11 for one and two megabits per second data rates. The operating transmit power of the WLAN is higher than that of CDMA which operates at noise level.

7.2.2 Channel Structure

ETSI defines 13 channels while ANSI defined 11 channels for Wireless LAN. Channel allocation without interference would mean three channels being allocated at a given location. Although the centre channel offset is 5 MHz, a 22 MHz channel separation is needed for interference-free operation. Thus, for an interference-free operation the possible combinations are $1 + 6 + 11$, $2 + 7 + 12$ and $3 + 8 + 13$.

7.2.3 Efficient Channel Sharing

The method to avoid collision and efficient channel sharing is specified in the IEEE 802.11 Standard. The carrier sense multiple access with collision avoidance (CSMA/CA) is used as the medium access method to define the method of accessing the shared medium. The Clear Channel Assessment (CCA) process is used for CSMA/CA in which the station before transmitting should listen if any other station is transmitting. Before a second frame is transmitted, the station listens to verify channel clearance. In the case of a busy channel, another back-off interval lesser than the first is selected and this continues until the waiting time reaches zero – then the station transmits again.

7.2.4 Parameters in WLAN Planning

Bit Rate

For an 11 Mbps WLAN system, the bit rate is 7.5 Mbps. The remainder is used as 'protocol overhead'. In some systems, the throughput is less than 7.5 Mbps (e.g. 6.5 Mbps) and the terminal throughput is even lower (e.g. approx. 5.5 Mbps).

Frame Error Rate

Frame Error Rate or FER is an indicator of link quality that is measured from the MAC layer, represented as a percentage. Not only is it indicative of the error frames in a given sequence, but it also indicates the amount of packets that do not reach the receiver. Higher FER indicates lower signal quality. Beyond a threshold level, the re-transmissions start taking place.

Channel-to-Interference Ratio

The ratio of channel signal strength and the interference signal is the 'channel-to-interference' ratio. In the case of WLAN, where does the interference signal come from? If two radio units are working in the same environment using the same channels, then interference will take place affecting the link quality. For optimal performance, the carrier should be 15 dB higher than the interfering signal.

Receiving Signal Strength Indicator (RSSI)

This is needed for calculation of the *C/I* and *C/N* ratios. As the name indicates, the RSSI gives the strength of the received signal.

Antenna Parameters

For WLAN, according to IEEE 802.11, the minimum transmitted power should be 1 mW. The EIRP is regulated to be 20 dBm (ETSI) while the maximum gain of the antenna is 7 dBi. However, in the USA, higher gain antennae are also used.

7.2.5 Coverage and Capacity in WLAN

IEEE 802.11 specifies the throughput as 11 Mbps. However, in practical cases, this throughput may vary from 1 to 11 Mbps. Tests have however indicated a throughput range between 4 and 6 Mbps. In a WLAN, the users' experiences are different at different locations. The differences are in terms of capacity, signal strength and quality. The area covered by the WLAN is also small, for example a few tens of metres. Sometimes signals are able to penetrate thin walls (e.g. in an office, people are able to connect their laptops in small rooms due to wireless connectivity but outside the office this is not possible) but is not able to penetrate thick walls. The factors that affect the throughput include range, multipath, obstacles, latency, quality/type of system used, etc. The user is able to move from one location to another in an office without any experience in network change – and even in this change, the user is moving from one access point to another. This is possible as the access points communicate directly with each other at the MAC layer level.

7.2.6 Security and Authentication

Security is a major issue in wireless transmission. Anyone with the right hardware (laptop, 802.11b card) and software ('Windows', etc.) can easily 'listen in' on the wireless conversation. Transmitted data are encrypted across by using the Wired Equivalent Privacy (WEP) protocol. This is available with different key lengths (40 and 120 bits) and is dependent on the amount of security needed. Thus, anyone 'listening in' would get only encrypted data. As the key is shared and limited, it makes WEP vulnerable. Given this, another security standard 802.1x is proposed that will solve most of the issues. The IEEE 802.1x Standard belongs to the IEEE802.1 group of protocols and is used for authentication for the network access control. It is based on the EAP (Extensible Authentication Protocol). It proposes to give a unique key to each user. For maintaining the security of a network with a large number of users in multiple locations, the correct authentication software and techniques are used, that is each user has a unique identification that allows only authorized users to log into the network.

7.2.7 WLAN Network Architecture

There are four fundamental terms in a WLAN Network.

7.2.7.1 Station

A typical example of a station can be a laptop (client) or a base station (access point). Any device that can connect to the network is called a station. This has a network interface card with which the device is capable to be used in the network. There are two types of stations: access points and clients. The former are base stations that transmit/receive radio frequencies for wireless devices. Clients are devices like laptops, PDA, etc., along with a network interface card.

7.2.7.2 Service Set

There are two types of service sets: Basic and Extended. All the stations that can communicate with each other constitute a basic service set. There are two kinds of BSSs: Independent and Infrastructure BSS. There are no access points in the independent BSS (i.e. it is not possible to connect to other BSSs) while the infrastructure BSS can communicate with other stations that are a part of a different basic service set. All the interconnected BSSs together form an Extended Service Set (ESS). These are connected through a distribution system (DS). This helps in extending coverage of the network.

7.2.8 WLAN Network Types

There are a few types of WLAN networks: Peer-to-Peer, Bridge and Wireless Distribution Systems. In a peer-to-peer network, wireless devices can communicate with each other directly without any base/central point or permissions to talk, for example two computers connected to each other. Bridge as the name suggests permits network connections; it also acts as the connection point to Wireless LAN. Interconnections between various access points will form a wireless distribution system. This helps in network expansion without the need of a wired backbone system to get them together.

7.2.9 Network Planning in WLAN

Planning a WLAN network generally starts on a wrong note with engineers focusing more on coverage than capacity. The fundamental aspect of WLAN planning is not only about understanding the coverage area but also the utilization of the WLAN. Another aspect is to understand any compliance issues that an organization may be having.

Site surveys as the name suggests are conducted to understand the site where the equipment will be located. This includes checking the suitability of the connection in terms of cabling, connections, power supply, etc. However, the site survey here is different from the one we have seen in previous chapters as the site survey would mainly be to understand the inside of the building. All this is done with a clear understanding of the building architecture such as high ceilings, outdoor coverage requirements, etc. The site survey should also include information of the other access points that will/might be a source of interference in the 'to-be-designed' network.

Site surveys are followed by pre-planning. This includes the identification of locations where the access points can be placed. It will also include some more detailed information about the interference sources. Generally every floor has its own plan. All important locations are covered (e.g. meeting rooms) while the unimportant ones (e.g. rest rooms) are least covered.

The planning goes hand-in-hand with measurements as WLAN is highly sensitive to the multipath effects. Usually the access point implementation and testing go together.

After this, the detailed planning phase starts, where configurations are done. The configurations such as SSIDs, VLAN assigned per SSID, transmit power, device naming and channel plans are carried out. The measurements done already are helpful in assigning of power and channels. Security and authentication setting (both of the network itself and integration with existing networks) are also a part of this process. This includes wireless client authentication, authorization, access control, encryption and resilience.

A coverage requirement in a WLAN is usually 100 %. As with most of the systems, with distance the throughput decreases. Factors such as interference have a big impact on the throughput and coverage. For increasing the coverage, diversity systems are used at the access points. In a high density network, frequency planning plays a very important role. The access point should not be working on the same frequency channel. An offset of 20–25 MHz would result in a higher performance network.

7.3 Wi-Fi Networks

Wi-Fi is the short name for Wireless Fidelity. It is the name of certification given by the Wi-Fi Alliance that was formerly called the WECA or Wireless Ethernet Compatibility Alliance. This group is responsible for the compatibility of the devices that are based in the IEEE802.11 standard. The logo for the Wi-Fi Alliance is shown in Figure 7.5. The Wi-Fi Alliance complies with the IEEE802.11 standard. The Wi-Fi Alliance promotes Wi-Fi networks, standardizes the Wi-Fi networks and tests/certifies the Wi-Fi products. Wi-Fi technology was designed and optimized for Local Area Networks.

7.3.1 Introduction to Wi-Fi Technology

We have seen various IEEE802.11 standards in Table 7.1. These different 802.11 standards have different modes of operation, making each of these standards offer different frequency, speed and ranges.

As shown in Table 7.2, the range of 802.11a has five times more speed than the 802.11b standard. However, the distance covered is much less (10 m as compared to 100 m of 802.11b). 80211a is based on OFDM technology (defined in Chapter 6). The operating frequency is 5 GHz with eight non-overlapping channels. 802.11b operates on 2.4 GHz at a speed of

Figure 7.5 Wi-Fi Alliance logo. Reproduced by permission of the Wi-Fi Alliance.

Table 7.2 Some IEEE 802.11 Standards for WiFI

Standard	Frequency	Speed	Range
WiFi a (802.11a)	5 GHz	54 Mbit/s	10 m
WiFi B (802.11b)	2.4 GHz	11 Mbit/s	100 m
WiFi G (802.11b)	2.4 GHz	54 Mbit/s	100 m

11 Mbps. Usually devices that operate 802.11a and 802.11b are incompatible with each other; however some dual-band devices operate in both the frequency bands.

7.3.2 Wi-Fi Network Architecture

The Wi-Fi network consists of three main units:

- Access point.
- Station (e.g. PC).
- PC cards.

A wireless access point is all what is needed to connect all the computers in a home. The whole system would consist of a DSL modem (or cable), wireless access point, firewall, router and an ethernet hub. The cable/modem brings the Internet line at home while the router allows inter-connection between various computers/devices. Many types of routers are used based on various IEEE 802.11 recommendations, but the 802.11g is used due to its speed and reliability. Once the router is established, it is succeeded by various settings as described in the following.

Service Set Identity (SSIDs)
This is the manufacturer's name and is usually default setting.

Channel Number
The routers use a single channel. However, if the channel is being used by another user, interference will be experienced. This problem can be mitigated by using another channel.

Security
As mentioned above, security is a key issue on these networks. Usually, having the individual user name and password is considered best from a protection perspective. Another one is WEP or Wired Equivalency Privacy. This makes a wireless network as secure as a wired network. However, as in most of the cases, it can be 'hacked into' as well. Taking this into consideration, a step forward from WEP is the WPA or the Wi-Fi protected Access (part of the 802.11 security protocol). This is used in most of the non-open public hotspots and it uses a TKIP or Temporary Key Integrated protocol. MAC address filtering is another method that uses computer hardware (and not software as with the WEP or WPA). The process involves specifying the individual MAC address which can be used. A new device would require a new MAC address to be specified. This is also not foolproof as it can also be 'hacked into'.

7.3.3 Wi-Fi Network Design

The network design involves looking into the following aspects:

- Coverage.
- Capacity.
- Frequency allocation.

Coverage and Capacity

The access point placement defines the coverage areas. This placement along with the availability of the required bandwidth would make sure that both the coverage and capacity requirements are met. The access points are assigned with the channels. However, at this point interference also needs to be considered as well. How are these access points placed? Let us consider the map of the area to be covered (e.g. a floor in the building). The area is divided into clusters and so we assess the traffic quantity in it. Also, other aspects such as installation facility, power connections, etc. are taken into account. Software tools are used to access the coverage plots (as in case of GSM, UMTS, etc. coverage planning) and hence decide the best location of the access points. As in GSM, etc., putting too many access points will increase the costs while too few will decrease the coverage. However, the area to be covered is much less and the probability of major infrastructure change, for example a thick wall getting constructed is also less. Thus, if planning is done with precision, the coverage, capacity and quality remain intact. The access points are placed in such a way that no gaps in coverage exist, at the same time making sure that high congestion on access points does not takes place. This is calculated by dividing the total traffic request on an access point by the bandwidth capacity of that access point. This process can make sure that the throughput across the network is better.

Frequency Allocation

After the coverage and capacity planning, the next step is to plan the frequency channels. The frequency is divided into 14 channels that are spread over the system. Both adjacent and co-channel interferences happen if the channel selection is not done correctly. More than three non-overlapping channels decrease the interference substantially.

7.4 WiMAX Networks

7.4.1 Introduction to WiMAX

As defined by the WiMAX forum (logo shown in Figure 7.6), the term WiMAX stands for Worldwide Interoperability for Microwave Access, introduced in 2001. The purpose was

Figure 7.6 Logo of the WiMAX Forum. Reproduced from the WiMAX Forum.

Table 7.3 Some IEEE 802.11 Standards for WiMAX

	802.16	802.16a	802.16e
Spectrum	10–66 GHz	2–11 GHz	< 6 GHz
Modulation	QPSK, 16 QAM, 64 QAM	QPSK, 16 QAM, 64 QAM, 256 QAM	QPSK, 16 QAM, 64 QAM, 256 QAM
Mobility	Fixed	Fixed	Mobile ($\leq$75 Mph)
Bit Rate	32–134 Mbps	70–100 Mbps	15 Mbps
Call Radius	1–3 miles	3–5 miles	1–3 miles
Bandwidth	20, 25, 28 MHz	1.25–20 MHz	5 MHz

to promote conformance and interoperability of the official IEEE 802.11 standard called 'Wireless MAN'. This technology may look similar to the Wi-Fi but offers some great differences/advantages over it:

- The Wi-Fi access is highly contended and the speed is slow as the back haul is mostly ADSL.
- The major advantage of WiMAX over Wi-Fi is range. WiMAX extends far beyond a few hundred metres and has a higher speed.
- LOS is not needed in Wi-Fi under good conditions.
- Interference is a problem in Wi-Fi as many users exist in urban environment unlike WiMAX that provides for many kilometres strong encryptions, less interference and symmetrical bandwidth.

WiMAX is based on the IEEE802.16 standards (as shown in Table 7.3). This technology is mainly associated as a high data rate 'last mile' connectivity technology. As mentioned above, the broadband technologies such as Wi-Fi have been using ADSL backbone network. However, WiMAX uses the *wireless* technology, thereby making it accessible even to areas that are not connected by cable or fibre.

Then, there is fixed wireless and mobile wireless, a summary of which is given in Table 7.4.

Table 7.4 Fixed versus mobile WiMAX

	Fixed WiMAX	Mobile WiMAX
Type of Network	Fixed	Fixed and Mobile
Modulation	OFDM	OFDMA
Frequency	2.5, 3.4–3.6, 5.8 GHz	2.3–2.4, 2.5–2.7, 3.3–3.4, 3.4–3.8 GHz
Duplexing	TDD, FDD	TDD, FDD
Handoff	No	Yes
Equipment	Outdoor/Indoor CPE	CPE, Embedded Modules

7.4.1.1 Fixed WiMAX

PMP (Point-to-Multipoint) broadband was developed over the wired infrastructure by the IEEE 802.16 working group. This was based on the TDMA protocol supported by the TDD and FDD on the air-interface. This works between 10 and 66 GHz with LOS and between 2 to 11 GHz with the NLOS condition.

7.4.1.2 Mobile WiMAX

Mobile WiMAX is based on the IEEE802.16-2004 and 802.16e-2005 standards that supported the mobile services with more than 120 km/h speeds. This is based on the OFDM access scheme, thus provisioning for the higher flexibility in assigning time and frequency resources. This section is dedicated to mobile WiMAX.

The main advantages of WiMAX over other systems are as follows:

- Better system performance due to advanced QoS mechanisms.
- Higher throughput due to robust adaptive technologies and spectral efficiency (almost three times better than current 3G systems).
- Capability to fill in the gaps (due to wireless) in a region having broadband due to cable/DSL.
- Interoperability (due to enforcing compliance).
- Has the ability of 'broadband on demand', with bandwidths up to 100 Mbps.
- System design does not need 'line-of-sight' conditions.

The performance of the WiMAX system is due to the usage of the OFDM and OFDMA air-interface technologies. Utilization of TDD also improves the system performance, though FDD can also be used. We have understood OFDM technology in Chapter 6, so will not go through its fundamentals again. Due to these multiplexing techniques, high data rates can be offered to the subscribers with a higher spectral efficiency. Due to 802.16, the quality of service is higher in the system. The MAC features allow an e2e IP based QoS as it provides flexibility in scheduling the resources over the air-interface. The WiMAX network is more interoperatable than the rest of the BWA networks. This is because global roaming among the WiMAX operators would allow subscribers to access various networks using the same devices. The choice of frequency is there as well, ranging from 2.3 GHz, 2.5 GHz, 3.3 GHz and 3.4–3.8 GHz in licensed and 5.8 GHz in unlicensed bands. Though the unlicensed bands seem to be a better option (no hassle to go through the rigorous frequency licensing process), there are problems associated with it, for example 5.8 GHz is universally unlicensed and hence the possibility to use it globally might not be there. But the biggest issue is interference – a high amount of interference can be expected from other technologies that may use this band, for example 'Bluetooth' or Wi-Fi, etc. There may be issues related to power control of unlicensed spectrums by the regulators, impacting the range of these networks. Latencies of less than 50 ms exist to provide optimal handovers for real time applications. Security is taken care of during these handover processes. The security features used are similar to the ones used in the Wi-Fi systems.

The key characteristics of the WiMAX system are:

- WiMAX is a Time Division Duplex System.
- Spectrum requirements are from 10 MHz to 30 MHz.
  - Frequency band – Licensed: 2.5/2.6 GHz; 3.5 GHz.
  - Frequency band – Unlicensed: 5.8 GHz.
- Throughput estimates.
  - Theoretical: 70 Mbps (for a 20 MHz carrier).
- Cell range.
  - For 2.5G Hz, 500 m–1.5 km.

7.4.2 OFDMA: Modulation in WiMAX

Mobile WiMAX adopts Orthogonal Frequency Division Multiple Access (OFDMA) for improved multi-path performance in 'non-line-of-sight' environments. Scalable OFDMA is introduced in the amendment to support scalable channel bandwidths from 1.25 to 20 MHz. The Mobile Technical Group in the WiMAX forum is developing Mobile WiMAX system profiles that will define the compulsory and optional features of the IEEE standard. This profile enables mobile systems to be configured based on a common base feature set. Release-1 Mobile WiMAX will cover 5, 7, 8.75 and 10 MHz channel bandwidths for licensed worldwide spectrum allocations for 2.3, 2.5, 3.3 and 3.5 GHz.

It is a multiplexing technique that subdivides the bandwidth into multiple frequency subcarriers. In this system, the input data stream is divided into several parallel sub-streams of reduced data rate which is modulated and transmitted on a separate orthogonal sub-carrier, as shown in Figure 7.7.

Here, the cyclic prefix can completely eliminate the Inter-Symbol Interference. This is the repetition of the last samples of data portions of the block that is attached to the beginning of the data payload. The OFDM has a very sharp spectrum, so a large fraction of the allocated channel bandwidth can be utilized for data transmission which compensates the loss in efficiency due to the cyclic prefix.

It uses the frequency diversity of the multipath channel and can be realized with efficient Inverse Fast Fourier Transform. In this system, resources are available in the time domain by means of symbols and in the frequency domain by means of sub-carriers. They can be organized into sub-channels for assigning to an individual user.

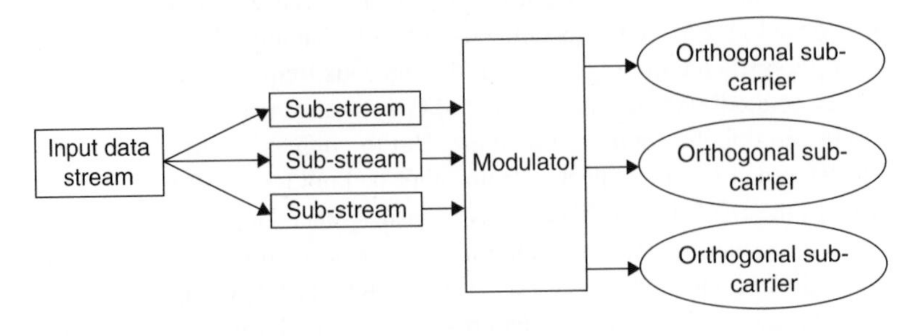

Figure 7.7 Architecture of OFDMA.

7.4.2.1 OFDMA Symbol Structure and Sub-Channelization

This has three types of sub-carriers:

1. Data sub-carriers: data transmission.
2. Pilot sub-carriers: estimation and synchronization.
3. Null sub-carriers: no transmission to guard bands and DC carriers.

Data and pilot sub-carriers are also called sub-carriers and their sub-sets are called sub-channels. It supports sub-channelization in both downlink and uplink. The minimum frequency-time resource unit of sub-channelization is one slot or 48 data tones.

7.4.2.2 Scalable OFDMA

This supports a wide range of bandwidths. The scalability parameters and their respective values for SOFDMA are shown in Table 7.5.

7.4.2.3 TDD Frame Structure

This is the preferred duplexing mode due to the following reasons:

- Enables adjustment of the DL/UL ratio for asymmetric traffic.
- Assures channel reciprocity for link adaptation, MIMO and closed loop advanced antenna technologies.
- Requires one channel for UL/DL leading to greater flexibility for adapting of various global spectrum allocations.
- Has easy and cheap transceiver designs.

In the frame structure, each frame is divided into DL and UL sub-frames separated by transition gaps. A frame uses the following control information:

- Preamble: provides synchronization.
- Frame Control Header: provides frame configuration information.

Table 7.5 Scalability OFDMA parameters

Parameters	Values			
System Channel Bandwidth (MHz)	1.25	5	10	20
Sampling Frequency (F_P in MH_Z)	1.4	5.6	11.2	22.4
FFT Size (N_{FFT})	128	512	1024	2048
Number of Sub-Channels	2	8	16	32
Sub-Carrier Frequency Spacing	10.94 kHz			
Useful Symbol Time ($T_b = 1/f$)	91.4 μsec			
Guard Time ($T_g = T_b/8$)	11.4 μsec			
OFDMA Symbol Duration ($T_s = T_b + T_g$)	102.9 μsec			
Number of OFDMA Symbols (5 ms Frame)	48			

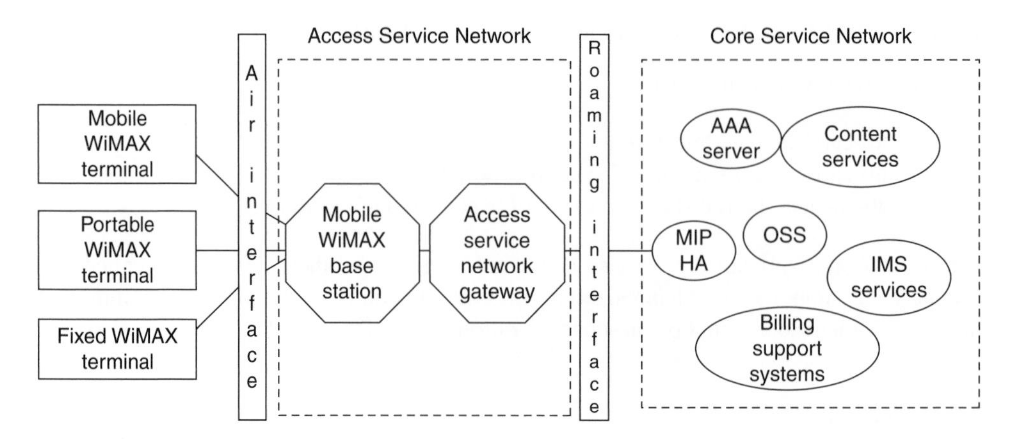

Figure 7.8 Block diagram of the WiMAX network architecture.

- Downlink and Uplink Media Access Protocol: provides sub-channel allocation and control information for DL and UL sub-frames, respectively.
- Uplink Ranging: allocated for mobile stations to perform closed-loop time, frequency and power adjustment and bandwidth requests.
- Uplink CQICH: allocated for the mobile station to feedback channel-state information.
- Uplink Acknowledgement: allocated for the mobile station to feedback down line HARQ acknowledge.

7.4.3 WiMAX Network Architecture

A simplified block diagram of the WiMAX network is shown in Figure 7.8. The WiMAX network consists of three main aspects:

- Mobile Station/ Equipment.
- Access Services Network (ASN).
- Connectivity Services Network (CSN).

7.4.3.1 Mobile Station

The mobile station as in other networks connects the user to the base station or the access network. The major difference with the current mobiles in the market (e.g. GSM or CDMA) is that this one needs to be IEEE802.16 compatible. Also, the mobile must have functionalities such as radio resource management (e.g. handover, paging, etc.), mobility management (e.g. location update), power saving capabilities (there are three power save classes), AAA (Authentication, Authorization and Accounting) and Session Management (e.g. triggering QoS based services).

7.4.3.2 Access Service Network

The access service network connects the mobile subscriber to the IP backbone using the OFDMA air-interface. It mainly consists of two parts: base station and ASN Gateway (ASN

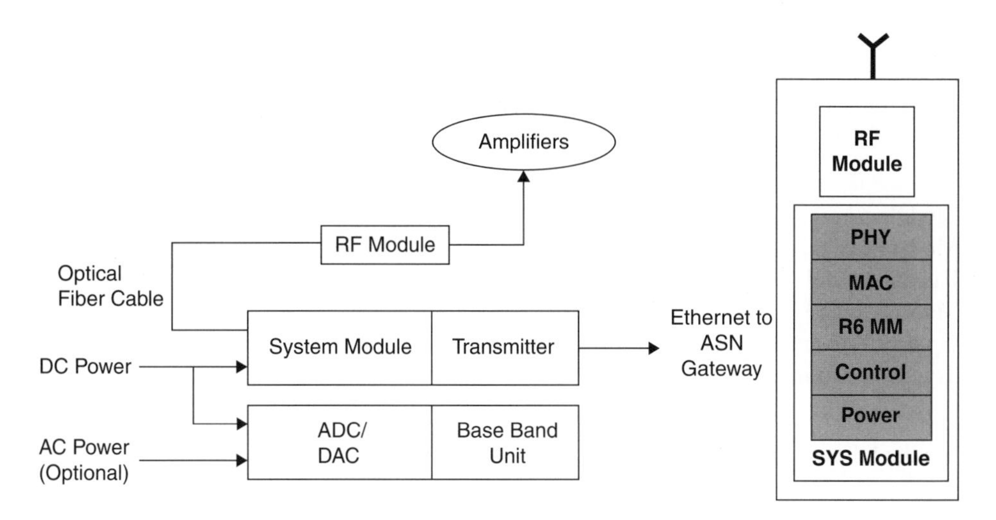

Figure 7.9 Simplified block diagram and protocol stack of WiMAX BS.

GW). One ASN may consist of one or more than one base station and one or more than one ASN gateway.

Base Station

A simplifies block diagram of the base station is shown in Figure 7.9. There is an RF module and system module. In practice, there are more than one RF modules in the base stations. Both these modules are connected by an OBSAI (Open Base Station Architecture Initiative) optical cable.

As the name indicates, the RF functions are performed by the RF module. The RF module has interfaces towards the system module, power and antenna system. The system module is responsible for the base station control and performs functions related to the physical layer (transmission and reception of OFDM symbols, power control, etc.) and the MAC layer (connection set-up, scheduling, mobility, etc.). It also performs the R6 mobility management procedures. These include storing required information related to the subscriber and terminal, service flow management (e.g. admission control), serving data path function (establishing and managing user plane between two peers), serving handover function (RRM) and sending the AAA information towards the gateway. Operations and maintenance, plus synchronization functions are also managed by this unit. Apart from this, power management and distribution are also parts of the functionalities of the system module. Thus, the system module has interfaces towards all radio modules, power, transmission, synchronization, external alarm and control, site support, etc.

The WiMAX base station works like the GSM base station – with towers standing high up in the air to broadcast radio signals and can cover up to 10 km radius. These can reach somewhere around 50 km or 30 miles but due to certain geographical limitations, they go only as far as 10 km or 6 miles. Any wireless connecting device for WiMAX will connect to the WiMAX network if it falls into this range.

The elements used here are as follows.

System Module
This contains the base band processing, operation and maintenance and synchronization, power distribution and transport functions. It is outdoor-capable and environmentally protected. It consists of the following main parts:

- Module core.
- Module casing.
- Fan sub-module.
- Module cables.
- Optical front and back covers.
- Transport sub-module.

It has the following external interfaces:

- Interfaces to radio modules which support optical interfaces for connection to the RF modules.
- Power input/output interfaces: these have DC power input connectors. The system module supports power output interfaces for distributing the DC current to the RF module.
- Transmission interfaces which are located on the front of the field-replaceable transport sub-module.
- Local management ports which are responsible for commissioning, operations, maintenance and testing of the base station.
- Site support interfaces which are used to carry initialization data, control data and signalling data from the base station to any auxiliary site equipment that has an ethernet port and an IP address.
- External alarms and control interfaces which are used to collect simple ON/OFF external alarms from any equipment that is external to the base station. They are also used to provide ON/OFF control signals for controlling external devices.
- External synchronization interfaces which are used to synchronize the base station with a GPS signal.

RF Module
This contains the radio frequency functions and is outdoor-capable and environmentally protected. It is designed to be mounted at pole and has the following interfaces:

- Optical interface.
- Power interface.
- Antenna feed.
- Remote antenna tilt interface.

WiMAX BTS Antenna
This is used to amplify the signals and also supports remote electrical tilt which enables the operator to optimize network performance remotely without a site visit.

GPS Antenna
This is installed outside for satellite visibility with an integrated receiver and is directly connected to the synchronization input of the system module.

Transmitter
This transmits the signals of the system module.

ADC or DAC
These are analogue-to-digital or digital-to-analogue converters and convert analogue signals into digital ones or digital signals into analogue ones.

There are various optional modules also which are described in the following.

Power Module
The BTS modules are designed for 45–50 V power supplies. They also have two modules, an AC–DC sub-module and a power battery sub-module.

Power DC–DC Module
This is an optional standalone DC/DC converter module and is capable to provide up power.

Alarm Extension Module
This can be installed in a stack or cabinet and is connected to the EAC interface in the system module.

Outdoor Cabinet
This can be used at outdoor and indoor sites where low lockable multi-purpose equipment is needed and includes a lock and a door alarm that is connected to the BTS system module EAC connector. It has the following modules:

- Cabinet site support module.
- Integrated battery backup unit.
- Long-term battery.

Indoor Cabinet
This is an optional indoor cabinet for new sites. It provides space for modules vertically.

Cabinet Accessories
They are an optional plinth, cabinet air filter and cabinet smoke detector.

7.4.3.3 ASN Gateway

As shown in Figure 7.8 above, the ASN Gateway is an interface between the ASN and CSN (e.g. home agent and AAA server). The ASN gateway is a logical entity that is responsible for functions such as mobility management, for example paging, QoS, lawful intercept interface, authenticator, data flow management, etc. The ASN gateway primarily performs one or many of the tasks such as control plane, user plane, serving and target gateways. Under the control plane anchor gateway role, the ASN GW performs functions such as location management, charging, mobility management, authentication, etc. The user plane anchor gateway performs functions related to service flow, tunnel anchor, foreign agent, subscription based rate limiting, etc. Functions such as relaying and forwarding messages between various elements in the network are performed by the serving ASN gateway while the target ASN gateway performs functions such as relaying/acknowledging the handover requests.

ASN Profiles

To manage the diversity in the ASN usage, three profiles are implemented in it. These three profiles are Profile A, Profile B and Profile C. Profile A includes Radio Resource Management (in ASN GW), Radio Resource Allocation (in the base station) and open interfaces R1, R3, R4, and R6 for Profile A. Profile B includes a 'no split if function', that is a distributed ASN solution with BS and ASN GW functionalities and no intra-ASN operatability. Profile C is similar to Profile A except that the base station has more functions (RRM is located in the base station).

7.4.3.4 Connectivity Service Network (CSN)

The Connectivity Service Network performs some key functions in a WiMAX network. As the name suggests, it provides connectivity to the external networks such as the Internet, PLMNs and others. Also it provides policy control for voice, VPN, etc. services. It is also responsible for IP address management (DHCP, AAA, etc.) and location, plus mobility management between the ASNs. The CSN is responsible for the QoS management as well. The Connectivity Service Network contains two main elements:

- Authentication, Authorization and Accounting (AAA) server.
- Home Agent (HA).

AAA Server

As the name suggest, AAA or the Authentication, Authorizing and Accounting server does these functions. Authentication and authorization procedures are based on the Extensible Authentication Protocol (EAP). The EAP runs on RADIUS between the AAA server and ASN. It is responsible for the security key generation and the mobile IP authentication. It is also responsible for charging, QoS control and bandwidth allocation.

Home Agent (HA)

'Mobile IP' is an international standard that allows users to seamlessly roam between various wireless networks. For this to take place, a mobile node in each mobile device and a home agent in each of these devices' networks are needed. The home agent intercepts the datagrams for the mobile node from a terminal and tunnels them using temporary IP addresses. The replies from the originating hosts can be sent from the mobile node directly or tunneled back to the home agent, which in turn send them to the host after unpacking them. Apart from this, other functions are to support the AAA server, inter-ASN mobility support, server load balancing, forward/reverse tunneling, etc. In short, the home agent makes it possible for the

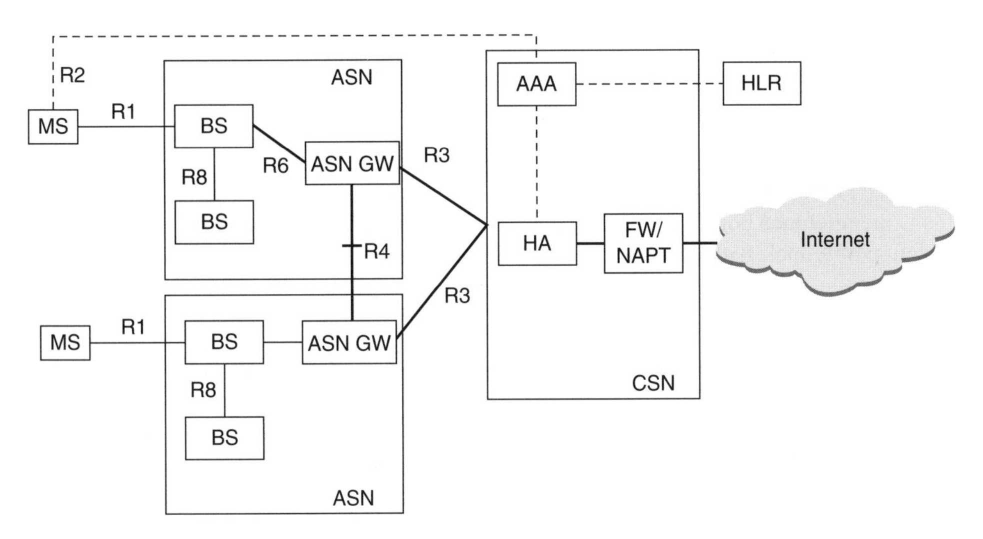

Figure 7.10 WiMAX network interfaces.

subscribers to be connected to the network and maintaining services even when roaming the WiMAX network.

7.4.3.5 Interfaces in WiMAX Network

Interfaces in the WiMAX network are called reference points (R). The following are the interfaces/reference points in the WiMAX network (shown in Figure 7.10).

Reference Point R1
This point is the interface between the mobile station and the base station. It implements the IEEE 802.16e-2005 standard including the physical and MAC layer features, carrying both the user traffic and user control plane messages.

Reference Point R2
The interface between the mobile station and ASN-GW or CSN is called the reference point R2. This is used for the protocols and procedures related to authentication, authorization and configuration management.

Reference Point R3
The interface between the ASN and CSN is called the reference point R3. This is used for implementing the tunnel between ASN and CSN, sending the data management/control plane messages, supporting the policy enforcement, AAA methods, etc.

Reference Point R4
This is the interface between two ASNs or two ASN-GWs. This is used for the implementation of data and control plane messages and mobility management.

Reference Point R5
This is the interface between two CSNs and is responsible for implementing the control and data plane methods between the home and visited networks.

Reference Point R6
This is the reference point between the base station and ASN. It is responsible for the implementation of tunnels between ASNs and for control plane signalling.

Reference Point R7
This is an informative reference point between the data and the control plane in an ASN gateway.

Reference Point R8
This is the reference point between two base stations. This is used for fast and efficient handovers between the base stations.

The additional elements used in the WiMAX network, based on the uses of the operator and existing architecture, are as follows.

7.4.3.6 Home Agent

This is a user plane element in the network that provides an interface between the WiMAX network and other internet protocol networks and services. It enables seamless service availability between different access networks.

7.4.3.7 Authentication, Authorization and Accounting Server

This covers the WiMAX policy and control function by interfacing with the operator's subscriber management and billing systems.

7.4.3.8 WiMAX Enabled End-User Devices

This allows the operator to offer consumer products to take advantage of WiMAX services.

7.4.4 Protocol Layers in WiMAX

The protocol layers in the WiMAX network are shown above in Figure 7.2. The IEEE802.16 standard however defines only two layers: physical Layer and MAC layer. As in any other protocol standard, the physical layer is responsible for data transfer between the two entities it connects. The MAC layer is that part of the data link layer which maps the IP/ATM packets into the protocol data units and subsequently onto the physical layer.

7.4.4.1 Physical Layer

The physical layer is responsible for transmitting/receiving the OFDM symbols between two entities and two directions (uplink and downlink). It is also responsible for the coding, modulation/de-modulation and power control. The OFDMA physical chain consists of the steps starting from randomization, FEC, interleaving and modulation. Randomization is used for protection and this is done by not using long sequences of ones or zeros. It is done on both the uplink and downlink directions. FEC or Forward Error Correction is done using the 'Reed Solomon Convolution Code (RS-CC), turbo codes and block turbo codes. The RS code is used in many systems and is done by adding some redundant bits into a digital sequence. At the receiver end, through sampling phenomena, the original signal is extracted (with fewer errors). In longer sequences, the errors are also long and may be quite difficult to correct. Here, bit interleaving is used in which by diversity introduction, error corrections can be done. The next step is repetition, on using which the signal margin improves further on top of the one facilitated by modulation and FEC. This is possible only on the QPSK modulations. 16QAM and 64QAM are necessary in DL while in UL, 64QAM is optional. For more details about the OFDM, please refer to the previous chapter. The physical layer in IEEE802.16 support both the TDD and FDD operations. Initial releases of WiMAX will see only TDD being implemented. In the TDD mode of operation, a single channel can be used for both the uplink and downlink unlike in FDD where two channels are needed. Adjustments are possible in the UL/DL traffic in TDD while in FDD, it is always fixed. Other features that are used on the physical layer are Hybrid ARQ (Automatic Repeat request), Fast Channel Feedback, Channel Quality Indicator (CQICH) and Adaptive Modulation and Coding (AMC). The Channel Quality Indicator, as the name suggests, provides the channel state information – starting from the mobile to the base station scheduler. HARQ provides faster response to packet errors, providing improvements to cell coverage. AMC supports in link adaptation in mobile environment (e.g. driving a car).

7.4.4.2 MAC Layer

The Medium Access Control or MAC layer is based on the DOCSIS standard and is able to simultaneously support the high data traffic, streaming video and voice over the same channel. The MAC layer consists of three parts: convergence sub-layer, common part sub-layer and security sub-layer. As mentioned above, MAC is responsible for the conversion of the IP/ATM packets onto the physical layer – this is the function of the convergence sub-layer. Functions such as scheduling and mobility are done by the common part sub-layer. Authentication, encryption are performed by the security sub-layer.

Quality of Service in MAC is provided by the service flow which is uni-directional in nature. These are defined by the set of QoS parameters, as shown in Table 7.6. The process of defining the QoS for a particular service is similar to that of UMTS. The two MAC layers interact with each other before data service is provided. Once the QoS parameter associated with the data service is defined, transmission and scheduling over the air-interface takes place. For the air-interface not to create hindrance with respect to, for example throughput, e2e QoS control is established. The MAC scheduler should be fast and efficient in allocating the resources in both the uplink and downlink directions. Scheduling is also dynamic – it change at frame level due to changes in the traffic/channel conditions.

Table 7.6 Mobile WiMAX applications and Quality of Service

Quality of Service Category	Applications	Quality of Service Specifications
Unsolicited Grant Services	VoIP	• Maximum Sustained Rate • Maximum Latency Tolerance • Jitter Tolerance
RT Polling Service	Streaming Audio or Video	• Maximum Reserved Rate • Maximum Sustained Rate • Maximum Latency Tolerance • Traffic Priority
Extended RT Polling Service	Voice with Activity Detection	• Minimum Reserved Rate • Maximum Sustained Rate • Maximum Latency Tolerance • Jitter Tolerance • Traffic Priority
NRT Polling Service	File Transfer Protocol	• Minimum Reserved Rate • Maximum sustained Rate • Traffic Policy
Best-Effort Service	Data Transfer, Web Browsing, etc.	• Maximum Sustained Rate • Traffic Policy

7.4.5 Security

We have seen some aspects of security in Wi-Fi networks before in this chapter. WiMAX also provides security aspects to protect the network. There are five key aspects of security in WiMAX networks, as shown in Figure 7.11.

Fast Handover Support

There is a 3-way handshake method to optimize the re-authentication mechanisms for supporting fast handshakes. This prevents the system from 'man-in-middle' attacks.

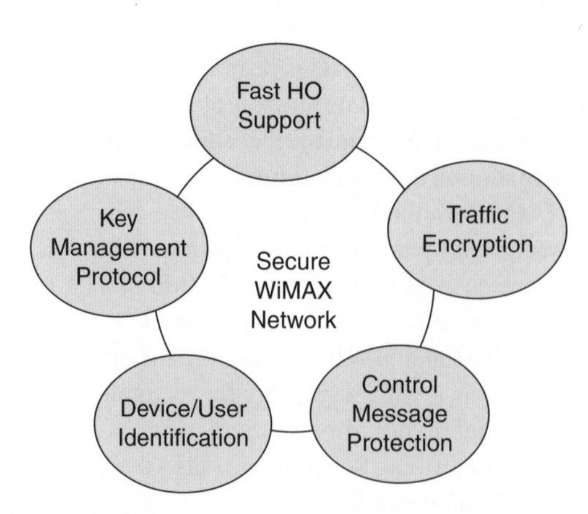

Figure 7.11 Security in WiMAX network.

Traffic Encryption

A cipher is used to protect all user data over the MAC interface. This is called the AES-CCM code (Advanced Encryption Standard – counter with Cipher block Chaining Message) authentication code and is generated from EAP authentication.

Control Message Protection

This is used to protect control data. This is done by using the AES based CMAC (Cipher based Message Authentication Code) or Message Digest-5 (HD-5)-based Hashed Message Authentication Code (HMAC).

Device/ User Identification

Both the device and user are protected in WiMAX. This is done by using the schemes based on SIM or username/password format. These are done by methods that are based on EAP protocol.

Key Management Protocol

Privacy and Key Management Protocol version 2 (PKMv2) is the basis of the security policy of WiMAX. This manages MAC security and major security messages (e.g. handover key messages, multicast/broadcast security messages) are based on this protocol.

7.4.5.1 Smart Antenna Technologies

They include the vector or matrix operations on signals due to multiple antennae and enhance system performance. They include:

- Beam forming. With this, the system uses multiple antennae to transmit weighted signals to improve coverage and capacity of the system and reduce outage possibility.
- Space-Time Code. Transmit diversity is supported to provide spatial diversity and reduce fade margin.
- Spatial Multiplexing. This is supported to take advantage of higher peak rates and increased throughput and using this, multiple streams are transmitted over multiple antennae. The user peak data rate and sector peak data rate for it in MIMO is just double that in the SIMO.

7.4.5.2 Fractional Frequency Re-Use

In this, all cells operate on the same frequency channel to maximize spectral efficiency. In Mobile WiMAX, the re-use is done through a sub-channel segmentation and permutation zone. The segment is a sub-division of the available OFDMA sub-channels and the permutation zone is a number of contiguous OFDMA symbols in DL/UL that use the same permutation.

7.4.5.3 Multicast and Broadcast Service

This combines the best features of DVB-H, MediaFLO and 3GPP E-UTRA and satisfies the following requirements:

- High data rate and coverage using a Single Frequency Network.
- Flexible allocation of radio resources.

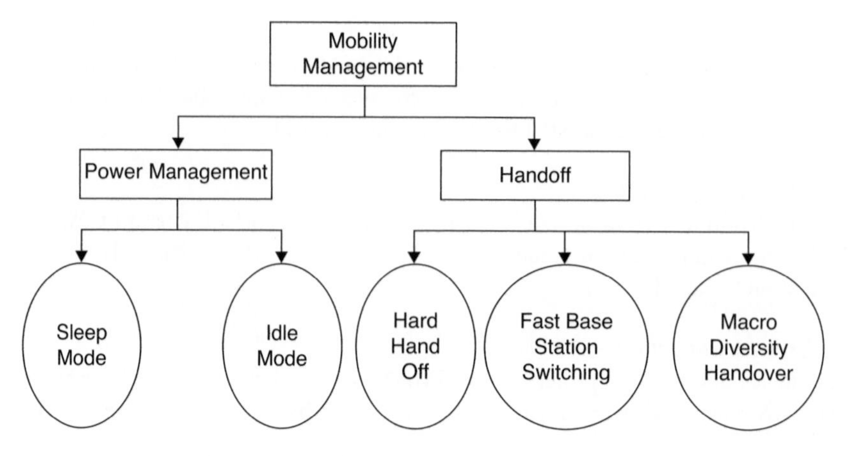

Figure 7.12 Mobility management in WiMAX.

- Low mobile station power consumption.
- Support of data-casting in addition to audio and video streams.
- Low channel switching time.

In this zone, the multi-base station mode is used by single frequency network operation and its flexible duration permits scalable assignment of radio resources to traffic.

7.4.6 Mobility Management

There are two critical issues for mobile applications which are described in Figure 7.12: power management and handoff.

The way that a mobile gets attached to the CSN is quite similar to the process in which the mobile gets attached in the GSM network. The process starts with the channel scanning in both the UL and DL directions along with relevant synchronization. The information is provided by the base station (and ASN). This is followed by the AAA, session establishment and capability management between relevant elements of the network. The mobile has four states: off, sleep, idle and normal modes. We have seen off, idle and normal modes in previous chapters, so let us understand the sleep mode. The sleep mode happens between bursts of data and helps make the mobile more power efficient. It is a state where pre-negotiated periods of absence are conducted by the mobile. Not only does it support power savings, but it also helps the mobile to scan the base stations for handoff processes. As mentioned above, there are three classes: Type 1, Type 2 and Type 3. Type 1 is for 'best effort' (NRT-VR), Type 2 for Unsolicited Grant Service (RT-VR) and Type 3 for multicast connections.

In the normal mode, there are three kinds of handovers/handoffs in a WiMAX network. These are hard handoff (HHO), macro-diversity handoff (MDHO) and fast base station switching handoff (FBSS). HHO is similar to the ones in GSM, that is the mobile needs to break off connection with one base station before getting connected to another base station and both the base stations are on different frequencies. Many techniques have been developed to bring down the delay to a level of 50 ms. MDHO is similar to the ones in UMTS – the mobile being 'in touch' with many base stations within a logical area and finally getting connected to the 'best' (e.g. based on signal strength) base station. In FBSS, the mobile is in the handover mode; it selects one base station that acts as an anchor for the traffic switching. This anchor is selected from the active set – a set of base stations involved in the FBSS with the mobile.

7.4.7 Network Design in WiMAX

The network planning process in a WiMAX network is quite similar to that of the planning process of GSM or UMTS networks. The stages are pre-planning, detailed planning and optimization. Nominal planning as the name says, will give out nominal plans that have some primary results on capacity, coverage planning, site surveys and model tuning. The detail planning exercise will give detailed capacity and coverage plans and parameter plans. Optimization is done to tune the network for best performance in terms of coverage, capacity and quality.

The initial network designing of WiMAX includes the study of a summary of the overall network objectives, goals, expectations, advantages and challenges from a user and network architecture perspective. It should include bandwidth considerations, link availability, frequency selection, number of sites and sectors, types of systems recommended, etc.

WiMAX works on both Internet and cellular networks. It has a speed of 2 Mbps and covers an area of 10 km. It has a larger range as compared to Wi-Fi. For laptops, its range can vary from 5 to15 km and for desktop computers its range is usually 50 km. WiMAX is a long range system, hence an area of several km can be covered using a few towers. It has a downloading speed of up to 20 Mbps. It provides facilities such as GPRS, gaming and downloading through mobile phones. It enhances the wireless coverage capacity by almost three times.

The designing of the network includes a detailed parts list showing the preferred vendors for access points, cabling, back-up power unit, antennae, switches, surge suppressors, connectors, jumpers, etc. The antennae allow a high degree of elevation and azimuth tilt and can be mounted on a mast or a wall. The antenna housing is also made available with a back plate that can be mated directly to an external CPE radio enclosure. In addition, the antenna can also be provided as a kit without the structure for integration into a CPE enclosure for a fully integrated CPE solution. It provides a gain of 18 dBi and covers the 3.4 to 3.6 GHz frequency range.

The network designing includes the total cost including final engineering, site acquisition, integration, training, testing, site optimization and rigger costs. The cost of installation and equipment of the base station of a WiMAX includes the total cost of the base station, sector antenna, total base station and total bandwidth. The designing also includes point-to-point backhaul link budgets and path profiles for each link in the network.

The backhaul network has the following features:

- Offers cost effective high-end IP services (e.g. video, gaming, etc.).
- High capacity bandwidth of up to 1.4 Gbps.
- Can carry mixed IP/TDM traffic or 100 % IP services.
- Extends network reach and expands the footprint.
- Easy failsafe redundancy for critical links.

A detailed path analysis and link budget for each link in the network includes calculations for free space losses, connector and cable losses, link availability and downtime in seconds per year. Free space loss is the loss that a radio signal experiences when travelling through free space. This calculation is important when determining the amount of power we will need to complete our 2.4 GHz WiFi point-to-point link. It is also very important for other frequencies as well. We have already seen the FSL calculations in previous chapters.

The adapter in WiMAX delivers up to 13 Mbps downlink and 3 Mbps uplink performance over WiMAX and up to 300 Mbps over WiFi.

During the course of network designing, it is important to understand the results of spectrum and noise analysis that were performed during the site survey. The designing also includes tower leg, water tank railing or ladder measurements. These measurements help in the procuring of appropriate mounting brackets and clamps.

Detailed hardware-specific multipoint propagation analysis with coverage estimates superimposed on geographical, political or satellite backdrops is done. Propagation as well plays an important role as in radio network design of any other wireless network. We have already seen the impact of propagation in previous chapters.

As with any network design, an optimal WiMAX network must be able to address specific market requirements, deployment geography, end-user demands and planned service offerings – both for today as well as for tomorrow.

8

Convergence and IP Multimedia Sub-System

8.1 Introduction to Convergence

The communications world is moving towards always-on and always-connected conditions for people who want to be connected to the world – real time. Both the means of communications, fixed and wireless, have undergone a sea-change. PSTN and ISDN are not only being used for voice but low-cost connectivity to the Internet world. Mobile communications have seen technology changing from first generation to the third generation and beyond (e.g. HSPA, LTE, etc.). And on top of this, we are witnessing a convergence of these two (fixed and mobile) worlds. Convergence is a word that may mean different things to different people. Convergence is happening on the service and application side, handset and communication devices, content and media, Internet and mobile; fixed and mobile networks, etc. But fundamentally, there are three types of convergence, as shown in Figure 8.1.

- Device Convergence.
- Network Convergence.
- Service Convergence.

A mobile device is no more just used for voice communication. It is used for data and can be used for viewing television programmes, etc. So also is the case with computers – no more just a data upload/download machine, it can be very much used for voice communications. Service industries are as well not just about using one technology for one device – they are about a seamless experience across multiple access technologies, for example browsing the Internet from a mobile phone or sending messages from a mobile to a PC. The world is moving towards mobiles – more and more people are subscribing to mobile services. This is a movement that is worrying for the fixed operators. Similarly, the mobile operators are using the fixed-mobile convergence to address the indoor voice market. Thus, both the mobile and fixed operators are moving towards network convergence. Many subscribers own more than one communicating device – each working on its own technology, having its own numbers (telephone numbers)

Cellular Technologies for Emerging Markets: 2G, 3G and Beyond Ajay R. Mishra

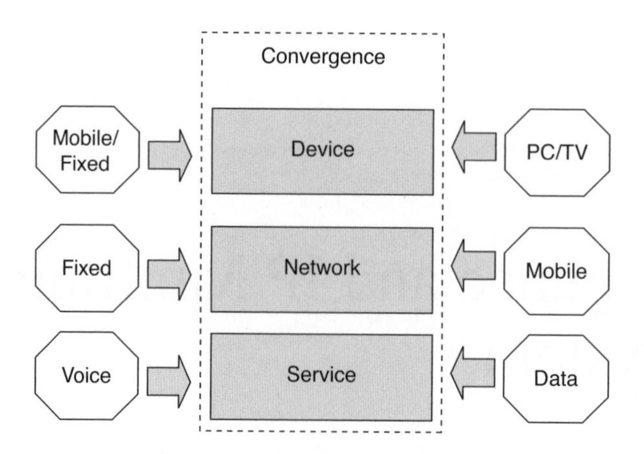

Figure 8.1 Types of convergence.

and different functionalities. This results in a need for convergence, resulting in, for example one contact number/address for all communicating purposes or one intelligent system that can detect the kind of traffic moving towards communicating devices and can control all the devices connected to the system.

A convergent system should be 'technology agnostic' and act as a bridge between the circuit- and packet-switched domains. It should be able to provide a centralized database that can give information about the subscribers to all the networks. This kind of system will be beneficial to all – subscribers will have only one device to maintain while having improved network coverage.

8.2 Key Aspects of Convergent Systems

8.2.1 Types of Convergence

As seen before, there are mainly three types of convergence. Each of these areas is further segmented to get the full picture of the FMC environment (shown in Table 8.1).

8.2.1.1 Device Convergence

A few years ago, each of the devices was used for a specific purpose, for example mobiles for voice (and SMS), and PCs (for the Internet). However, converged devices are the ones that can be used for multi-purposes – having the capability to provide more than one type of service

Table 8.1 FMC environment

Element/ Component	Type/Layer	User
Device	Phone, PDA, Headset, Laptop	All Individual/Group Users/Subscribers
Network	Access and Core	CSPs (Communications Services Provider)
Service	Voice, Messaging, VPN, Multimedia	All Individual/Group Users/Subscribers

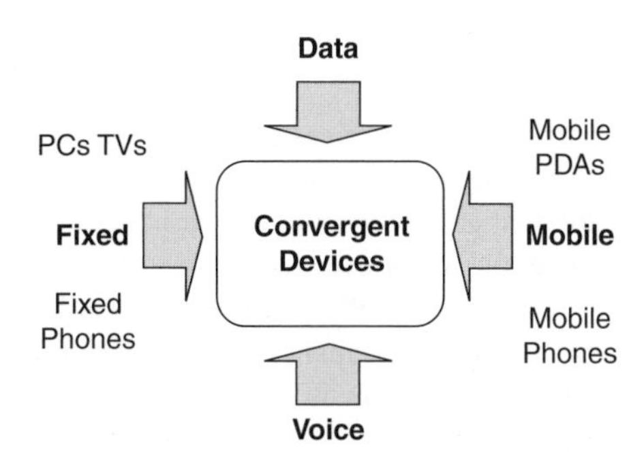

Figure 8.2 Device convergence.

and get connected to more than one type of network. The development of mobile phones has been phenomenal in the last few years with each new device being more convergent than the previous one (as shown in Figure 8.2). The advances in the areas of processing speed and memory has made it possible for multiple applications to run at the same time. The subscribers look for two main features: Quality of Services and Mobility in these convergent devices. Even the PCs are getting smaller with higher memory and capacity. As the PCs have more capability to handle the multiple applications and have multiple interfaces to connect to various networks, they have been fundamentally more of convergent devices than the mobiles. Also PDA and other devices such as Wi-Fi's VOWLAN (Voice Over Wireless LAN) are being used as convergent devices.

The rapidly increasing capability of handsets in terms of memory, screen space, speed to downloading and streaming capabilities has made applications like watching a football game or listening to a music video – something that was possible only on television a few years back – or downloading emails, something that was possible only on a computer. One device that is catching up is the 'Smartphone'. 'Smartphones' are cellular phones that are built on an open operating system, for example Symbian, palm, OS etc. Applications can be downloaded into these phones including convergent applications. These phones are able to provide both the voice and multimedia services in this environment. An open IP network will bring in an unprecedented number of devices and their inter-operability success will play a key role in the open and convergent world – as smart devices mean smart connectivity wherein the complexities are hidden from the subscriber so as they get a seamless experience. Some of the devices that offer these kinds of services are the Blackberry, Nokia 9500, Motorola MPx range, NEC's 900iL, etc.

8.2.1.2 Network Convergence

The applications and types of devices that are available make sure that the network operators gear up to handling more than one hundred times of the traffic in the next few years. Plus traffic has different types of variations, for example real time, non-real time, etc. Providing connectivity to the new sets of *converged* devices, providing applications and services means a network that can handle both the IP and Internet would be needed – on top of handling voice traffic. The access network is to cater to various technologies such as GSM, EDGE, WCDMA, WiMAX, UMA, etc. The core network is moving towards THE All-IP format. Implementation of IMS enables real convergence. We will look into the details of IMS in the latter part of this chapter.

As mentioned before, the fixed network operators are moving towards mobiles while the mobile network operators are moving towards indoor coverage. Since mobile technology has come into being, the fixed subscribers are moving towards being mobile subscribers. In the developing world, the phenomenon looks even better – for example in China, during this decade the fixed subscribers increased by 50 % while the mobile penetration increased by 100 %. In advanced countries, for example Japan, fixed line penetration has reached its peak. Fixed operators are facing trends, for example market saturation, migration from PSTN to IP, call price depreciation, etc. that is forcing them to look into other options such as migration to broadband network operations. This will enable them to provide a new range of services to their customers and thereby increase the revenue potential. Also, mobile operators are looking towards the option of routing the mobile traffic to the indoor fixed network. Quite a high percentage of calls are made inside a building – home, office, mall, etc. Most of these are already having some networks, for example Wi-Fi, installed. Routing the traffic through this network would not only reduce the costs and increase quality, but also utilize the fixed networks, for example PBX. Also, fixed networks are quickly moving towards broadband service providers (ISPs/ Internet Service Providers). Though mobiles are capable of carrying data, most of the developing world (and also the developed world) still relies on wired broadband networks and computers as the device to access the Internet. Networks such as CDMA2000 or WCDMA are making inroads, but still a PC is the preferred device for accessing the Internet.

As shown in Figure 8.3, an access network would be an aggregation of technologies that will be able to handle both the voice and data applications. The networks are moving towards the All-IP domain making the architecture a 'flat one'. Transport networks would be more unified – integrating the latest network technologies with that of aggregation technologies. The services provided would have a seamless connectivity making sure that subscribers are able to use their devices, anytime and anywhere, with a high quality of service. Convergence with IT is the main feature of the application layer.

With the availability of Docsis technology, a higher bandwidth is possible for the cable operators. It offers data rates of at least 160 Mbps in the downstream and 120 Mbps in the upstream. This gives the possibility for cable operators to compete with the fibre-based telecommunication operators.

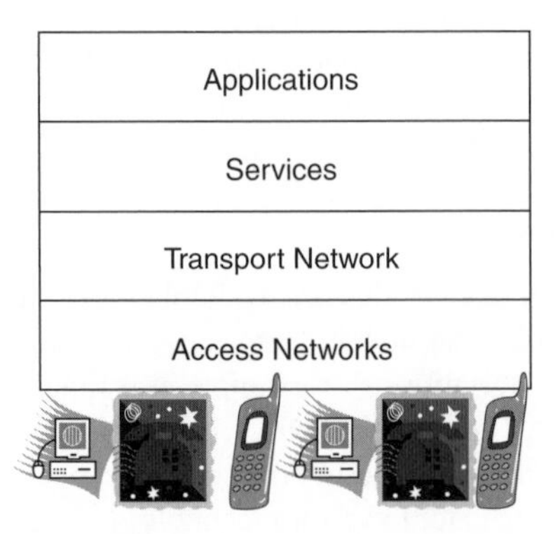

Figure 8.3 Converged networks.

In the coming years, it is expected that there will be no more fixed or mobile network operators but only network operators. With these networks it would be possible to access different technologies that would make the applications and services access agnostic.

8.2.1.3 Service Convergence

Service convergence is the most advanced – it has been happening for most of the time and has been independent of both the devices and the network advancements. Some of the key examples of service convergence are:

- One number service: single number for more than one subscription.
- One E-mail directory service: one e-mail directory on more than one terminal.
- Telephony convergence: one telephone can access many networks.
- Corporate VPN: extending the VPN services in remote/wireless locations.
- Push-to-talk: instant messaging across service boundaries.

Though multiple services are being offered, service convergence has still a long way to go. Some services, for example dual mode voice such as UMA, are being seen. Service 'bundling', such as voice, data, Internet, TV, etc. is being offered in the form of triple or quadruple plays, as shown in Figure 8.4. And to implement these services, the fixed operators will be tying up with the mobile operators and other service providers. This not only gives convenience to the subscriber, for example in the form of a single bill, but also to the operators as a bundled service offering will land them with higher revenues.

8.2.1.4 Quality of Service

Circuit-switched networks such as PSTN have evolved and so have the PLMNs. Both these networks have provided voice and data communications and both are utilized by the subscribers to fulfil their communication needs. With IP entering the scene the communication scenarios have changed drastically with voice, real time data, streaming data, etc. The VoIP applications are becoming increasing popular and preferred by individual/business clients for their low costs. However, all this leads to the ever important question of quality of services in convergent networks. The packet-switched networks provide only 'best effort quality of service' without any guarantees on bandwidth, delays, etc. making the quality vary during the course of the

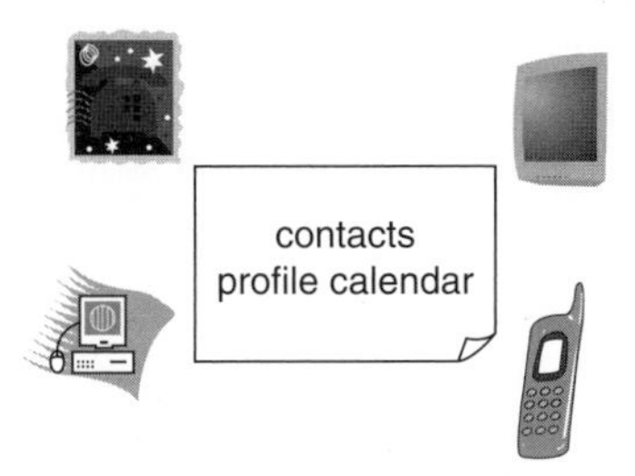

Figure 8.4 Service convergence.

connection (voice/data). Convergence takes care of aspects such as synchronization to have better user experience. QoS helps in more predictable user behaviour. It also allows efficient utilization of resources, that is utilization of the resources in the best possible way, allows application of grades of services to subscribers, etc.

8.2.2 Applications

Convergence provides an ecosystem for creating applications with IT and separation from access and transport layers. Due to open systems and APIs, rich and dynamic combinations of application and services are possible, making various permutations with respect to services very much a possibility. Faster, richer service development at lower costs, combining the CS and PS traffic and providing services like Web 2.0 to the service providers, is the hallmark of convergence. Some of the key applications are:

- Multimedia services, for example advertising.
- 'Push-to' services, for example push to talk, push to video, etc.
- Gaming, for example multi-party gaming.
- Personalized Information Services, for example alerts or calendars.
- Combined voice, video and data calls.
- Video streaming, interactive voice response.
- Messaging – unified, instant.

Let us discuss some of the above mentioned applications.

8.2.2.1 Location Based Services

The location based services combine the required information to a geographical position, for example an address along with the directions. There are three kinds of location based services: emergency services, operator services and commercial services. Emergency calls are tracked along with the location of the call. Operator based services includes services like network management systems providing network performance statistics at the cell level. Some examples for commercial services include services like navigation, tracking, etc.

8.2.2.2 Push-to Services

In this, the content is pushed from the server directly to the subscriber terminal. This is done through an SMS, MMS, or e-mail. This is opposite to the usual methodology employed by the mobiles in pulling the information from the server. Unlike the push services, these do not use the bandwidth during the downloading of information as in push-to services; the information is send out automatically.

8.2.2.3 Video Streaming

In streaming, the end user is able to view the content almost as soon as the downloading starts. In this methodology the waiting time is negligible. Live events can as well be broadcasted through this technique – these are called webcasts (or net casts).

8.3 Architecture in Convergent Networks

8.3.1 Business and Operator Support Networks

The focus now in business has shifted from providing an expected bandwidth to providing expected services. The business support deals with the customer side, including aspects such as bill processing, payment collections, etc. The operator support system includes the network itself and its associated aspects such as provisioning, inventory, etc.

The initial systems were quite complex, supporting the operations as shown in Figure 8.5. Due to the high level of complexity, minor system level changes affect all of the interfacing systems. Similar is the case with the logical coupling with the software systems. Any changes to it may end in overhauling of the whole system. Also, these systems are so large and complex integration with external systems is almost impossible so that any change that needs to be done has to be done in multiple systems. To tackle these problems, the next generation systems will be managing IP networks. These systems will hold the key to operator promise of delivering any service to any device over any network, anytime, anywhere. Key components of the next generation systems (as shown in Figure 8.6) are as follows.

Business Process Management Layer

This layer is responsible for automated business processes. The tools that offer the automated business processes are usually deployed on the application servers and have an access to all the features such as security, remote connectivity, etc.

Application Queue

The initiation of the business processes is done through message sending. These messages are sent on the application queue and are deployed to provide synchronous communications.

Connector

This is initiated through the request in the queue or web services and is a composition of various elements such as in-adapter, out-adapter, connector queue, etc. Through this, only the business process management interacts with the external systems.

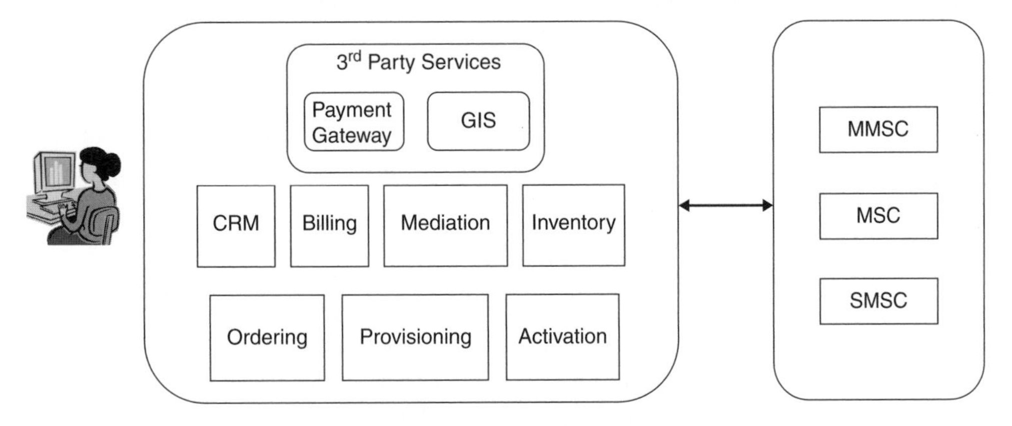

Figure 8.5 Traditional OSS-BSS systems.

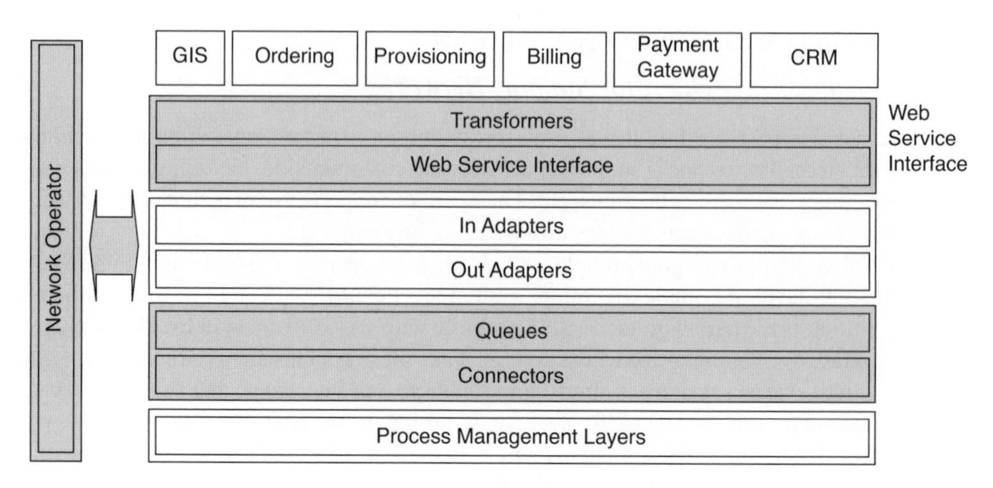

Figure 8.6 Block diagram of NGN converged OSS-BSS system.

In-Adapter

This is used for placing the request in the application queue. In simpler terms, it is responsible for placing the update requests with other parts of the systems when an update has taken place in one part of the system.

Out-Adapter

Once the requests have been placed through the in-adapter, the out-adapter is used to process the requests. The out-adapter reads the requests and takes the necessary steps for fulfilling these requests.

Connector Queue

This takes in all the asynchronous requests from the out-adapter.

Web-Service Interface

This helps in converting the application specific parameters and converting them to external system specific parameters and vice versa.

Transformer

As the name suggests, it transforms, but only with only transformation logic and not with business logic, for example Java to XML or vice versa. This is used for third party systems.

8.3.2 Technology

The key access technologies that are used in convergence are the likes of DSL, WiMAX and Ethernet from the fixed side and GPRS, HSDPA, EV-DO, UMA, WiMAX, OFDM, etc. on the mobile side. We have seen the details for most of these technologies earlier in this book. From the core perspective, all technologies – fixed and mobile – get converged into one network;

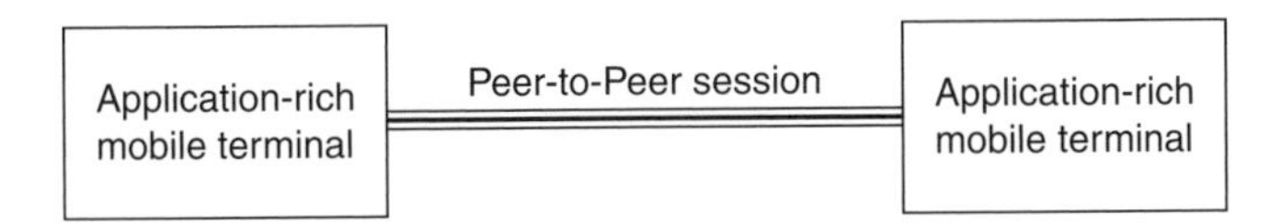

Figure 8.7 SIP based peer-to-peer session.

technologies such as SIP, MPLS, IMS, etc. However, because of its exciting features and its most relevance to the convergence theme, we will look into details of IMS in the next section.

8.4 IMS

8.4.1 Introduction to IMS

IMS or IP Multimedia Sub-system is a next generation network architecture that is used for providing multimedia services – both fixed and mobile. IMS started off as a technology under 3GPP, but is now a part of the wired world as well. This technology is built on SIP (Session Initiation Protocol), a technology that is used for controlling the communications in IP based NGNs (Next Generation Networks). Session Initiation Protocol (SIP) enables clients to invite others into a session and negotiate control information about the media channels needed for the session. A peer-to-peer session between application-rich terminals is established (as shown in Figure 8.7) and services like sharing real time video, MP3-coded music stream, real time game data, presence information, conferencing or shared two-way radio sessions (i.e. push-to-talk) can be realized. IMS provides real time services on top of the Universal Mobile Telecommunications System (UMTS) packet-switched domain. The skill to combine mobility and the IP network will be crucial to the success of any service in the future.

The Internet Protocol Multimedia Sub-system defines the complete class architecture and framework that enables the convergence of voice, video, data and mobile network technology over an IP-based infrastructure. It was originally designed by 3GPP and is now being adopted by other standard bodies like ETSI. IMS is 'access-independent' as it supports the following multiple access types (as shown in Figure 8.8):

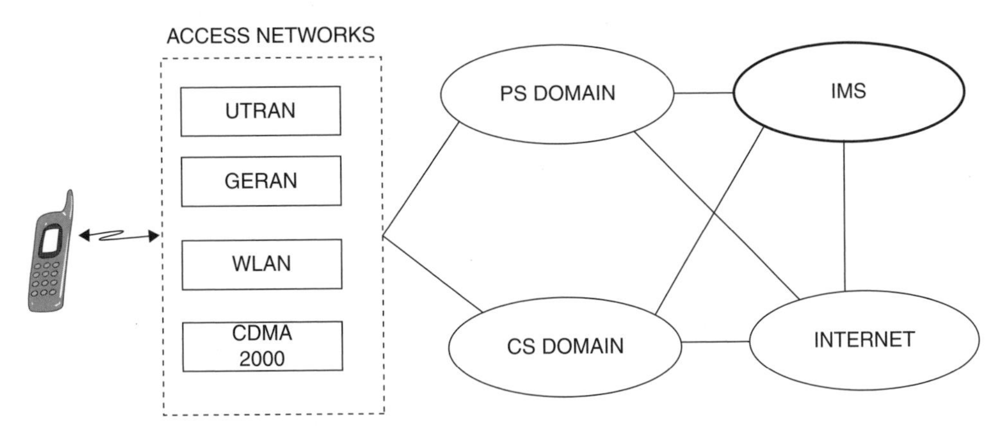

Figure 8.8 Position of IMS in a cellular network.

- GSM.
- WCDMA.
- CDMA2000.
- WLAN.
- Wire line broadband.
- Other packet data applications.

To ease the integration with the Internet, IMS uses IETF protocols wherever possible. IMS will make Internet technologies such as web browsing, e-mail, instant messaging and video conferencing available to everyone from any location. It is also intended to allow operators to introduce new services such as web browsing, WAP and MMS at the top level of their packet-switched networks.

What is in for subscribers? IMS-based services enable person-to-person and person-to-content communications in a variety of modes including voice, text, pictures and video or any combination of these in a highly personalized and controlled way.

And operators find that IMS takes the concept of layered architecture one step further by defining a horizontal architecture where service enablers and common functions can be re-used for multiple applications.

Due to these reasons, IMS is becoming a preferred solution for fixed and mobile operators' multimedia businesses. It promises to provide seamless roaming between mobile, public Wi-Fi and private networks for a wide range of services and devices.

IMS is a standardized reference architecture and consists of session control, connection control and an applications services framework along with subscriber and services data. It enables new converged voice and data services, while allowing for the interoperability of these converged services between Internet and cellular subscribers. Effectively, IMS provides a unified architecture that supports a wide range of IP-based services over both packet- and circuit-switched networks, employing a range of different wireless and fixed access technologies. A key point of IMS is that it is intended as an open-systems architecture, that is, services are created and delivered by a wide range of highly distributed systems, real-time and non-real-time, possibly owned by different parties, cooperating with each other.

IMS enables services to be delivered in a standardized, well-structured way that truly makes the most of the layered architecture. At the same time, it provides a future-proof architecture that simplifies and speeds up the service creation and provisioning process, while enabling legacy interworking. For fixed and mobile operators there are benefits of introducing the IMS architecture today. In the longer term, IMS enables a secure migration path to an all-IP architecture that will meet end-user demands for new enriched services.

8.4.2 IMS Development

IMS was first introduced in 3GPP Release 5, where SIP, defined by IETF, was chosen as the main protocol for IMS. It has been further enhanced in Releases 6 and 7 of 3GPP to include additional features like presence and group management, interworking with WLAN and CS based systems and Fixed Broadband access. 3GPP2, also standardized their own IMS. The initial release of the 3GPP2 specifications on IMS is largely adopted from 3GPP Release 5. The two IMS networks defined by the two organizations are fairly similar but not exactly the same. 3GPP2 added appropriate adjustments for their specific issues. Nevertheless, the purpose of both organizations is to ensure the IMS applications will work consistently across different network infrastructures. In addition to 3GPP and 3GPP2, OMA plays an important role on specifying and developing IMS service standardization. The services defined by OMA are built on top of the IMS infrastructure, such as IM, PS and GMS.

8.4.3 Applications of IMS

IMS is used in the following fields:

- Presence services.
- Full duplex video telephony.
- Instant messaging.
- Unified messaging.
- Multimedia advertising.
- Multiparty gaming.
- Video streaming.
- Web/audio/video conferencing.
- Push-to services, such as push-to-talk, push-to-view, push-to-video.

It delivers person-to-person real-time IP-based multimedia communications as well as person-to-machine communications and enables applications in mobile devices to establish person-to-person connections. QoS provides some type of guaranteed level transmission. Easy user set-up of multiple services in a single session or multiple simultaneous synchronized sessions are possible.

8.5 IMS Architecture

The architecture of IMS is split into three main layers or planes which are as follows (Figure 8.9).

8.5.1 Service or Application Layer

The application layer comprises application and content servers to execute value-added services for the user. Generic service enablers in the IMS standard, such as presence and group list management, are implemented as services in an SIP application server. The service layer is where all of the actual services 'live'. This includes traditional voice services as well as new applications built on the IMS architecture. This layer performs the following functions:

- Configuration storage, identity management and user status which is held by the HSS.
- Billing services provided by a charging gateway function.
- Control of voice and video calls and messaging provided by the control plane.

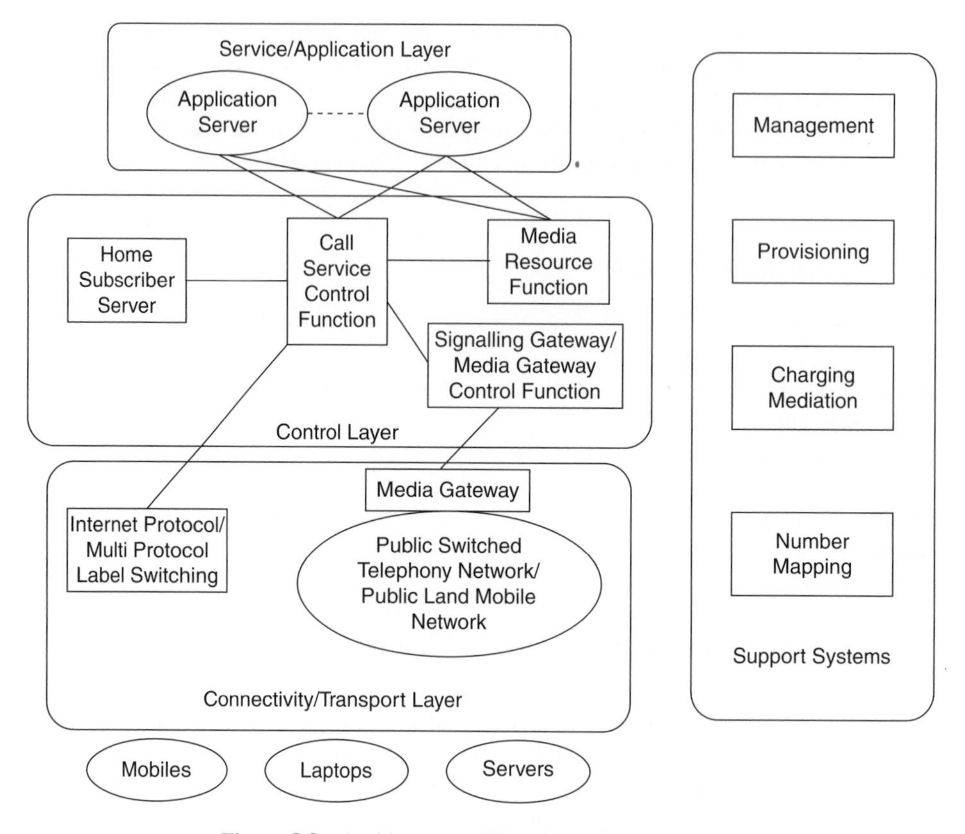

Figure 8.9 Architecture of IP multimedia sub-system.

8.5.2 Control Layer

The control layer comprises network control servers for managing call or session set-up, modification and release. The most important of these is the CSCF, also known as the SIP server. This layer also contains a full suite of support functions such as provisioning, charging and operation and management. Interworking with other operators' networks and/or other types of networks is handled by border gateways. This layer performs the following functions:

- Routes the call signalling and tells the connectivity layer what traffic to allow.
- Generates billing information for the use of the network.

8.5.3 Connectivity or Transport Layer

The connectivity layer comprises routers and switches, both for the backbone and the access network. The transport layer is responsible for the abstraction of the actual access networks from the IMS architecture. In essence, this layer acts as the intersection point between the access layers and the IP network above it. It is responsible for doing initial IP provisioning as well as facilitating the registration of devices with the higher layers. This layer performs the following functions:

- The I-BCF controls transport level security.
- The BGF provide media relay for hiding endpoint addresses with managed pinholes to prevent bandwidth theft.

As the main impact happens in the core site, let's try to understand the core site of the IMS Network.

8.5.4 IMS Core Site

An IMS Solution may consist of Connection Processing Servers (CPSs) and IP Multimedia Register (IMR) products. They may be co-located with other core systems like the Media Gateway and Mobile Switching Centre. An example of an IMS Core site is shown in Figure 8.10.

Many protocols are used in implementation of the IMS site and are used for providing features such as resilience, efficiency, security, etc. Initially 3GPP Rel-5 defined that the terminals shall use IPv6 (exclusively) when communicating with the IMS. Later on, this specification was changed to allow also IPv4 to communicate with the IMS. In some IMS interfaces, for example towards legacy network elements, support for IPv4 is in practice mandatory at the moment. Also IPv4 interfaces are most likely to be used for network management, charging, etc. interfaces for some time still. Hence IMS network elements should support dual stack, that is both IPv4 and IPv6. Some key elements in the IMS core site are as follows.

Domain Name Server (DNS)
This is the hierarchical, distributed database. It stores information for mapping Internet host names to IP addresses and vice versa and other data used by Internet applications. In the

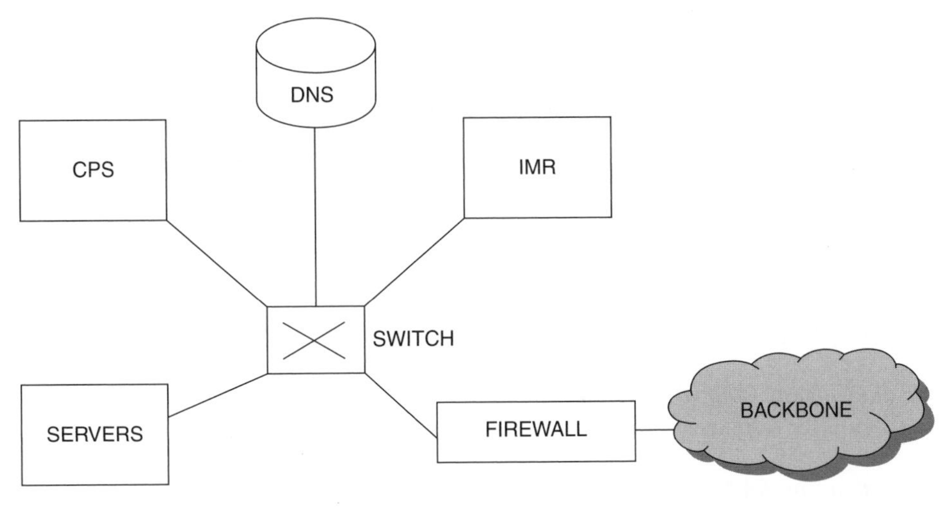

Figure 8.10 IMS core site.

IMS solution a DNS-cache should be implemented in every node to reduce Ethernet traffic and minimize DNS query time in the server. Slave DNS servers for IMS elements should be located inside server clusters. A slave DNS server answers to cluster nodes requests and receives zone transfers from the master DNS server. The master server is the ultimate source of information in an IMS domain. This DNS server is a stand-alone element. The primary master is an authoritative server configured to be the source of zone transfer for one or more secondary slave servers. The slave servers load the zone contents from the master server using a replication process known as a 'zone transfer'. With DNS server caches, resolution services to local clients are offered by running a named process on the local host. DNS queries are performed with DNS protocol to a local name server. If the answer is not in the cache, the query will be forwarded to a specified (slave) DNS server and received answers will be cached.

Firewall

Firewalls provide the most effective means for controlling the flow of IP traffic between two networks or servers. The principle of firewall operation is that all traffic across an interface is vetted in the firewall and only allowed to pass if explicitly permitted by the defined security policy for that interface. The firewall can apply different security rules for incoming and outgoing traffic on an interface. Firewalls are generally used at the boundaries of a network. They can also be used to implement network 'islands' and security domains within a wider network. This is the case, for example in the server site solution. The network islands need not be in one place and can be interconnected securely using IPSec encrypted VPN tunnels. For the IP Multimedia Subsystem IPv6, firewall capabilities are required. A IPv6 firewall protection is needed for the following applications:

- Protection of mobile users from other mobile users as peer-to-peer traffic has to be generally allowed.
- Protection of the IMS elements from the mobile users. Generally only SIP signalling should be allowed.
- Protection of the IMS elements at interconnects to other networks.

Later protection of the mobiles from the public Internet is also needed. In practice the initial functionality may consist of filtering of unwanted services.

Other network elements in IMS network are as follows.

Access Gateway

This network element provides an interface between the radio network and the IP-based network.

Access Network

This is the radio portion of the network.

Breakout Gateway Control Function

This controls the resources allocation to IP sessions.

Call Session Control Function

This provides control and routing function for IP sessions.

Foreign Agent
This 'advertises itself' to mobile stations in the serving area and provides registration information to the home agent. It also forwards packets from the mobile to the home agent.

Home Agent
This tracks the current FA serving the mobile and forwards packets to the current FA.

Home Subscriber Server
This can take the place of a HLR in an all-IP network and contains AAA functions and other databases.

Media Gateway
This provides an interface for bearer traffic between IP and PSTN.

Media Gateway Control Function
This provides signalling interoperability between IP and PSTN domains.

Policy Decision Function
As IP networks, unlike TDM networks, assign network bandwidth and resources in real-time, its role is to assign resources according to demand and quality of service requirements.

Position Determining Entity
While some mobiles can determine position independently, the PDE can provide assistance by way of location determination algorithms.

SIP Application Server
This represents a platform for SIP application development and operation

8.5.5 Functions and Interface in IMS

The functions and interfaces of IMS are shown in Figure 8.11.

Call Session Control Function (CSCF)
This provides the session control for both applications and terminals. This includes securing SIP message routing, monitoring and communicating, plus it interacts with the HSS as well.

Serving Call Session Control Function (S-CSCF)
This is the heart of an IMS system. This provides provision for the SIP signalling, routing, translation, maintenance of sessions, interaction with other services and charging. It also provides the information to the end points involved in sessions through associated P-CSCFs and also supports in authentication of users after registrations.

Interrogating Call Session Control Function (I-CSCF)
A peer IMS network will first contact the I-CSCF of the home network. It helps in routing the SIP request to an assigned S-CSCF.

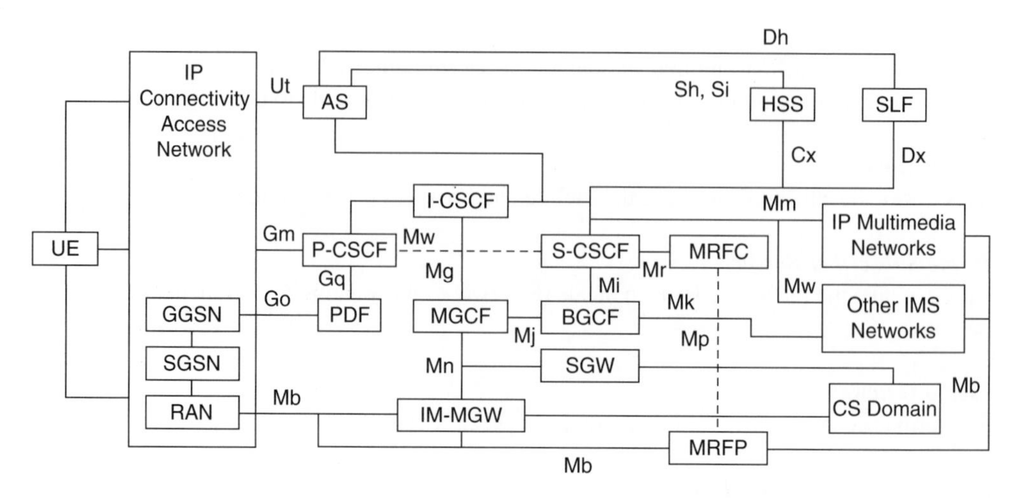

Figure 8.11 IMS functions and reference points.

Proxy Call Session Control (P-CSCF)

This is a SIP proxy that is the first point of contact for the IMS terminal. It can be located either in the visited network or in the home network. Some networks may use a SBC for this function. The terminal discovers its P-CSCF with either DHCP or it is assigned in the PDP context. It is assigned to an IMS terminal during registration and does not change for the duration of the registration. It sits on the path of all signalling messages and can inspect every message. It also authenticates the user and establishes an IP security association with the IMS terminal. This prevents 'spoofing' attacks and 'replay' attacks and protects the privacy of the user. Other nodes trust the P-CSCF and do not have to authenticate the user again. It may include a PDF, which authorizes media plane resources and generate the charging records as well. Also it compresses and decompresses SIP messages using signalling compression, which reduces the round-trip over slow radio links.

Media Gateway Control Function (MGCF)

This controls the connection for media channels in an IMS-MGW mode and communicates with the CSCF. It also supports in protocol conversion between ISUP and IMC call control protocols.

IP Multimedia Sub-system-Media Gateway Function (IMS-MGW)

This provides support to bearer control, media conversion and payload processing.

Media Resource Function (MRF)

This comes into play when the IMS application needs to provide media services from the network and implements functionality to manage and process media streams.

Multimedia Resource Function Controller (MRFC)

As the name suggests, this is responsible for the controlling of media stream resources of the MRFC.

Multimedia Resource Function Processor (MRFP)

The MRF processor provides functions and provisions for multimedia resources. These resources are controlled by the MRFC.

Subscription Locator Function (SLF)

In response to the queries from the I-CSCF or AS, this is located in the subscriber database.

Breakout Gateway Control Function (BGCF)

This is responsible for the routing of the telephony sessions that are initiated in the IMS moving towards the circuit-switched network. The BGCF selects the break out operator for outbound session – routing it to the MGCF (Media Gateway Control Function), BGCF to BGCF of another operator.

Application Server (AS)

This is placed in the user's home network (or third party location) and provides value added IMS services.

Home Subscriber Server (HSS)

This is the master database that contains all the user and subscriber information, including identification handling, authentication, authorization, mobility management, etc. In the case of more than one HSS, the SLF is used to detect the HSS where the subscriber information is being stored.

Signalling Gateway Function (SGF)

This is responsible for signalling conversion (in both directions) between SS7 and IP networks.

Policy Decision Function (PDF)

This allocates (or does not allocate) the IP bearer resources entering the packet-switched network.

8.5.6 Reference Points

There are many reference points in the IMS network. A brief description is given in Table 8.2.

8.5.7 Protocol Structure in IMS

SIP (Signalling Inititation Protocol) is the main protocol that is used in the IMS (as defined in RFC 326). SIP has been chosen as the main IMS protocol because it comes closest to the IMS requirements such as flexibility and security. This protocol is the function of establishment, modification and termination of the multimedia sessions between two devices/terminals. Some principles in SIP are similar to HTTP and SMTP protocols. Other functions handled by SIP include QoS authorization, subscriber management, billing, service control, billing, resource management, etc. Under AAA authentication, authorization and accounting protocols exist, replacing the RADIUS protocol (defined in RFC 3588). Diameter security is provided by IPSEC or TLS. This is used by I-/S-CSCF, AS functions, RACS and CLF. Another protocol

Table 8.2 IMS reference point descriptions

Interface Name	IMS Entities	Description	Protocol
Cr	MRFC, AS	Used by MRFC to fetch documents like scripts and other resources from an AS	HTTP over dedicated TCP/SCTP channels
Cx	I-CSCF, S-CSCF, HSS	Used to communicate between I-CSCF/S-CSCF and HSS	Diameter
Dh	SIP AS, OSA, SCF, IM-SSF, HSS	Used by AS to find a correct HSS in a multi-HSS environment	Diameter
Dx	I-CSCF, S-CSCF, SLF	Used by I-CSCF/S-CSCF to find a correct HSS in a multi-HSS environment	Diameter
Gm	UE, P-CSCF	Used to exchange messages between UE and CSCFs	SIP
Go	PDF, GGSN	Allows operators to control QoS in a user plane and exchange charging correlation information between IMS and GPRS networks	COPS, Diameter
Gq	P-CSCF, PDF	Used to exchange policy decisions-related information between P-CSCF and PDF	Diameter
ISC	S-CSCF, I-CSCF, AS	Used to exchange messages between CSCF and AS	SIP
Ma	I-CSCF, AS	Used to directly forward SIP requests which are destinated to a Public Service Identity hosted by the AS	SIP
Mg	MGCF, I-CSCF	MGCF converts ISUP signalling to SIP signalling and forwards SIP signalling to I-CSCF	SIP
Mi	S-CSCF, BGCF	Used to exchange messages between S-CSCF and BGCF	SIP
Mj	BGCF, MGCF	Used to exchange messages between BGCF and MGCF in the same IMS network	SIP
Mk	BGCF	Used to exchange messages between BGCFs in different IMS networks	SIP
Mm	I-CSCF, S-CSCF, external IP network	Used for exchanging messages between IMS and external IP networks	SIP
Mn	MGCF, IM-MGW	Allows control of user-plane resources	H.248
Mp	MRFC, MRFP	Used to exchange messages between MRFC and MRFP	H.248
Mr	S-CSCF, MRFC	Used to exchange messages between MRFC and S-CSCF	SIP
Mw	P-CSCF, I-CSCF, S-CSCF	Used to exchange messages between CSCFs	SIP

Table 8.2 (*Continued*)

Interface Name	IMS Entities	Description	Protocol
Rf	P-CSCF, I-CSCF, S-CSCF, BGCF, MRFC, MGCF, AS	Used to exchange offline charging information with CCF	Diameter
Ro	AS, MRFC	Used to exchange online charging information with ECF	Diameter
Sh	SIP AS, OSA-SCS, HSS	Used to exchange information between SIP AS/OSA-SCS and HSS	Diameter
Si	IM-SSF, HSS	Used to exchange information between IM-SSF and HSS	Mobile Application Part (MAP)
Sr	MRFC, AS	Used by MRFC to fetch documents like scripts and other resources from an AS	HTTP
Ut	UE, AS (SIP-AS, OSA-SCS, IM-SSF)	Facilitates the management of subscriber information related to services and settings	HTTP, XCAP

that is related to the policy support is the Common Open Policy Services (COPS) protocol. It is used for supporting the policy control over the Quality of Service (QoS) signalling protocol (e.g. RSVP), for conveying policy requests and decisions between PDPs (policy decision points) and PEPs (policy enforcement points). Real-Time Protocol (RTP) provides transport functions for transmitting real-time data while MeDaCo or H248 are used for controlling media serving functions. Monitoring of real-time delivery, provision for minimal control and identification functions are defined by the Real-Time Control Protocol (RTCP).

As mentioned above, 3GPP has chosen SIP for signalling between UE and the IMS as well as between the components of IMS in order to facilitate maximum interoperability with existing Internet systems, devices and protocols. SIP is an application layer control protocol based on a request-response paradigm for creating, modifying and terminating multimedia sessions with one or more participants. SIP is a signalling protocol, widely used for controlling multimedia communication sessions such as voice and video calls over IP. Other feasible application examples include video conferencing, streaming multimedia distribution, instant messaging, presence information and online games.

There are five types of logical entities in an SIP network, namely user agents, proxy servers, redirect servers, location servers and registrars. SIP works in conjunction with the SDP to initiate multimedia sessions and describes multimedia session functions such as session initiation and session announcement.

SIP is used to establish sessions according to the following general steps:

- **Session Initiation:** The user's device signals the need for a session and the user's network location is identified and a unique session identifier (SIP URI) is assigned.
- **Session Description:** Delivers a description of the session to the user device. The SDP protocol is used for this function.

- **Session Management:** Once the session is accepted by the user device, media streams or other content are then directly exchanged between the end points. RTP or RTSP are commonly used.
- **Session Termination:** Either party in the session can request session termination once the data or media exchange is complete.

Some of the key aspects of IMS are as follows.

Resilience

Resilience is the ability of the network to withstand faults and failures and to provide a high degree of availability for services. Resiliency is required at network element level, site level, transport level and disaster recovery level. Resilience mechanisms at site level are network element redundancy, element internal hardware redundancy, network element recovery mechanisms, redundancy between network elements and flexible re-routing between elements. The last two are achieved with the Hot Standby Router Protocol/Virtual Router Redundancy Protocol. The target of site resilience is to avoid single point of failure at site. One way to achieve high availability in site solution is to use the Virtual Router Redundancy Protocol (VRRP) which provides network redundancy for IP networks, ensuring that user traffic immediately and transparently recovers from 'first-hop' failures in network devices. In VRRP, two or more routers can act as a single 'virtual' router by sharing an IP address and a MAC (Layer 2) address. The members of the virtual router group continually exchange status messages. This way, one router can assume the routing responsibility of another, should it go out of commission for either planned or unplanned reasons. Hosts continue to forward IP packets to a consistent IP and MAC address and the changeover of devices doing the routing is transparent.

Virtual LAN

Normally in Ethernet network all nodes see each other. In a Virtual Local Area Network (Virtual LAN or VLAN) nodes are grouped together to virtual LANs, so that the nodes in different VLANs do not see each other. For example, if all O and M interfaces of different nodes are assigned to a single VLAN and charging interfaces are assigned to another VLAN. The O and M and charging traffic are logically separated and they cannot reach each other, even though they share the same L2 network. This logical grouping of network nodes helps to free site managers from the restrictions of their existing network design and routing infrastructure. It offers a fundamental improvement in the ease with which LANs can be designed, administered and managed. And since VLANs are logical, they allow the network structure to quickly and easily adapt to the addition, relocation or reorganization of nodes when L2 topological changes are needed. VLANs address the limitations of standard switch segmentation by containing broadcast as well as node-to-node traffic. This helps to eliminate router bottlenecks and reduces the danger of broadcast storms. Since all packets travelling between VLANs pass through a router, standard router-based security measures can be implemented to restrict access as needed. One VLAN can have multiple IP sub-nets, but one IP sub-net can be only in one VLAN; this is for IP routing reasons. Usually in practice each VLAN is one IP sub-net, which helps maintenance and configuration. There are two ways to use VLANs, port- and IP-based. In port-based implementation, the administrator assigns each port of a switch to a VLAN. For example, ports 1–3 might be assigned to the O andM VLAN, ports 4–6 to the charging VLAN and ports 7–9 to the SIP signalling VLAN. The switch determines the VLAN membership of each packet by noting the port on which it arrives. In the IP-based method, the VLAN membership of a packet is based on IP addresses. This usually provides more flexible traffic grouping possibilities. Instead of the switch, the sending host adds a tag to the packet to indicate VLAN membership. This eases the switch configuration and improves its performance. It is recommended that at least four VLANs are used with IMS for the following:

- O and M traffic.
- Charging traffic.
- Lawful interception traffic.
- All the other signalling traffic.

Multiprotocol Label Switching (MPLS)

Multiprotocol Label Switching (MPLS) is a technology for speeding up network traffic flow and making it easier to manage. MPLS involves setting up a specific path for a given sequence of packets, identified by a label put in each packet, thus saving the time needed for a router to look up the address for the next node to forward the packet to. MPLS is called a 'multiprotocol' because it works with the Internet Protocol (IP), Asynchronous Transport Mode (ATM) and frame relay network protocols. MPLS allows most packets to be forwarded at the layer-2 level rather than at the layer-3 level. In addition to moving traffic faster overall, MPLS makes it easy to manage a network for Quality of Service (QoS). MPLS-based Virtual Private Networks (VPNs) can be used for carrying the different types of traffic separated across the IP/MPLS backbone. In the PE device, different VLANs are connected to different MPLS VPNs.

IPSec

IPSec provides security for transmission of sensitive information over unsecured networks. IPSec operates in one of two modes:

- Transport mode places the IPSec header after the original outer IP header and before the upper layer protocol.
- Tunnel mode encapsulates the entire IP header and datagram, presents an IPSec header and then creates an outer IP header to tunnel the packet.

Two main protocols provided by IPSec are the Authentication Header (AH) and Encapsulating Security Payload (ESP). The AH protects a packet against modification during transit. The AH functions in transport mode by inserting an AH header into the datagram after the IP header which increases the packet size. The AH functions in tunnel mode by placing an AH header in front of the entire original datagram and adding another IP header on the outside. ESP is the encryption part of IPSec. In transport mode ESP encrypts only the payload leaving the header unmodified. In tunnel mode the whole packet is encrypted and encapsulated in another IP packet.

In the operator backbone some applications may require enhanced security in addition to the basic traffic separation provided by the backbone itself, for example MPLS VPNs. For example, signalling and charging may need encryption because of regulation or operator security policy. If needed, there are two separate ways to use IPSec in an IMS network. In the IMS, a solution using direct IPSec connection in tunnel mode between the IMS elements is recommended in the case where network elements are on the same site and the traffic between them needs to be protected with IPSec. In 3GPP specifications this interface is referred to as Zb. In the 3GPP security model all secure communication between security domains takes place through Security Gateways. This interface is referred as to Za in the specifications.

8.6 IMS Security System

In the multimedia domain, service is not provided until a security association is established between the mobile equipment and the network. IMS is essentially an overlay to the PDS and has a low dependency on the PDS. PDS can be deployed without the multimedia session capability. Consequently a separate security association is required between the multimedia client and the IMS before access is granted to multimedia services.

IMS authentication keys and functions at the user side may be stored in some secure memory location on an UE. It shall be possible for the IMS authentication keys and functions to be logically independent to the keys and functions used for PDS authentication.

There are five different security associations and different needs for security protection for IMS and they are numbered 1, 2, 3, 4 and 5 in Figure 8.12. Their functions are as follows:

1. This association provides mutual authentication between the UE and the S-CSCF. The HSS collective delegates the performance of subscriber authentication to the S-CSCF. However the HSS is responsible for generating keys and challenges. The long-term key in the secure memory of the UE and the HSS is associated with the user private identity. The subscriber will have one user private identity and at least one external user public identity.
2. This association provides a secure link and a security association between the UE and a P-CSCF for protection of the Gm reference point. Data origin authentication is provided, that is the corroboration that the source of data received is as claimed.

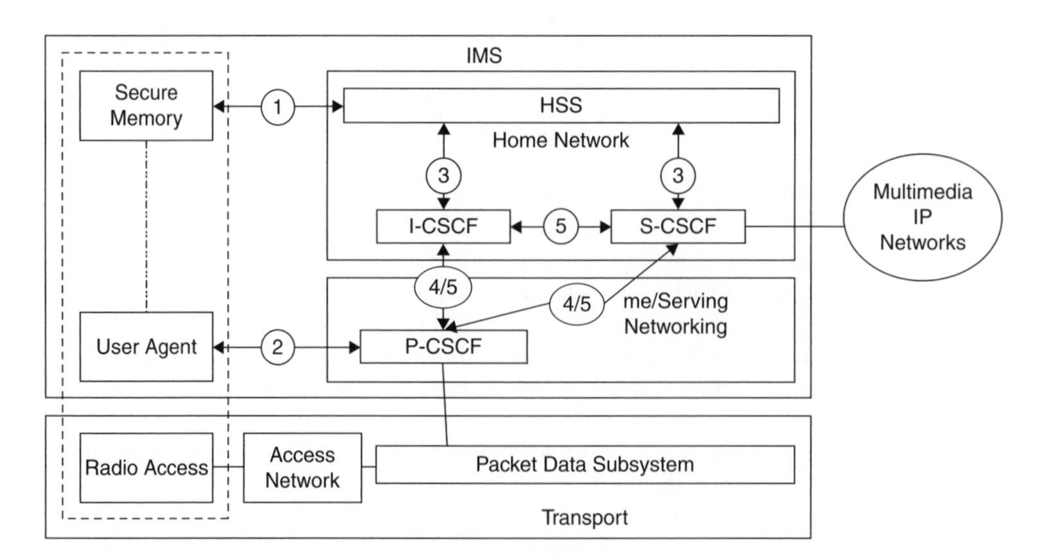

Figure 8.12 IMS security system.

3. This association provides security within the network domain internally for the Cx-interface.
4. This association provides security between different networks for SIP capable nodes. This security association is only applicable when the P-CSCF resides in the virtual network.
5. This association provides security within the network internally within the IMS sub-system between SIP capable nodes. This security association is also applied when the P-CSCF resides in the home network.

8.7 IMS Charging

Offline charging is applied to users who pay for their services periodically, whereas online charging, also known as credit-based charging, is used for prepaid services or real-time credit control of postpaid services. Both may be applied to the same session.

8.7.1 Offline Charging

All the SIP network entities involved in the session use the Diameter R_f interface to send accounting information to a CCF located in the same domain. The CCF will collect all this information and build a CDR, which is sent to the billing system of the domain. Each session carries an ICID as a unique identifier. IOI parameters define the originating and terminating networks. Each domain has its own charging network. Billing systems in different domains will also exchange information, so that roaming charges can be applied. This is shown in Figure 8.13.

8.7.2 Online Charging

S-CSCF talks to an SCF which looks like a regular SIP application server. The SCF can signal the S-CSCF to terminate the session when the user runs out of credits during a session. This is shown in Figure 8.14.

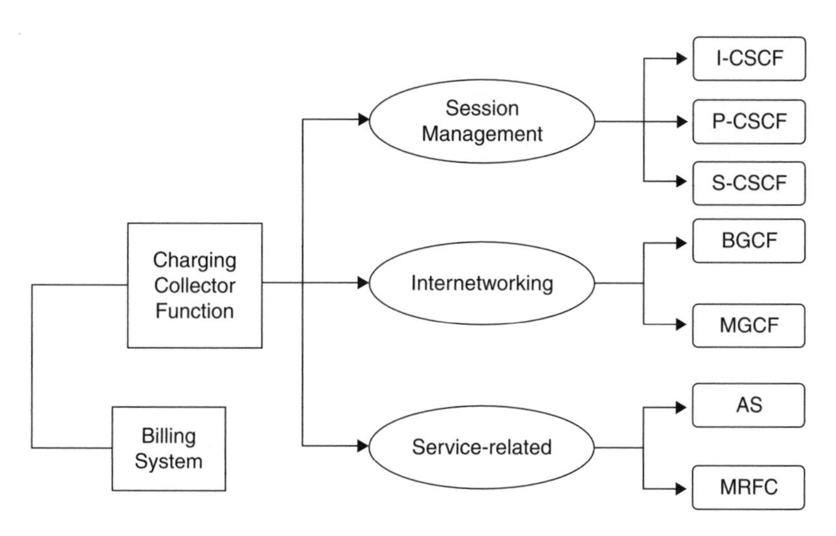

Figure 8.13 Offline charging.

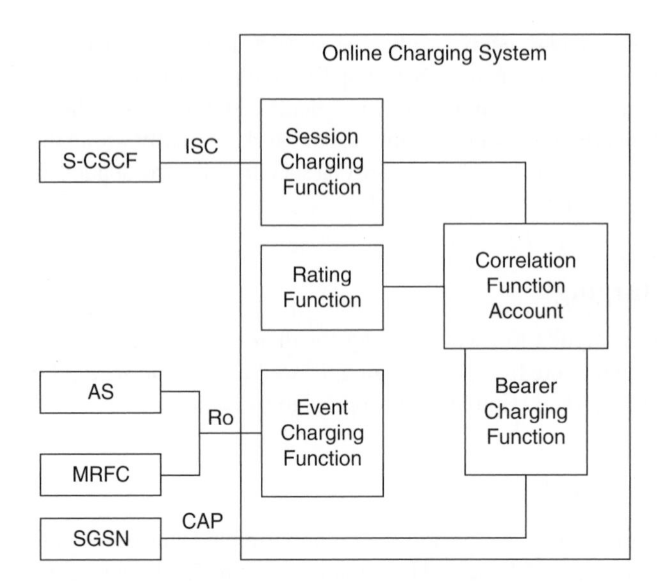

Figure 8.14 Online charging.

The AS and MRFC use the Diameter Ro interface towards an ECF:

- When IEC is used, a number of credit units are immediately deducted from the user's account by the ECF and the MRFC or AS is then authorized to provide the service. The service is not authorized when not enough credit units are available.
- When ECUR is used, the ECF first reserves a number of credit units in the user's account and then authorizes the MRFC or the AS. After the service is over, the number of spent credit units is reported and deducted from the account and the reserved credit units are then cleared.

8.8 Service Provisioning in IMS

As service providers develop their IP service provisioning strategies, in the core, at the edge and even in the data centre itself, economic and technological baselines continue to move. Customers are demanding more capability, at lower cost, delivered almost instantaneously. It is a whirlwind of possibility, potential and problems. As we have understood, IMS provides the provision to enable services. This includes defining the service, creating related service data and subsequently passing an incoming request to application servers.

8.8.1 Registration in IMS

The following points are considered as requirements for the purpose of the registration procedures:

- The architecture shall allow for the S-CSCFs to have different capabilities or access to different capabilities.

- The network operator shall not be required to reveal the internal network structure to another network. Association of the node names of the same type of entity and their capabilities and the number of nodes will be kept within an operator's network. However, disclosure of the internal architecture shall not be prevented on a per-agreement basis.
- A network shall not be required to expose the explicit IP addresses of the nodes within the network.
- It is desirable that the UE will use the same registration procedure within its home and visited networks.
- It is desirable that the procedures within the network are transparent to the UE, when it registers with the IP multimedia core network sub-system.
- S-CSCF is able to retrieve a service profile of the user who has IMS subscription. The S-CSCF shall check the registration request against the filter information and if necessary inform application servers about the registration of the user and it shall be possible for the filter information to allow either just the initial registrations of the user or also subsequent re-registrations to be communicated to the application servers. S-CSCF knows how to reach the P-CSCF currently serving the user who is registered.
- The HSS shall support the possibility to bar a PUI from being used for IMS non-registration procedures. The S-CSCF shall enforce these barring rules for IMS.
- The HSS shall support the possibility to restrict a user from getting access to an IP multimedia core network sub-system from unauthorized visited networks.
- It shall be possible to register multiple public identities through a single IMS registration procedure from the UE.
- It shall be possible to register a PUI that is simultaneously shared across multiple contact addresses through IMS registration procedures. However, each registration and each de-registration process always relate to a particular contact address and a particular PUI. The number of allowed simultaneous registrations is defined by home operator policy. It shall be possible for the UE to indicate to the network whether the registration adds a new contact to an existing registration from the same UE.
- Registration of a PUI shall not affect the status of already registered PUIs, unless due to requirements by the IR set.
- When multiple UEs share the same public identity, each UE shall be able to register its contact address with IMS.
- The UE may indicate its capabilities and characteristics in terms of SIP user agent capabilities and characteristics during IMS registration. The UE may also update its capabilities by initiating a re-registration when the capabilities are changed on the UE.
- If a UE supports GRUU, the UE shall indicate its support for GRUUs and obtain a P-GRUU and a T-GRUU for each registered PUI during IMS registration.
- The P-CSCF may subscribe to notifications of the status of the IMS signalling connectivity after successful initial user IMS registration.
- When the access network type information is available from the access network, the P-CSCF shall ensure that the IMS registration request received from the UE to the SIP server contains the correct information. The P-CSCF may subscribe to notification of changes in the type of access network.
- P-CSCF shall cancel any active subscription.

When a user has a set of PUIs defined to be implicitly registered through single IMS registration of one of the PUIs in that set, it is considered to be an implicit registration.

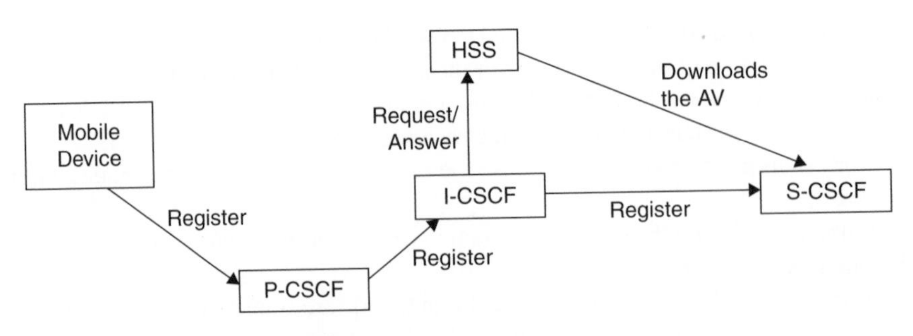

Figure 8.15 IMS registration.

The process of registration can be explained from Figure 8.15.

The SIP protocol is mostly used to establish sessions, which means a context, which is used for further exchange of the data. First the client has to register to some server. In the IMS terms the entry point for communication is called P-CSCF. During the registration phase, the client is first authenticated and then based on triggers the register is send to various application servers in the IMS to let them know that a client is online.

After the PDP context is established, the client sends a register request to the P-CSCF. The P-CSCF normally selects/resolves the I-CSCF node address in the client home IMS. I-CSCF is the entry point in a home IMS network and it selects the S-CSCF node, which asks the client to authenticate itself. S-CSCF communicates with HSS to retrieve the authentication parameters. This results in sending the second register request with authentication parameters.

8.8.2 De-Registration in IMS

De-registration from IMS may be mobile-initiated or network-initiated. The network-initiated de-registration may also be initiated by registration timeout or by a network administrative function such as HSS or S-CSCF.

8.8.2.1 Mobile Initiated De-Registration

When the UE wants to de-register from the IMS then the UE shall perform application level de-registration. De-registration is accomplished by a registration with an expiration time of zero seconds. It follows the following procedure:

- The UE decides to initiate de-registration. To de-register, the UE sends a new register request with an expiration value of zero seconds. The UE sends the register information flow to the proxy.
- Upon receipt of the register information flow, it shall examine the home domain name to discover the entry point to the home network. The proxy does not use the entry point cached from prior registrations. The proxy shall send the register information flow to the I-CSCF. A name–address resolution mechanism is utilized in order to determine the address of the home network from the home domain name. The P-CSCF network identifier is a 'string' that identifies at the home network, the network where the P-CSCF is located.

- The I-CSCF shall send the Cx-Query information flow to the HSS.
- The HSS shall determine that the PUI is currently registered. The Cx-Query is sent from the HSS to the I-CSCF.
- The I-CSCF, using the name of the S-CSCF, shall determine the address of the S-CSCF through a name–address resolution mechanism and then shall send the de-register information flow to the S-CSCF.
- Based on the filter criteria, the S-CSCF shall send de-registration information to the service control platform and perform whatever service control procedures are appropriate. The service control platform removes all subscription information related to this specific PUI.
- Based on operator choice, the S-CSCF can send either Cx-Put and the PUI is no longer considered registered in the S-CSCF. In the case where the user has services related to an unregistered state, the S-CSCF sends Cx-Put in order to keep the S-CSCF name in the HSS for these services. The HSS then either clears or keeps the S-CSCF name for that PUI according to the Cx-Put request. If the S-CSCF name is kept, then the HSS shall be able to clear the S-CSCF name at any time.
- The HSS shall send Cx-Put to the S-CSCF to acknowledge the sending of the Cx-Put.
- The S-CSCF shall return the '200 OK' information flow to the I-CSCF. The S-CSCF may release all registration information regarding this specific registration of the PUI after sending the information flow '200 OK'.
- The I-CSCF shall send the information flow '200 OK' to the P-CSCF.
- The P-CSCF shall send the information flow '200 OK' to the UE. The P-CSCF releases all registration information regarding this specific registration of the PUI after sending the information flow '200 OK'. If the P-CSCF has an active subscription to notifications of the status of the IMS signalling connectivity, the P-CSCF shall cancel the subscription.

8.8.2.2 Network Initiated De-Registration

If an ungraceful session termination occurs, when a stateful proxy server is involved in a session, memory leaks and eventually server failure can occur due to hanging state machines. To ensure stable S-CSCF operation and carrier grade service, a mechanism to handle the ungraceful session termination issue is required. This mechanism should be at the SIP protocol level in order to guarantee access independence for the IM CN sub-system. The IM CN sub-system can initiate network initiated de-registration procedures for the following reasons:

- **Network maintenance:** Forced re-registrations from users, cancelling the current contexts of the user spread among the IP multimedia core network sub-system network nodes at registration and imposing a new IM registration solves this condition.
- **Network/traffic determined:** The IM CN sub-system must support a mechanism to avoid duplicate registrations or inconsistent information storage. This case will occur when a user roams to a different network without de-registering the previous one. This case may occur at the change of the roaming agreement parameters between two operators, imposing new service conditions to roamers.
- **Application layer determined:** The service capability offered by the IP multimedia core network sub-system to the application layers may have parameters specifying whether sub-system registrations are to be removed or only those from one or a group of terminals from the user.

- **Subscription management:** The operator must be able to restrict user access to the IP multimedia core network sub-system upon detection of contract expiration, removal of IM subscription, fraud detection, etc. In the case of changes in the service profile of the user, it may be possible that new S-CSCF capabilities, which are required from the S-CSCF, are not supported by the current S-CSCF which has been assigned to the user. In this case, it shall be possible to actively change the S-CSCF by using the network initiated de-registration by a HSS procedure.

9

Unlicensed Mobile Access

9.1 Introduction to UMA

Unlicensed Mobile Access (UMA) technology helps mobile operators in completing the cost and performance advantages of IP access technologies of fixed networks to deliver good-quality, low-cost, mobile voice and data services in locations where subscribers spend most of their time in homes and offices and to extend the technology beyond homes and offices, precisely to 'hot-spot' areas. As the name suggests, this technology promotes the use of unlicensed spectrum technologies as a part of an integrated communication network. The need to deliver high performance and low cost services (voice and data) to home/office subscribers gave rise to UMA technology. This technology promotes the use of unlicensed spectrum technologies such as WiFi/WLAN/802.11b/g and 'Bluetooth' as part of an integrated communication network.

UMA is also known as a Generic Area Network (GAN). It is a telecommunication system that allows seamless integration of the cellular network with unlicensed wireless technologies. A GAN/UMA system headset system can operate in the following four modes:

- Cellular network.
- Cellular network where available and WiFi otherwise.
- WiFi where available and cellular network otherwise.
- WiFi only.

UMA is a decentralized, diverse, seamless connection and is simple for the user. UMA-enabled dual mode handsets enable operators to provide high-performance, low-cost mobile services to subscribers when in the range of a home, office or public Wi-Fi network. With this, the subscribers can automatically roam and handover between cellular and Wi-Fi access, receiving a consistent set of services as they transmit between networks.

How does UMA works? UMA provides alternative access to GSM and GPRS core network services through IP-based broadband connections. In order to deliver a seamless user experience, the specifications define UNC (UMA Network Controller) and associated protocols that provide for the secure transport of GSM/GPRS signalling and user plane traffic over IP.

Cellular Technologies for Emerging Markets: 2G, 3G and Beyond Ajay R. Mishra

9.1.1 History and Evolution of UMA

The UMAC (Unlicensed Mobile Access Consortium) was formed by leading companies within the wireless industry to promote UMA technology and to develop its specifications. The UMAC worked with 3GPP through the collaborative agreement between different telecommunication standards bodies to develop the formal standards of UMA. The initial set of UMA specifications was published in September 2004, and details the use of the same device over a licensed radio spectrum connection when users are outside the UMA coverage and using an unlicensed radio spectrum when being inside the UMA coverage. In UMA there exists a parallel radio access network known as UMAN (UMA Network) that interacts with the mobile core network using existing GSM-defined standard interfaces.

UMA technology delivers a number of key service advantages. With UMA, mobile operators can allow millions of subscribers to securely access the mobile core service network over an IP access network. Through UMA, all services available over GSM networks are available over IP access networks in a transparent manner. Services can also be provided at different environments such as home, office, 'hot spot', coffee shop, campus and airport.

9.1.2 Benefits of UMA

UMA technology allows mobile operators and service providers to maximize their revenue potential and improves subscriber retention by increased use of mobile phones. It provides the following advantages to mobile operators, service providers, as well as clients:

- Optimizes the use of GSM radio network resources by using an alternative lower-cost and higher-bandwidth access network.
- Reduces capital and operational expenditure on radio networks by using an alternative low-cost access network.
- Provides advanced and consistent services over both fixed and mobile networks.
- Offers 'bundled' fixed and mobile services and makes the mobile handset the customer's only phone, thereby increasing their share of the customer's total expenditure.
- Greatly increases the use of mobile voice and data services in locations where usage was discouraged due to cost or network coverage.
- Delivers enhanced reach as well as improved voice quality.
- Brings increased usage and allows new services to be offered and delivers the broadband data rates to the handsets.
- The same terminal can be used everywhere by the clients.
- Delivers mobile voice and data services over unlicensed wireless networks seamlessly.
- Provides the same mobile identity on cellular RAN and unlicensed wireless networks.

9.2 Working on UMA Network

UMA technology allows mobile subscribers to seamlessly roam between mobile and home wireless networks or WLAN 'hot-spots'. As subscribers move between networks, they continue to receive mobile voice and data services in a consistent manner. In fact subscribers within buildings can obtain good quality voice datas due to improved signal strength. Through UMA, mobile users can take advantage of potentially faster data services through avoiding the bandwidth constraints of the GSM.

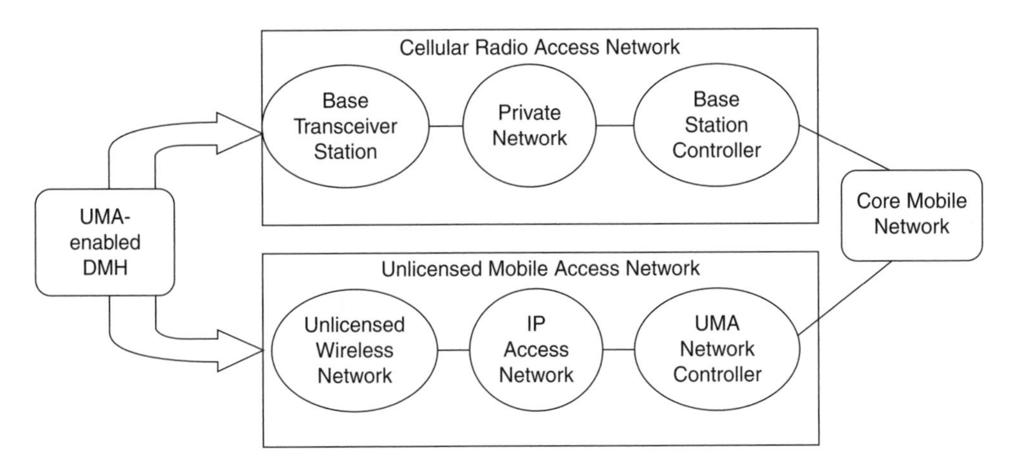

Figure 9.1 Working of UMA.

The connection to a fixed network occurs automatically when a mobile subscriber with UMA-enabled dual-mode mobile handsets moves within the range of an unlicensed wireless network to which the handset is allowed to connect. Upon connecting, the handset contacts UNC over the broadband IP access network to be authenticated and authorized to access GSM voice and GPRS data services through the unlicensed wireless network. If approved, the subscriber's current location information stored in the core network is updated and from this point on, all mobile voice and data traffic is routed to the handset via the UMAN rather than the cellular RAN. This is shown in Figure 9.1.

9.3 Architecture of UMA

UMAN consists of more than one AP and one or more UNCs interconnected through a broadband IP network. The functional architecture of UMA is shown in Figure 9.2.

The main features of the UMAN architecture are as follows:

- New entities and entities with enhanced functionality.
- Mobile station and access points.
- UNC which appears to the core network as a GERAN BSS and includes a security gateway that terminates secure remote access tunnels form the MS, providing mutual authentication, encryption and data integrity for signalling, voice and data traffic.
- A broadband IP network which provides connectivity between the AP and the UNC. The IP transport connection extends all the way from the UNC to the MS, through an AP. A single interface, the U_p interface, is defined between the UNC and the MS.
- Co-existence with GERAN and interconnection with the GSM core network via the standardized interfaces defined for GERAN.
- interface for circuit-switched networks.
- Gb-interface for packet-switched networks.
- In this architecture, the principle elements of transaction control and user services are provided by the network elements in the core network, namely the MSC/VLR and the SGSN/GGSN.

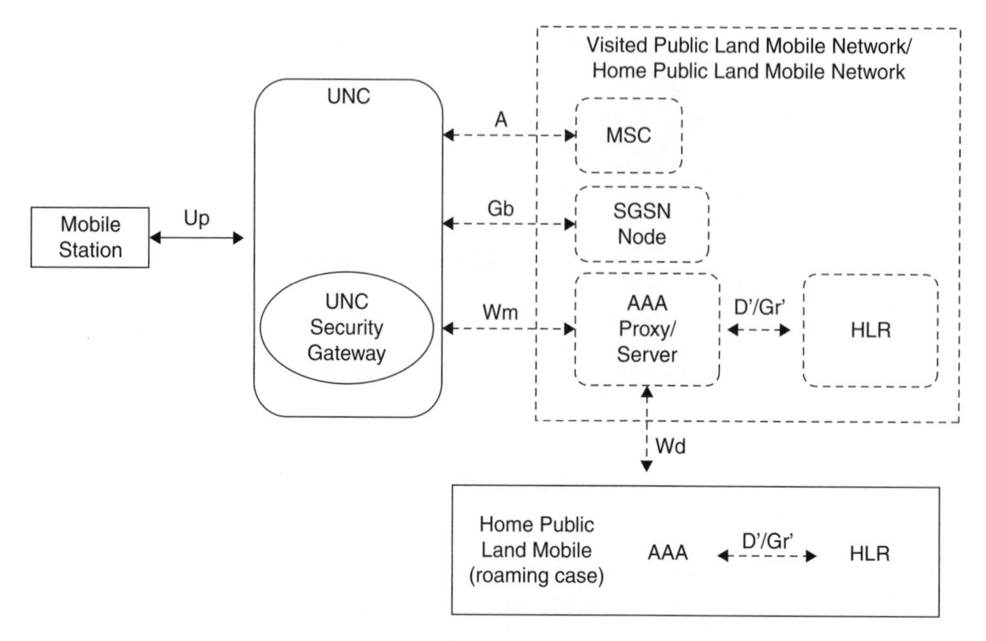

Figure 9.2 UMA functional architecture.

- Use of an AAA server over the Wm-interface as defined by the 3GPP. The AAA server is used to authenticate the MS when it sets up a secure tunnel.
- The UMAN shall support simultaneous CS and PS services. Indication of support of DTM shall be provided through appropriate signalling to the MS.

Key elements of the UMA Network are as follows:

Mobile Station

The MS shall include dual mode radios and the capability to switch between them. The MS supports an IP interface. In other words, the IP connection from the UNC extends all the way to the MS.

Access Point

The AP provides the radio link towards the mobile station using an unlicensed spectrum. It connects through the broadband IP network to the UNC and does not provide any UMA-specific gateway functions. Any generic AP can be used to interconnect the MS to the UNC via the broadband IP network.

UMA Network Controller

The UNC connects to a unique MSC and SGSN via the A-interface and G_b-interface, respectively. It provides functions equivalent to that of a GERAN BS controller and connects via an IP transport connection to an AP. The UNC interfaces to the MS using the U_p-interface. It maintains end-to-end communication with the MS and relays GERAN signalling to the A or G_b-interface towards the core network. The following functions are supported by UNC:

- U_p user plane speech services: inter-working speech bearers over the U_p interface to speech bearers over the A-interface including transcoding voice to/from the MS to PCM voice when some of the features are not being utilized from/to the MSC.
- U_p user plane data services: inter-working data transport channels over the U_p interface to packet flows over the Gb-interface.
- U_p security gateway to terminate secure remote access tunnels from the MS.
- U_p control functionality.
- Registration for UMA services.
- Set-up for UMA bearer paths for CS and PS services. These include participation in establishment, management and 'tear down' of secure signalling and user plane bearers between the MS and the UNC.
- UMA functions equivalent to GSM RR and GPRS RLC such as for paging and handovers.
- Transparent transfer of L3 messages between the MS and core network.

9.4 U_p Interface in UMA

This is the interface between the UNC and MS and it operates over an IP transport network and relays GSM/GPRS signalling between the PLMN core network and the MS. The features of U_p protocol architecture in support of CS domain signalling as well as UMA specific signalling are given below.

GSM protocol is carried transparently between the MS and MSC. This allows the MS to obtain all GSM services that it can receive through a GSM BSS, through a UMAN.

GSM RR protocol is replaced with a UMA RR protocol. The unlicensed radio link presents different characteristics from that of the licensed radio GSM link, so the UMA RR protocol is customized to take advantage of these characteristics. As in a GSM BSS, the UNC, acting like a BSC, terminates the UMA RR protocol and inter-works it.

The features of U_p protocol architecture in support of GSM voice transmission are given below:

- Audio flows over the U_p interface according to the RTP framing format.
- Support for GERAN.
- AMR FR is the preferred codec type when operating in the UMA mode.

The features of U_p protocol architecture in support of GPRS signalling are given below:

- GPRS LLC PDU for signalling and higher layer protocols is carried transparently between the MS and SGSN. This allows the MS to obtain all GPRS services in the same way as if it were concerned to a GERAN BSS.
- GPRS RLC protocol is replaced with an equivalent UMA RLC protocol. Given the transport characteristics over U_p, interface with the GPRS TBF abstraction is not applicable and reliability is ensured by TCP. Therefore, the UMA RLC is significantly lighter than GPRS RLC. As in a GERAN BSS, the UNC acting like BSC, terminates the UMA RLC protocol and inter-works it to the G_b-interface using BSSGP.

9.5 Protocols in UMA

The UMA architecture uses the following standard GERAN protocols:

- GSM MM, CM and higher layer protocols are used without any changes in the MS or the MSC.
- GSM voice encoding carried over IP between the MS and UNC.
- GPRS LLC and higher layer protocols are used without any changes on the MS and SGSN.
- A-interface protocols are used between the MSC and UNC.
- G_b-interface protocols are used between the SGSN and UNC.
- Wm-interface protocols are used between the UNC and AAA server.

9.5.1 *Standard IP-Based Protocol*

The UMA architecture uses the following standard IP-based protocols:

- IP over standard lower layers.
- TCP to provide a tunnel for GSM/GPRS signalling and SMS.
- IPsec ESP to provide a secure tunnel for the GERAN user and control plane traffic.
- IKEv2 and EAP SIM for authentication and establishing and maintaining a security association between MS and UNC.
- UDP for IPsec NAT traversal.
- UDP for GPRS data transfer.
- RTP/UDP for transfer of GSM frames over IP transport.

9.5.2 *UMA Specific Protocols*

It includes the following types of protocols.

UMA RR

This protocol provides a radio resource management layer, which is the peer of GSM RR, in the MS. It is designed to take advantage of the characteristics of the unlicensed radio link as it is quite different from that of the GERAN radio link. It provides the following functions:

- Registration with UNC.
- Setup of the bearer path for CS traffic between the MS and UNC.
- Handover support between GERAN and UMA-functions and support for identification of the AP being used for UMA access.

UMA RLC

This protocol provides the following services:

- Delivery of the GPRS signalling and SMS messages over the secure tunnel.
- Paging, flow control and GPRS transport channel management.
- Transfer of GPRS user plane data.

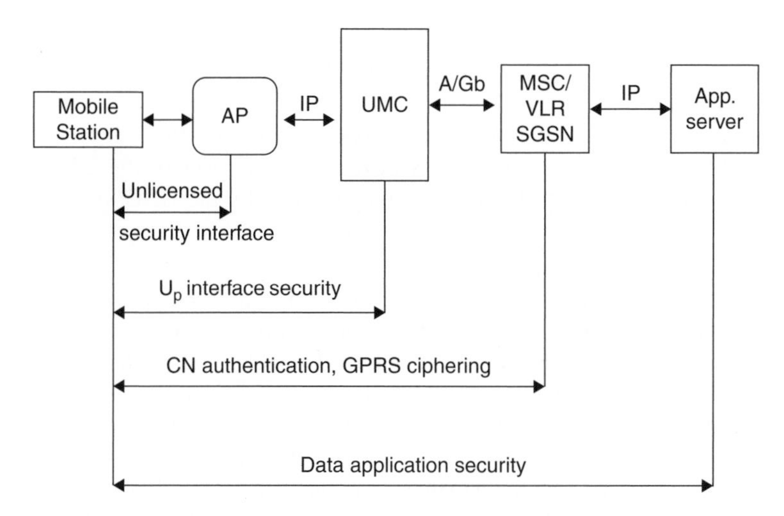

Figure 9.3 Security flow in UMA.

9.6 Security Mechanism of UMA

The security mechanisms applied over the unlicensed radio interface are the authentication and encryption functions defined for the unlicensed mode radio interface protocols applied between the MS and AP. These mechanisms are applied to voice, data and signalling over the radio interface.

The security mechanisms over the U_p interface protect signalling, voice and data traffic flows between the MS and the UNC from unauthorized use, data manipulation and eavesdropping, that is, both authentication and encryption mechanisms are supported (shown in Figure 9.3). Authentication of the subscriber by the core network occurs between the MSC/VLR or SGSN and the MS and is transparent to the UNC. There is, however, cryptographic binding between the MS core network authentication and the MS UNC authentication. GPRS ciphering is the standard LLC layer ciphering that operates between the MS and the SGSN.

Additional application level security mechanisms may be employed to secure the end-to-end communication between the MS and the application server or gateway.

9.7 Identifiers and Cell Identifiers in UMA

The following are the key MS and AP addressing parameters.

- **The IMSI Associated with the SIM in the Terminal**
  This identifier is provided by the MS to the UNC when it registers to a UNC. The UNC maintains a record for each registered MS.
- **Public IP Address of the MS**
  The public IP address of MS is the source IP present in the outermost IP header of packets received from the MS by the UNC security gateway. If available, this identifier may be used

by the UNC to support location services and fraud detection. It may also be used by service providers to signal managed IP networks IP flows that require QoS treatment.

- **The AP Identifier**
  This is the MAC address of the unlicensed mode access point through which the MS is accessing UMA service. This identifier is provided by the MS to the UNC via the U_p interface, when it requests UMA service. It may be used by the UNC to support location services.

A single UNC represents a single cell and is referred to as a UMA cell, for the purpose of handover from GERAN to UMAN. The cell is identified by the CGI. The CGI assigned to the UNC is configured as the target handover cell in all neighbouring GERAN cells which are those cells whose service area overlaps the UNC service area, for the purpose of handover. For the purpose of providing location services and emergency services an operator may create cell identification contexts locally to a UNC. These cell identification contexts are not visible to the MS. These cell identifiers are additional to the one assigned to the UNC for handover purposes.

The GERAN to UMAN handover method makes use of ARFCN and BSIC parameters to identify the UMA target cell. Selection of ARFCN should follow the following guidelines:

- ARFCN should not be allocated from the operator's existing BCCH pool so that a 'scarce' BCCH is not used.
- ARFCN is desired to be the same unique number across the whole operator network so that the BSS configuration effort can be minimized.

The following are the available options for the selection of the AFRCN:

- Ideally, UMAN is assigned an AFRCN value which is not in the frequency bands currently used by the operator.
- Typically, different PLMNs in the same country have disjoint frequency allocations. For each PLMN, some of the frequencies are reserved for BCCH. It will be transmitted with constant maximum power on time slot zero. Other frequencies are dedicated as traffic channels. The UMAN AFRCN could use any non-BCCH frequency from the carrier's existing frequency pool. Standard GSM MSs will be able to tune onto this channel but will not be able to find a FCCH 'burst'.
- Alternatively, in a 1900 MHz network, ARFCN can be any value falling within the GSM or DCS band. Standard GSM handsets operating in the 1900 MHz frequency will ignore this ARFCN. Tri-band handsets supporting automatic band change will not be able to find a BCCH on the associated frequency.

9.8 Mode and PLMN Selection

9.8.1 Mode Selection

At any given time, the MS is in one of the two operating modes:

- **GSM mode:** in this mode, GSM RR is the serving RR entity.
- **UMA mode:** in this mode, UMA RR is the serving RR entity.

On power up, the MS starts in the GSM mode and executes the normal GSM power-up sequence. After that, it might switch into the UMA mode based on mode selection preference determined by user preferences and services provider configuration. Emergency calling procedures have certain exceptions which are as follows:

- **GERAN-only:** the MS shall stay in the GSM mode and never switch to the UMA mode.
- **GERAN-preferred:** the MS shall stay in the GSM mode, as long as a PLMN is available via GERAN. If no PLMN is available via GERAN, the MS may search for UMA coverage, and if it is detected, shall execute the registration procedure, thereby switching to the UMA mode. At any time, when in the UMA mode, if a GSM PLMN becomes available or the MS leaves UMA coverage, the MS shall execute the 'rove-out' procedure, thereby switching to the GSM mode.
- **UMAN-preferred:** at any time, when in the GSM mode and a PLMN search is not in progress, if UMA coverage becomes available, the MS shall execute the 'rove-in' procedure per sub or handover to the UMAN procedure.
- **UMAN-only:** the MS shall stay in the UMA mode and never switch to the GSM mode.

9.8.2 PLMN Selection

There shall be no change from the GERAN procedures for PLMN selection in the NAS layers in the MS. The PLMN available via the serving RR entity shall be made available to the NAS layers for PLMN selection.

GSM mode
In this mode, only the PLMNs available via GERAN can be selected and the PLMNs available via UMAN are not reported.

UMA mode
In this mode, only a PLMN associated with the successfully registered UMAN can be selected and the PLMNs available via GERAN are not reported.

Normal PLMN selection rules shall be applied based on the available networks reported by the serving RR entity.

Manual PLMN selection mode
In this mode, the available PLMNs are displayed to the user for selection.

Automatic PLMN selection mode
In this mode, the PLMN is selected by the network itself for the user.

9.9 UMAN Discovery and Registration Procedures

9.9.1 Registration

The discovery procedure is performed by the MS when first attempting to obtain UMA service in order to determine the identity of the default UNC which may also serve as the serving

UNC for that connection. The UMA registration procedure is performed between the MS and UNC. It serves the following purposes:

- It informs the UNC that a MS is now connected through a particular AP and is available at a particular IP address. The UNC keeps track of this information for the purposes of providing service.
- It provides the MS with the operating parameters associated with the UMA service. The 'GSM System Information' message content that is applicable in the UMA mode is delivered to the MS during the UMA registration process.

These procedures are applicable for the following conditions and mode selection preferences:

- UMAN-only.
- UMAN-preferred.
- GERAN-preferred and no GSM PLMN available.

When an MS supporting UMA first attempts to connect to a UNC based on a UMA subscription, it needs to identify the serving UNC. In order to do this, it first connects to a provisioning UNC and then discovers a default UNC, which in turn can redirect the MS to a serving UNC. A MS supporting UMA may be provisioned with the FQDN or IP address of the provisioning UNC and the associated security gateway.

In case the SIM is not provisioned with the FQDN or IP address, the MS shall derive a FQDN for the provisioning UNC and the secure gateway, based on its IMSI. The MS shall set up a secure tunnel using the provisioned or derived address and connect to the provisioning UNC. It shall then obtain the FQDN or IP address of the default UNC and the associated security gateway, through the discovery procedure. The default UNC serves as the primary registration destination address for the MS when it fails to register on an alternate serving UNC which is stored in the MS and associated with previously joined WLAN APs if there is no GSM coverage or with GSM cell identities if there is GSM coverage.

Following the discovery procedure the MS shall establish a secure tunnel with the secure gateway of the default UNC and attempt to register with the default UNC. The default UNC network may also serve as the serving UNC for that connection. The procedure may result in the MS getting redirected to a different serving UNC. UNC redirection refers to the capability of a UNC to redirect an MS to a UNC distinct from the one it initially requests access to, based on MS provided information and operator chosen policy. If no GSM coverage is available when an MS connects to the UNC for UMA service, then the UNC cannot necessarily determine the location of the MS for the purposes of assigning the MS to the correct serving UNC. The UNC shall permit the operator to determine the service policy in this case.

The MS shall store the address of the provisioning UNC and of the default UNC. The MS shall also store serving UNC information for serving the UNCs with which the MS was able to complete a successful registration procedure. If there is no GSM coverage by the AP, the stored serving UNC information shall be associated with the AP identifier. If there is GSM coverage by the AP, the stored serving UNC information shall be associated with the GSM CGI. The stored serving UNC information is as follows:

- Serving security gateway FQDN or IP address following successful registration.
- Serving UNC FQDN or IP address following the successful registration.

The number of such entries to be stored in the MS is implementation-specific. For a particular AP or GSM cell, only the last successfully registered UNC association shall be stored. An MS may preferentially join a WLAN AP whose association with a serving UNC has been stored in memory. If the CGI value cached for that AP does not match the current camping GSM cell, the MS may invalidate the cache entry. On joining a WLAN if the MS has a stored serving UNC for the joined AP or the GSM cell, the MS shall attempt to register with the associated serving UNC in its memory. The UNC may still reject the MS for any reason even though it may have served the MS before. The MS shall delete from its stored list the address of the serving UNC on receiving a registration reject.

If the MS does not receive a response to the registration request sent to the serving UNC, it shall attempt to register with the default UNC in order to obtain a new serving UNC for the joined AP. If the MS does not receive a response to the registration request sent to the default UNC, it shall attempt the discovery procedure with the provisioning UNC in order to obtain a new default UNC. When the MS joins a WLAN, for which it does not have a stored serving UNC in its memory, it shall attempt to register with the default UNC. The discovery and registration procedure consists of the following steps:

- Joining a WLAN.
- Discovery of default UNC through the provisioning UNC.
- Registration with the default UNC.
- Potential redirection to a serving UNC or registration.
- Registration with a serving UNC.

Through the registration procedure the MS may get redirected to another serving UNC. A successful registration procedure results in the UNC establishing a context for the MS. The MS obtains the necessary system information for the UMAN it has registered on and can trigger a normal location/routing area update procedure with the CN.

The process of registration and discovery is shown in Figure 9.4.

9.9.2 De-Registration

The UMA de-registration procedure allows the MS to explicitly inform the UNC that it is leaving the UMA mode, allowing the UNC to free the resources that it assigned to the MS. The UNC may also implicitly de-register the MS when the TCP connection to the MS is abruptly lost. The UMA can be de-registered in the following modes:

- **De-registration Initiated by the MS**
  In this mode, the MS sends the URR DEREGISTER to the UNC which removes the MS context in the UNC. This is shown in Figure 9.5.
- **De-registration Initiated by the UNC**
  In this mode, the UNC sends the URR DEREGISTER to the MS. This is shown in Figure 9.6.

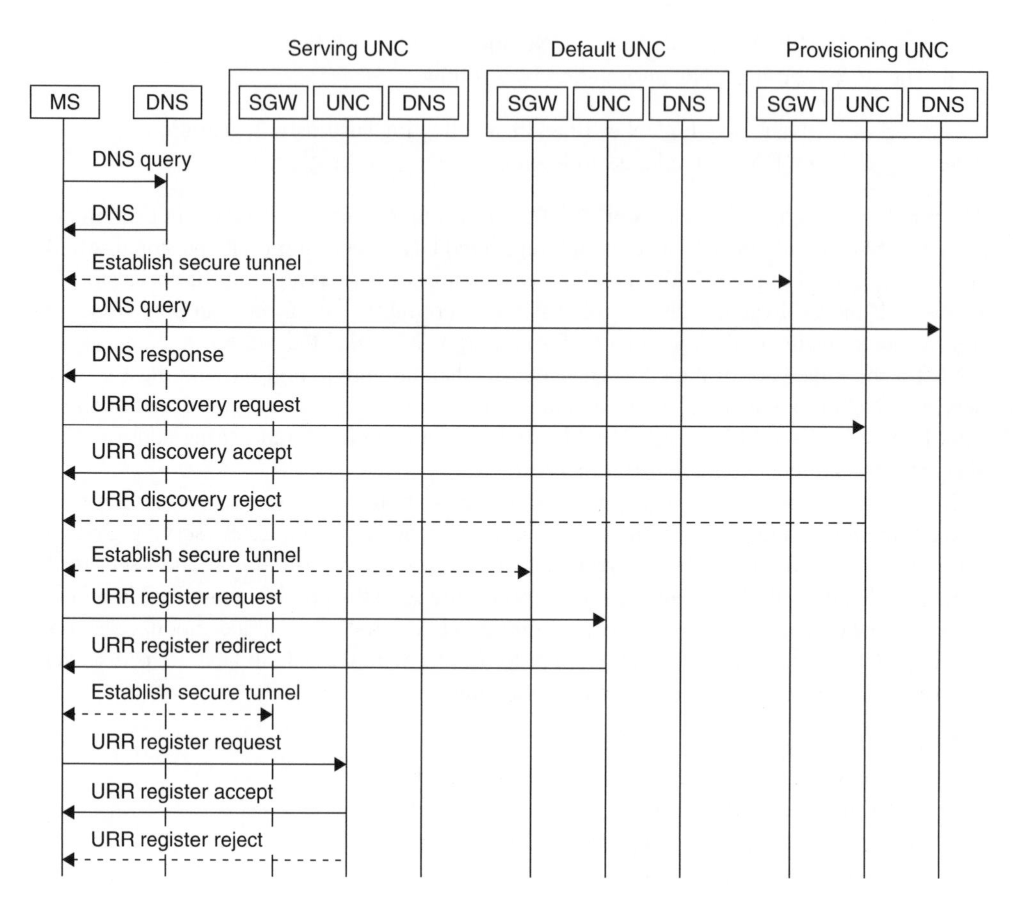

Figure 9.4 Registration and discovery process.

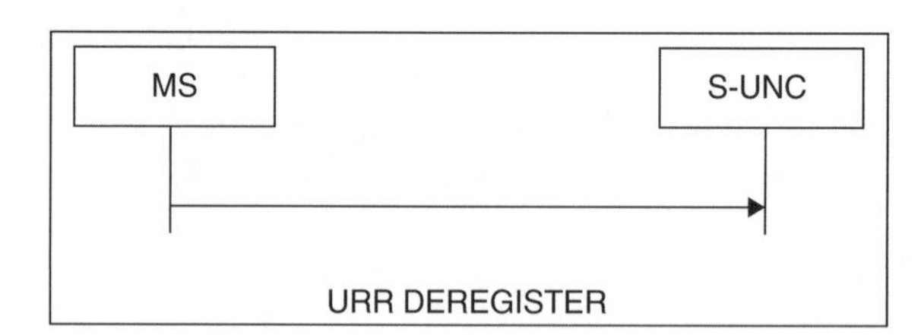

Figure 9.5 De-registration initiated by a mobile.

Figure 9.6 De-registration initiated by UMA.

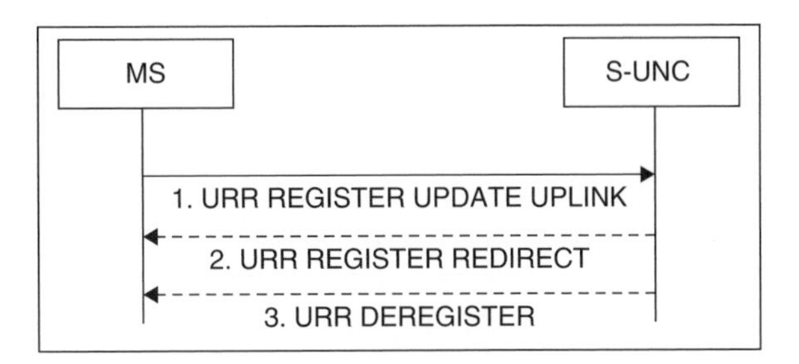

Figure 9.7 Registration update.

9.9.3 Registration Update

This can be done by using the following procedures.

- **Registration Update Uplink**
  This procedure is used to indicate to the UNC that information pertaining to the joined AP or the identity of the overlapping GSM cell has changed. This is shown in Figure 9.7.

  The procedure is explained below:

  - When the MS detects GSM coverage after reporting no coverage during registration, it shall send the URR REGISTER UPDATE UPLINK to the UNC with the updated information. Whenever the MS changes AP it shall send a URR register update uplink to the UNC with the updated AP identity information.
  - The UNC may optionally send the URR REGISTER REDIRECT when it wants to redirect the MS based on updated information.
  - The UNC may also optionally de-register the MS on receiving an update by sending URR DEREGISTER to the MS.

- **Registration Update Downlink**
  This procedure is used by the UNC to update information in the MS such as system information or status of location services. This is shown in Figure 9.8.

  In this, the UNC sends URR REGISTER UPDATE DOWNLINK with updated information.

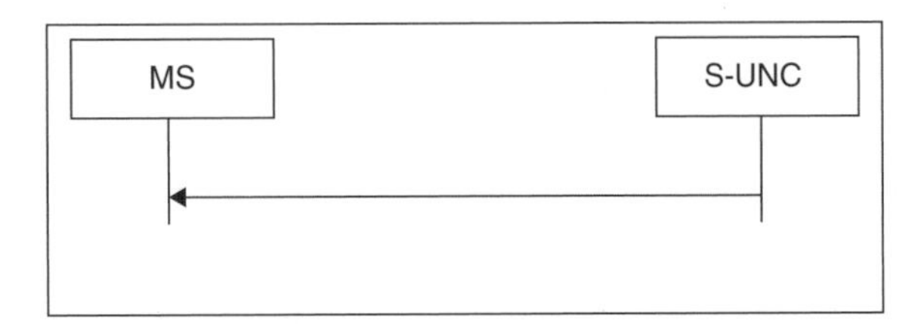

Figure 9.8 URR register update downlink.

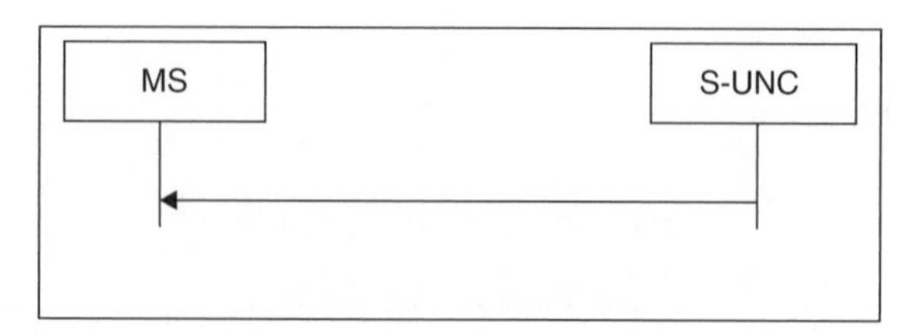

Figure 9.9 URR 'keep alive'.

9.9.4 'Keep Alive'

This process is a mechanism between the peer URR entities to indicate that the MS is still registered to the UNC. Using periodic transmissions of the 'keep alive' message the MS in turn determines that the UNC is still available using the currently established lower layer connection. This is shown in Figure 9.9.

9.10 UNC Blocks

- **Access Point**
  This provides the radio link towards the mobile station using an unlicensed spectrum. It connects through the broadband IP network to the UNC. The AP provides 'Bluetooth' or 802.11 access point functionality. The AP does not provide any UMA-specific gateway functions and any generic AP can be used to interconnect the MS to the UNC via the broadband IP network.
- **Security Gateway**
  This terminates the secure remote access tunnels from the MS, providing mutual authentication, encryption and data integrity for signalling, voice and data traffic.
- **Home Base Station Controller**
  This remains in contact with the base stations of the service providing company.
- **Home Mobile Support Node**
  This facilitates the manageability of the GAN network in terms of validation, access control, centralizing of common data and management of the MS. The HSN node also handles additional functionality to enable compliancy to international emergency call handling and routing as well as lawful interception.
- **AAA Server**
  The AAA server in CDMA data networks is an entity that provide IP functionality to support the functions of authentication, authorization and accounting. The AAA server in the CDMA wireless data network architecture is similar to the HLR in the CDMA wireless voice network architecture.
- **A-and Gb-Interface**
  The A-interface is for circuit-switched services and the Gb-interface is for packet- switched services.

Table 9.1 Comparison between femtocells and UMA

Type	Gain a foothold 'in home': to ensure a place within the connected home future	Maintain and increase the mobile AR per unit: in the face of downward pressure on mobile voice revenues	Optimize network resources: to avoid macro network overload due to increasing mobile data usage
UMA	• UMA fixed telephony launched by T-Mobile US • Broadband and UMA bundles launched by Orange • Takes advantage of installed base of WiFi devices in the 'connected home'	• HD mobile TV on UMA now available • Unlimited Surf offer launched by Orange • T-mobile US have launched both mobile and fixed telephony on UMA	• Takes advantage of existing WiFi router installed based • No WiFi management required • Laptops use WiFi when 'at home' to avoid mobile core network overload • No impact on RAN evolution
FETMO	• All this is possible, but not yet proven	• All this is possible, but not yet proven	• No installed base of fetmo-compatible CPE • Additional CPE to manage • Fetmocell upgrades required to match macro RAN evolution

9.11 Comparison between Femtocells and UMA

The comparison can be seen in Table 9.1.

The consumers have the following benefits from UMA:

- **Unlimited Tariffs:** These give the following price benefits:
  - Cheaper mobile calls.
  - No fixed line monthly fee.
  - Get more for less with multi-play bundles.
- **Improve Indoor Coverage:** This gives the following usage benefits:
  - No missed calls on the mobile.
  - Make voice calls at home.
  - Faster mobile broadband at home.
- **Coverage Benefits:** These give the following service benefits:
  - One provider.
  - One bill.
  - Single customer service.

9.12 Conclusion

The level of mobile use has grown rapidly around the globe. Particularly in Western countries where users have access to both fixed and mobile services there has been a growing desire to use just one phone for all their services. With access to wireless connectivity rising from the increased use of DSL lines to the home as well as wireless coverage in the workplace and in WiFi 'hot-spots' there is the possibility of using a variety of short-range wireless connectivity.

An additional factor is that IMS and VoIP services are now becoming far more widespread and this raises the possibility of using a single user interface and connecting the phone either over the cellular network, or by using wireless connectivity through WiFi access points, with the system choosing the optimum service.

With users providing a definite push to generate the technology for this fixed-mobile convergence, UMA is bound to be a growing technology. UMA promises to offer new opportunities to users, mobile operators and manufacturers alike. Thus, UMA could be one of the major drivers for the mobile telecommunications industry in the near future, providing revenue from new sources as well as providing an upgrade path for existing users.

10

DVB-H

10.1 Mobile Television

Watching television on handheld devices has caught the imagination of many in the industry. As the name suggests, mobile TV is all about transferring multimedia content to handheld devices or in simpler terms television on a mobile device (as shown in Figure 10.1).

Some technologies that are being used in this space are ISDB-T, DMB, Media FLO and DVB-H. ISDB-T stands for Integrated Services Digital Broadcasting and is a terrestrial television standard. It has been developed in Japan and 1/13 of the DTV networks have been allocated for broadcasting towards handheld devices. DMBs or the Digital Media Broadcasts use the Eureka-147 DAB standard for transmitting to handheld devices. MediaFLO has been developed by Qualcomm and is a proprietary system that uses OFDM for transmitting to the handheld devices. Another technology is MBMS (Multimedia Broadcast/Multicast Service) which is based on the3GPP release 6 and uses the EDGE/UMTS infrastructure. Let's have a brief look at some of the technologies used in Mobile TV.

10.1.1 Bearer Technologies for Handheld TV

Digital Audio Broadcasting (DAB)

This was initially developed for digital audio but can now encompass the whole range of multimedia, including audio broadcasting services (video, audio, data, text, etc.). Though there are many DAB technology variants, there are two main technology variants: DAB-IP and T-DMB. The former one is about video and audio being delivered directly over IP using the packet mode while the latter one uses the stream-mode-to-transfer combined video and audio. T-DMB was previously adopted as a standard by the WorldDAB (a consortium that developed DAB standards in 1995). T-DMB uses the OFDM transmission technology. It can be deployed both in single and multiple frequency networks.

Digital Video Broadcasting for Handheld (DVB-H)

This is a new standard built on top of the DVB-T (terrestrial). We will look into this technology more in detail later in this chapter.

Cellular Technologies for Emerging Markets: 2G, 3G and Beyond Ajay R. Mishra

Figure 10.1 Television on a handheld device.

Forward Link Only (FLO)
This is the bearer technology of the MediaFLO system. It is designed for efficient transmission of the multimedia multicast stream to handheld devices. This technology is 'backward compatible' as well as it has been developed to enable the broadcast network overlaid onto a cellular network.

There are some viewing issues with handheld televisions though! The quality of viewing as perceived might not be an issue. User-perceived picture quality is defined in two ways: the number of picture elements and the number of colours rendered. As compared to standard television, 'high mobile phones' can render 1/16 and PDA devices 1/4 of the picture elements. And as the screen of handheld devices is scaled down, the perceived quality is similar to that of a standard television. The picture element density can be increased two-fold using the current technologies, thereby increasing the picture quality. Complications of the handheld television technology are much more than the terrestrial television, as seen in Table 10.1.

10.1.1.1 Broadcast and Uni-Cast Technologies

Broadcast technology is one wherein the same content is used for million of users while unicast technology is one wherein the same content is used for selected subscribers. While radio and television are examples of broadcast technology, streaming videos of a football match is an example of unicast technology. In broadcast technology, there is no limit on the number of viewers. DVB-T and DVB-H are examples of broadcast technology while 'video-on-demand' is an example of unicast technology.

Table 10.1 Handheld TV versus terrestrial TV

Handheld TV	Terrestrial TV
Limited power supply (battery)	Unlimited power supply
Due to mobility, robust transmission of signals needed	Due to stationary nature, robust transmission is not needed
Microscopic antennae needed (more complicated)	Normal antennae are OK

10.1.2 Service Technology for Handheld TV

The service technology that is used for handheld TV is defined for some bearer technologies only. Operators should be careful in selecting the bearer technologies as they impact the service technology selection. These service technologies specify many of the following content, description of service, security, roaming, subscription management, etc. Some of these service technologies are described in the following.

Service technology related to DAB
DAB was originally only an audio service but has since changed into an interactive, on-demand service. DAB permits many services to run over it. Services such as broadcast websites, slide shows and electronic programme guides are possible over the DAB. IP datacasting is also possible in DAB as the system is capable of carrying the IP packets using the IP/UDP protocols. These packets flow in one direction only (IP multicasting) from one service provider to many users through the DAB packet mode service component, through the encapsulation process

IP Datacast (IPDC)
IP Datacasting over DVB-H involves specifications that are necessary to deploy the commercial mobile TV. This is based on the IP abstraction layer and involves the electronic programme guide, use cases, architecture, content delivery protocols, security, etc. IPDC was first used for the DVB-H physical layer but now is being considered to be used for other technologies (e.g. DAB).

Multicast Device Network Interface (MDNI)
The Multicast Device Network Interface (or MDNI) specifications are used, delivering services over the FLO bearer technology. MDNI is almost coupled with the FLO and has no clear abstraction layer. It is not an open standard.

Open Mobile Alliance Broadcast (OMA BCAST)
Open Mobile Alliance (OMA) BCAST standards are for mobile services. Unlike MDNI, it is technology agnostic although the IP abstraction layer presence is a definite requirement. Streaming, transfer of data files, etc. services are possible in this service technology.

10.2 Introduction to DVB

Digital Video Broadcasting for Handheld or DVB-H is the latest technology developed by the Digital Video Broadcasting (DVB) project. DVB technology is related to digital television. The DVB project maintains a suite of internationally accepted open standards for digital television. These are published by a JTC (Joint Technical Committee) that compromises ETSI (European Telecommunications Standards Institute), CENELEC (European Committee for Electro-technical Standardization) and EBU (European Broadcasting Union). The DVB-H standards are based on the DVB's terrestrial (television) standards known as DVB-T. A comparison with other technologies is shown in Table 10.2.

There were some disadvantages associated with the existing technologies:

- UMTS: not possible to scale it for huge content delivery.
- DAB: it has a narrow spectrum (not possible to transmit data).

Table 10.2 Mobile TV technology comparison

Technology	Data Rate	Frequency Band
DVB-H	5–31 Mbps	UHF
DAB	1.15 Mbps	VHF
MBMS	300 kbps	WCDMA frequency band
Korean DMB	1.15 Mbps	VHF

As the name suggests, DVB broadcasts digital video data. The transfer is done through various methods such as satellite (called DVB-S), Cable (DVB-C), terrestrial methods (DVB-T), through handheld devices (DVB-H), microwave (DBV-MT), etc. Both the DVB-S and DVB-C standards were ratified in 1994 and 1996, respectively.

10.2.1 Digital Video Broadcasting – Terrestrial

DVBT technology allows 'one-to-many' broadband wireless data transport (IP packets, video and audio). The capacity is quite high, for example 54 channels, each 5–32 Mbps and cell sizes going up to 100 km. This technology is developed for MPEG-2 (Moving Pictures Expert Group) but has the capability to carry any data. The modulation scheme COFDM with 2k and 8k modes is applied, having a bandwidth of approximately 8 MHz. Some other technical specifications are:

- Many transmission modes, 4.98–31.67 Mbps @ $C/N = 25$dB.
- Carrier modulation: QPSK, 16 QAM or 64 QAM.
- In 2k mode: 1705 sub-carriers with spacing of 4464 Hz.
- In 8k mode: 6817 sub-carriers with spacing of 1116 Hz.
- Error correction: convolution code and Salomon–Reed.

The whole idea of DVH-T transmission (see Figure 10.2) is to streamline the various data sources at the physical level – providing flexibility for network and service providers. This requires hierarchal scheming in DVB networks. There are three streams of audio, video and data that are multiplexed. There are several such streams that are multiplexed to make a single transport stream. The splitter does the function of splitting the streams into one of the two hierarchical schemes. DVB-T has two hierarchical modes wherein two transport streams can be sent simultaneously with different priorities – high and low. The two transport schemes are fed to the modulator – one controls the constellation placement in the quadrant while the other defines the 16QAM signal. The former is with high priority while the latter is of low priority. There are defined three possible constellation offsets, $\alpha = 1, 2$ and 4, but the principle

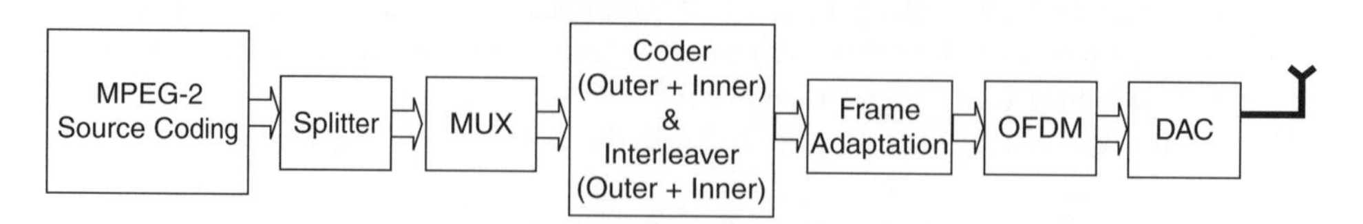

Figure 10.2 Simplified block diagram of a DVH-T transmitter.

is the same. The benefit of this encoding is that the system allows one program for mobile or portable applications together with service supporting; typically three programs to be picked up in the stationary application. We need to here remember that in non-hierarchical modes that all the transport streams have the same coding and mapping in the physical layer while in the hierarchical mode simulcast and multicast modes are used. In the simulcast mode, the program is split into two streams – each being high and low priority, respectively. It is possible to receive the high priority stream even in difficult conditions while the low priority has a high bit rate that is responsible for better receiver quality. In multicast mode, the two programs are mapped onto different hierarchies – each on high and low priority, respectively. The receiver is able to distinguish between the hierarchies by TPS (Transmission Parameter Signalling). This contains the information related to receiving and demodulating the signal.

10.2.2 Digital Video Broadcasting – Handheld

With both 3G and DVB-H being in the implementation stage, the data transfer for mobile TV is one on 3G networks as well. As this kind of data transfer is unicast (bi-direction point-to-point) in nature resulting in inefficiency, a point-to-multipoint DVB-H broadcast network is used for delivering the same. High data rate broadcast services for handheld terminals are possible with the DVB-H networks. However, hybrid networks (unicast and broadcast networks) are used for providing these services to users. The planning and optimization of the hybrid networks is more difficult than the single network. DVB-H standards specifications are for converging the cellular radio networks and the broadcasting networks. The specifications for DVB-H were released by the DVB project in 2004. These specifications were published by ETSI (EN 302304). Subsequently, the implementation guidelines TR 102377 were also published.

DVB-H has two main advantages over the DVB-T systems:

- It has lower battery power consumption.
- It has improves robustness in difficult atmospheric conditions due to in-built antennae.

10.2.3 History of DVB-H

The history of DVB started with the DVB project in 1998, including research on terrestrial DVB and its first commercial deployment in Europe. The study resulted in conclusions such as the possibility of reception of DVB-T but implying dedicated broadcast networks. A few years down the line, it was possible to permit mobile broadcast services on DVB-T. However, with it there was an increase in the requirements of the users – they needed rich multi-media content on a handheld device like television. In the year 2002, the work related to such systems started and it led to DVB-H where 'H' stands for 'Handheld'. These standards were published in EN 302304 in November 2004.

10.3 DVB-H Ecosystem

The mobile phone industry fundamentally comprises the cellular operators and vendors. But with the advent of mobile television, broadcasters will come to play a very important role in the mobile ecosystem (as shown in Figure 10.3). In some countries, it is the broadcasters that

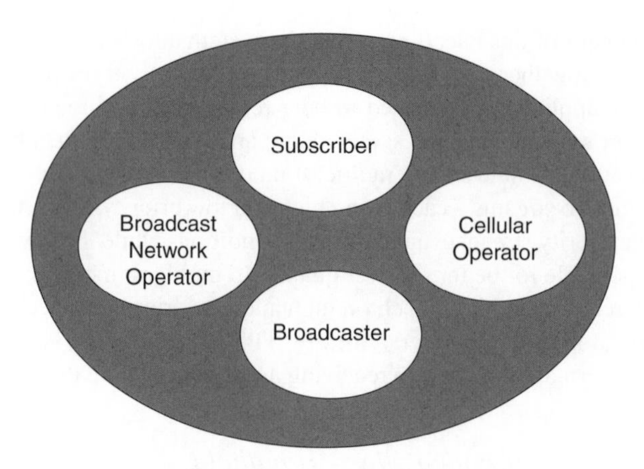

Figure 10.3 DVB-H ecosystem.

have taken a lead role in driving the mobile television services. However, in some cases, the mobile operators may as well don the role of broadcaster to provide the television services to its handheld subscribers. Broadcast network operators already have the infrastructure in place to support the DVB-H services. What advantages are there for broadcasters to enter this field? The primary one is that they can reach their customers/viewers without any limitation of location and time – viewers can watch their favourite football match they missed last evening when going into work the next morning on the train to the office. Not only will the broadcasters have new audiences watching television when they want to, plus they will focus more on content providers/aggregators and less on transmission methodology. Another factor that will play its role is costs. The cost of reception on the handheld device through the UMTS network might be much higher than that of the broadcaster's network.

10.4 DVB-H System Technology

As mentioned before, the DVB-H standards have been evolved from the DVH-T standards. There are three major additions to the physical layer of existing standards that make the DVB-H standards (ETSI EN 300744). Combination of the physical layer, link layer and service information together will define the DVB-H system (shown in Figure 10.4). The technologies involved in the link layer and physical layer are mentioned (these technologies are explained below).

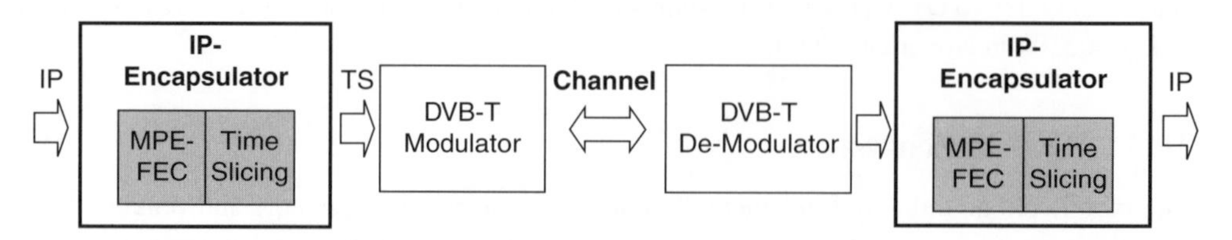

Figure 10.4 DVB-H system.

Link Layer

- Time-slicing – this reduces the average power consumption of the terminal and enables smooth and seamless frequency handover.
- Forward error correction for multiprotocol encapsulated data (MPE-FEC). This improves the *C*/*N*-performance and Doppler performance in mobile channels. It also impacts tolerance to impulse interference positively.

Physical Layer

Apart from the DVB-T (Specifications EN 300 744 -1) the following technical elements are added, targeting for DVB-H use:

- 4K mode: this is useful for trading off between the mobility and SFN (Single Frequency Network) size.
- DVB-H signalling: the TPS-bits are used to enhance and speed up service discovery.
- Cell identifier: this is carried on TPS-bits to support quicker signal scans and frequency handover on mobile receivers
- In-depth symbol interleaver: this is used to improve robustness of the system.

Both the terrestrial and handheld versions of DVB use the same physical layer. DVB-H is downward compatible with DVB-T. However, there are some enhancements that make the DVB-H system more robust than the DVB-T system. These enhancements are related to time slicing, error protection and signal modulation.

10.4.1 Time Slicing

This is an important feature that is related to prolonging the battery life of the handheld device (shown in Figure 10.5). For users to be able to use their devices longer without having to re-charge the batteries, DVB-H uses the time slicing technique. In time slicing, data are delivered to the handheld device in the form of 'bursts' at some given time intervals. The IP datagrams are transmitted as data bursts that may contain up to 2 Mbps data. These data

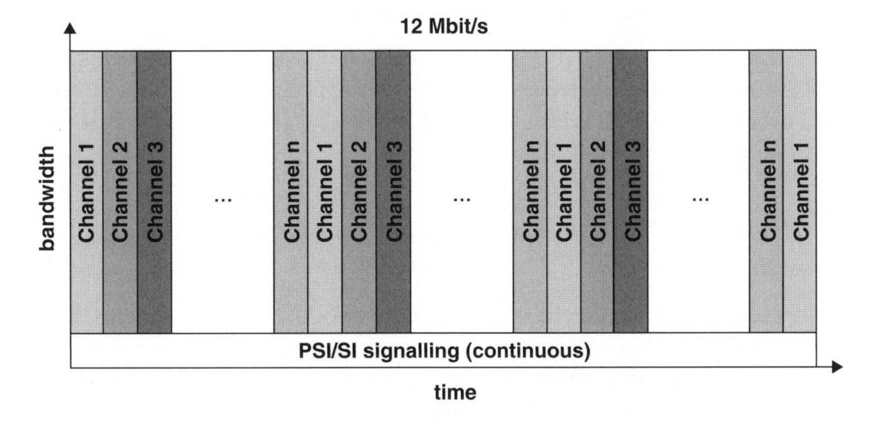

Figure 10.5 Time slicing principle.

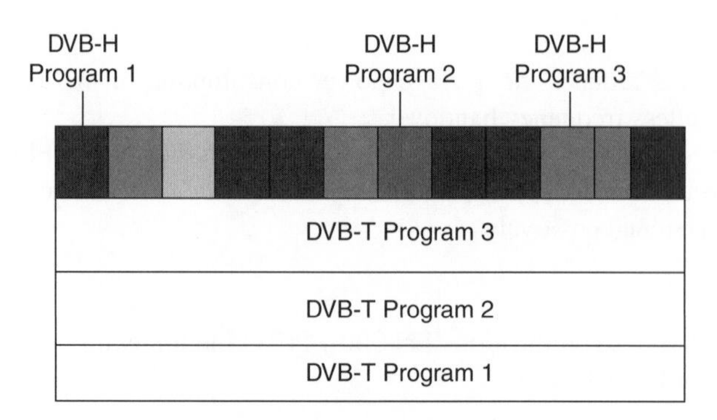

Figure 10.6 DVB-H frame structure.

bursts are in small time slots. During the non-receiving time, the tuner is inactive and thereby in a 'less-power-using' mode. This saves around 90–95 % of the power as compared to the DVB-T tuners. The time between the bursts is responsible for the power savings. During the 'off-time', the same receiver can be used to monitor the neighbouring cells.

Principle of Time Slicing

Power saving is possible due to time slicing. Instead of sending an MPE section of a particular ES with constant low bit rate, they are sent in high bit rate bursts. No sections of the ES are transmitted, allowing the receiver to completely power-off during that period. How will the receiver know when to 'power-on'? This is done by the delta_t parameter in the header of all sections of the burst making signalling very robust against transmission errors.

DVB-H Frame Structure

The frame structure for the DVB-H is more efficient than the DVB-T due to the time slicing technique. As seen in Figure 10.6, due to time slicing, the power savings take place as compared to the DVB-T.

10.4.2 IPDC (Internet Protocol Datacasting)

How is data transferred in DVB-H? The transmission of audio and video content is done through IP datacasting. This is a process in which data are encapsulated in the IP packets and transmitted over the Internet. It is also used for file delivery. This enables DVB-H technology to use the existing protocols for its data transmission.

10.4.3 MPE/FEC (Multiple Protocol Encapsulation/Forward Error Correction)

This provides robust transmission to the DVB-H system. As in mobile systems, reception is from multiple directions. To make sure that the data reception is good/error free, a strong error protection is as well needed. In DVB-H systems; this is done through the addition of FEC to the MPE layer. With MPE-FEC the IP datagrams of each time sliced burst are protected by Reed–Solomon parity data (RS data), calculated from the IP datagrams of the burst (as shown

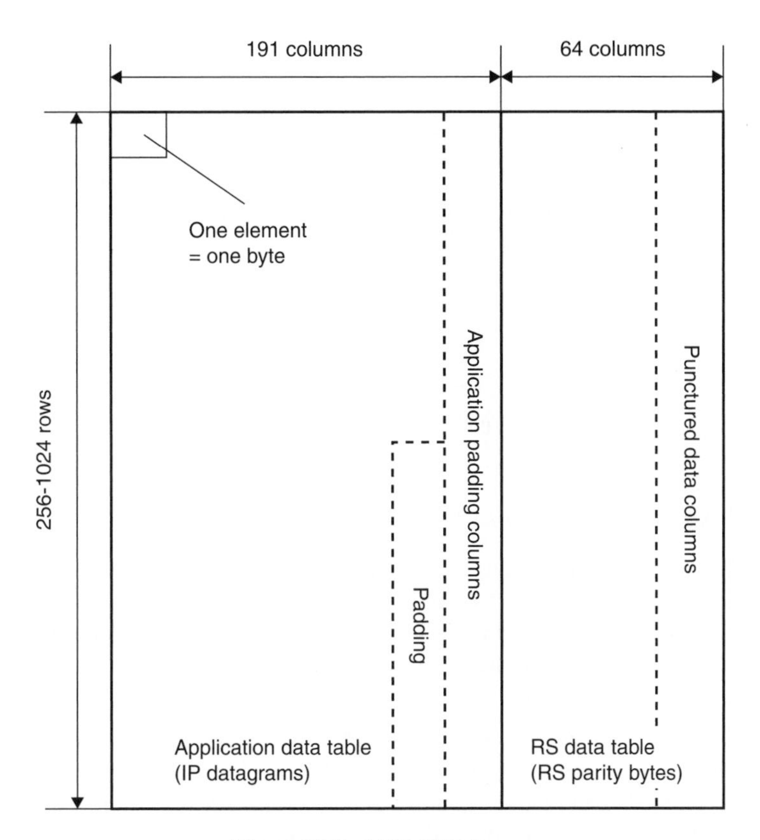

Figure 10.7 MPE-FEC frame.

in Figure 10.7). The RS data are encapsulated into MPE-FEC sections, which are also part of the burst and are sent immediately after the last MPE section of the burst, in the same ES, but with different table_id than the MPE sections, which enables the receiver to discriminate between the two types of sections in the ES. Each burst is of 2 Mbps and there are 64 parity bits for each 191 data bits. These are protected by the FEC. Different frames are needed for optimal planning (e.g. 256, 512, 1024, etc.) that includes capacity of services, number of services, buffer size, etc. It should be noted that though MPE/FEC reduces the *S/N* and improves Doppler performance in mobile channels, it also requires extra power consumption for transmitter and receiver. It also lowers the throughput due to parity overhead.

10.4.4 Protocol Stack for DVB-H

The payloads of the DVB-H network are the IP datagrams. These are encapsulated in the MPE sections (as shown in Figure 10.8). FEC provides improvement in the Doppler performance and the *C/N* performance, although this is optional. This is then exchanged through video coding standards, for example MPEG, and then transmitted over the DVB-T physical layer. The receivers of the DVB-T are not influenced by the DVB-H signals. As mentioned before, the time slicing or MPE-FEC does not impact the DVB-T physical layer as this is implemented on the link layer.

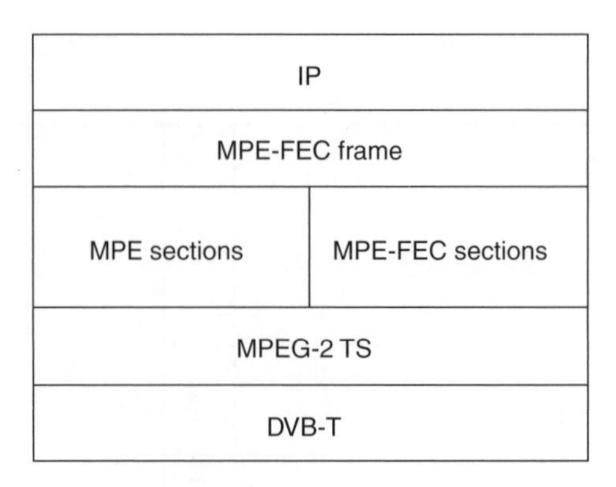

Figure 10.8 Protocol stack for DVB-H.

10.4.5 4k Mode and In-Depth Interleavers

To provide the network optimization, a 4k mode has been added in the DVB-H standard. The advantage of the 4k mode is that it gets 'immunity' against signal variations and interference. It also introduces a trade-off between the 2k (high receiving speed) and 8k (large transmission cell size and low receiving speed) modes. The trade-offs are:

- The DVB-T 8k mode can be used both for single transmitter operation and for small, medium and large SFNs. It provides a Doppler tolerance allowing high speed reception.
- The DVB-T 4k mode can be used both for single transmitter operation and for small and medium SFNs. It provides a Doppler tolerance allowing very high speed reception.
- The DVB-T 2k mode is suitable for single transmitter operation and for small SFNs with limited transmitter distances. It provides a Doppler tolerance allowing extremely high speed reception.

The SFN size of the 4k mode network is twice that of the 2k mode networks.

10.4.6 Multiplexing and Modulation

OFDM (Orthogonal Frequency Division Multiplexing) is the multiplexing scheme used in DVB-H. In OFDM, data are transmitted in parallel using several carriers in the same time. These are modulated using the modulation schemes of the DVB-H network. OFDM has been explained in detail before in Chapter 6.

The modulation schemes that are used in DVB-H networks are as follows:

- QPSK (Quadrature Phase Shift Keying)
  - o 2 bits/symbol.
- 16QAM (Quadrature Amplitude Modulation)
  - o 4 bits/symbol;
  - o ~6 db stronger signal needed compared to QPSK.

64QAM is used in DVB-T. It is not used in DVB-H networks as it is not feasible for mobile reception.

10.4.7 DVB-H Signalling

Signalling in DVB-H is used for enhancing and increasing the service delivery speed. The TPS channel allows a faster way to access the signalling than demodulating and decoding the Service Information (or the MPE-section header) and also allows the TPS-lock in a demodulator with low *C/N* values. Two TPS bits are used – one for time slicing and other for MPE-FEC. Other signalling includes that of the 4k mode and symbol 'interleavers'.

10.4.8 SFN

SFN or the Single Frequency Network, as the name indicates, uses only one frequency. The network itself consists of several transmitters. All these transmitters are accurately synchronized. An area of around 60 km is easily covered without using any 'high towers'. These types of networks are called Single Frequency Networks. SFN is more like a distributed transmitter and not a network. As all transmitters send exactly the same signal, hence overlapping signals result in gain (also called SFN Gain). As the signal arrives from many locations simultaneously, 'slow fading' does not have much impact on it.

10.4.9 Power Consumption

As mentioned above, the time slicing technique was introduced so as to adapt the DVB-T stream to handheld devices. A reduction of 5 % power consumption would mean that a receiver can be turned on for 500/10 000 (= 5 %), when it is receiving 500 kbps. On top of this, synchronization is added as well. However, the total power saving can reach 10 %. One of the most recent front-end technologies consumes approximately 500 mW of power that makes possible the reception of 500 kbps. It should be noted that the DVB-T MPEG-2 program stream can utilize up to 10 M bps which is much more than a handheld device can display.

10.4.10 Signal Quality in DVB-H Networks

In signal transmission of terrestrial networks, 'echo' is one of the problems faced. The impact of this echo is mitigated by the use of a guard interval between each modulated symbols. This reduced the inter-symbol interference and saves degradation of signal quality. Also the antennae facing the direction of the signal broadcast station prevent the echo to happen. However, in handheld devices this is an issue. The signal quality becomes worse with the increase in speed as the frequency shift affects the echoes. Echoes due to the Doppler shift are perceived as noise that increase the inter-carrier interference. Experiments have found that the acceptable level of *C/I* remains until the maximum tolerable Doppler frequency is reached. By using the dedicated signal processing techniques, the impact of inter-carrier interference can be mitigated.

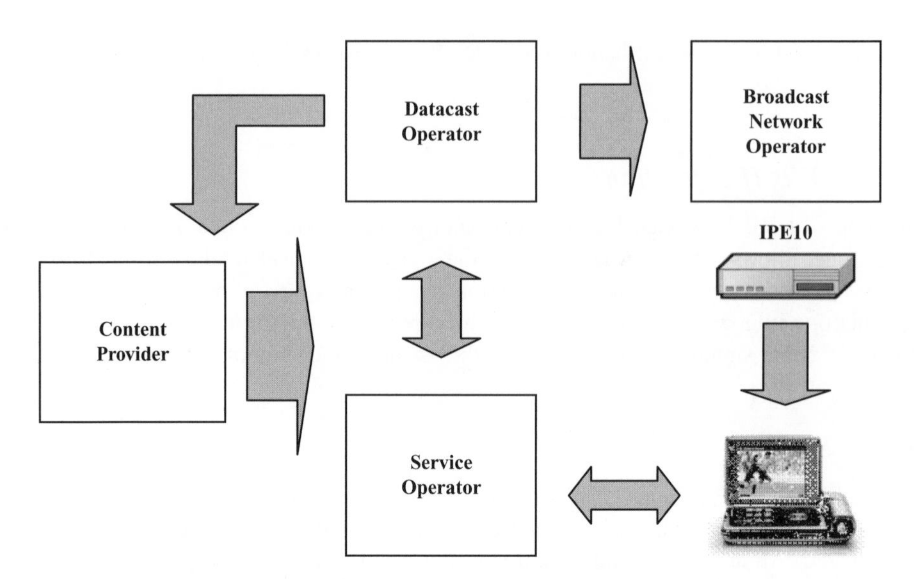

Figure 10.9 Simplified block diagram of a DVB-H network.

10.5 DVB-H Network Architecture

A simplified block diagram of the DVH-H network architecture is shown in Figure 10.9. Let us now understand the major network elements/blocks and their functionalities.

10.5.1 Content Provider

This is responsible for providing the content to the solution. The content includes sports, music, weather, news, etc. This content is provided to both the datacast operator and the service (cellular) operator.

10.5.2 Datacast Operator

This block contains quite a few elements including ones that control the broadcast services, for example encryption, multicast routing, encapsulation, etc. The planned broadcast content of content providers can be stored in repositories. IP encapsulation is controlled by it for the execution of the stored plans. The service guides (also called ESGs or Electronic Service Guides) are transmitted for each service area and are controlled by the datacast operator. The pricing and service packaging are stored in a central repository as it allows the service operators to configure their own repository.

10.5.3 Service Operator

This is the cellular network operator. It contains the core network and the access network. The traditional GSM network can be used to provide the broadcast services delivered by DVB-H. Separate devices are used for DVB-H related charging in cellular networks – and these are capable of supporting both the pre-paid and post-paid customers.

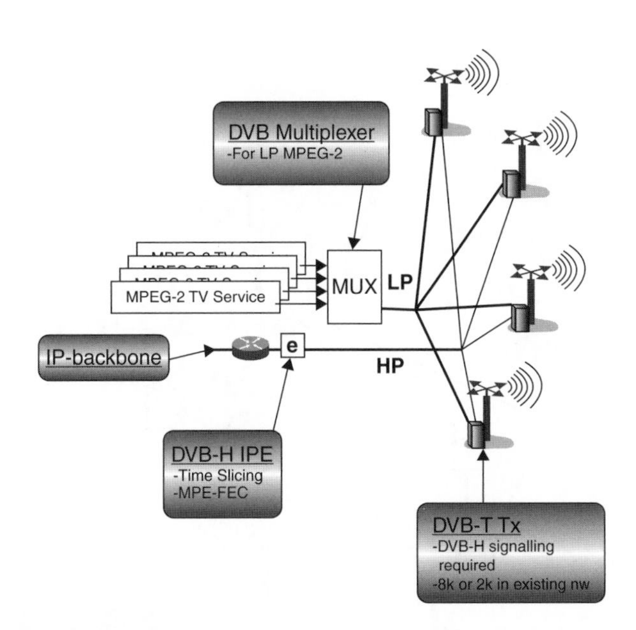

Figure 10.10 DVB-H and DVB-T multiplexing. © European Telecommunications Standards Institute 2009. Further use, modification, copy and/or distribution are strictly prohibited. ETSI standards are available from http://pda.etsi.org/pda/.

10.5.4 Broadcast Network Operators

To make the actual broadcast possible, a broadcast encapsulator is used by the broadcast network operator. It consists of an IPE (IP encapsulator) and an IPE manager through which IP forwarding and encryption settings are managed. The broadcast networks receive the content from the content providers through datacast operators. The IPE is used as a gateway between the IP multicast network and DVB-H transmitter.

10.6 DVB-H Network Topologies

10.6.1 Multiplexing – DVB-T and DVB-H Networks

Figure 10.10 shows the multiplexing of the DVB-H and DVB-T networks. The DVB-T transmitters serve both the terrestrial and the handheld devices. This is done by using the DVB-H IP encapsulator that is located in each of the coverage area (both SFN and MFN) and is done at multiplex level. The equipment in the DVT-T network is upgraded for better DVB-H support. The DVB-T network should be upgraded to be able to support indoor coverage for handheld receivers.

10.6.2 Dedicated DVB-H Networks

In this kind of network, the sharing of TS between the DVB-H and DVB-T services does not take place. A dedicated multiplexer is used to carry the DVB-H services. There are some changes that need to take place in the DVB-T network, such as upgrading the TPS to enable the DVB-T modulators to handle DVB-H services, some changes in the IFFT (to support

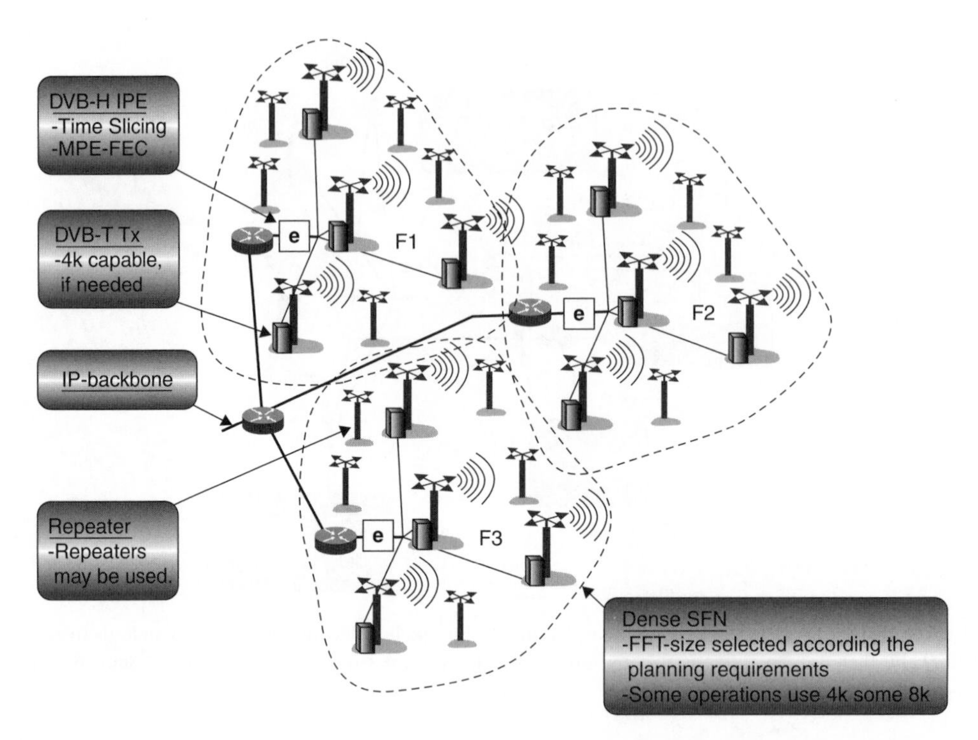

Figure 10.11 Dedicated DVB-H networks. From the Institute of Electrical and Electronic Engineers (IEEE) from G. Faria, J.A. Henriksson, E.Stare and P. Talmola, 'DVB-H: Digital Broadcast Services to Handheld Devices', *Proceedings of the IEEE*, **94** (1), January 2006.

the 4k mode) and changes in symbol interleavers. As shown in Figure 10.11, such networks consist of several SFNs where the size of this SFN is dependent on the guard interval, FFT, etc. The power transmitted in the DVB-H network is higher and so is the number of transmitters. However, the transmitter power and antennae heights are lower than the ones used in the DVB-T network.

10.6.3 Hierarchal DVB-T and DVB-H Networks

This is another way to share the DVB-H networks. The advantage of using this method of sharing is the optimal usage of bandwidth (separate sets of modulation parameters for terrestrial and handheld are used) and no 'jitter' and ID management are needed as in the case of multiplexing related sharing. However, the disadvantage is that there is no flexibility in terms of bandwidth usage – a fixed amount of bandwidth is needed. The DVB-H IP services are inserted in the high-priority stream of the DVB-T modulator, increasing the robustness over the low-priority streams. A new TS distribution network for the HP stream is needed as well as the IP-encapsulator with DVB-H capability (see Figure 10.12).

10.7 Network Design in the DVB-H Network

As in the design of any network, there are a few factors that are necessary to be kept in mind when planning a DVB-H network. Customer requirements are coverage area (along with the

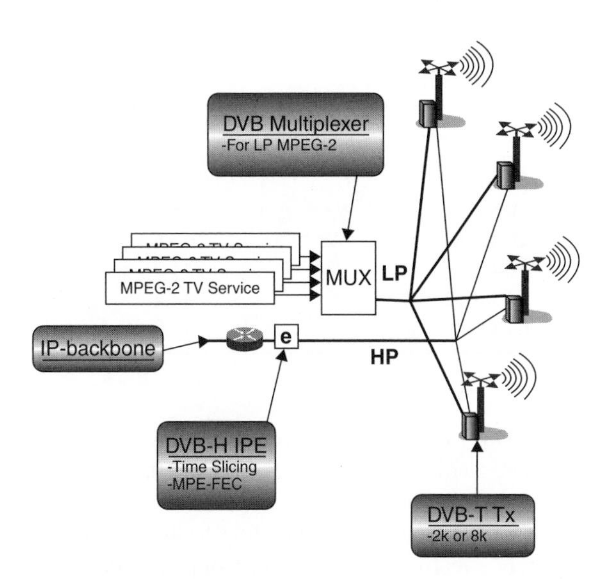

Figure 10.12 DVB-T and DVB-H hierarchal transmission. © European Telecommunications Standards Institute 2009. Further use, modification, copy and/or distribution are strictly prohibited. ETSI standards are available from http://pda.etsi.org/pda/.

subscriber base), available sites and frequencies available. Network characteristics that will be important are costs, transmitter size and SFNs. The DVB-H network design fundamentally focuses on the coverage. Let us now understand a few parameters before going into the details of planning.

Impact of 4k mode on Network Planning

As mentioned before in this chapter, the 4k mode provides a more flexible architecture and a more acceptable performance of the mobile receiver. This mode provides two times a better Doppler performance than the 8k mode. When designing the SFN, the acceptable inter-transmitter distance is proportional to the maximum echo delay acceptable by the transmission that is dependent on the guard interval value. In the 4k mode, this distance is larger than the 2k mode and smaller than the 8k mode.

MFER (MPE-FEC Frame Error Rate)

Degradation of the MPE-FEC Frame affects the service reception. This is because this erroneous frame would impact the service reception of the whole interval between the bursts negatively. Thus, the MFER or MPE-FEC Frame Error Rate would define degradation points to the frequency of the 'lost frames'. A small change in *C/N* will result in a large change in MFER. MFER is agreed to be 5 % to mark the degradation of the DVB-H network.

C/N

The C/N (dB) values are calculated using the theoretical C/N figures given in EN 300 744. The performance in Table 10.3 gives the expected values of the DVB-H receiver where Noise (*N*) is applied with the carrier (*C*) with a signal bandwidth of 7.61 MHz (with a degradation MFER of 5 %).

Table 10.3 *C/N* (dB) for 5 % MFER in the Gaussian channel

Modulation	Code rate	Gaussian
QPSK	1/2	3.6
QPSK	2/3	5.4
16-QAM	1/2	9.6
16-QAM	2/3	11.7
64-QAM	1/2	14.4
64-QAM	2/3	17.3

The performance of the DVB-H receiver is given by two figures: C/N_{min} and F_{d3dB}. C/N_{min} gives the minimum *C/N* for 5 % MFER while the F_{d3dB} gives the Doppler frequency where the *C/N* requirement is raised by 3 dB from the C/N_{min} value. The required *C/N* would increase at a lower speed (i.e. a lower Doppler) corresponding to the portable reception case as the MPE-FEC is not completely effective due to a lower Doppler frequency.

Noise

Typically, when the DVB-H receiver is added to the GSM mobile, a GSM reject filter is added in front of the receiver to prevent the high power from the GSM transmitter entering the DVB-H receiver. This adds an insertion loss of about 1 dB, making a total loss to 6dB at 700 MHz.

Antennae in Handheld Devices

There are two types of antenna solutions: integrated and external antennae. The mobile terminals have integrated antennae and are small as compared to the wavelength. When an antenna needs to cover a wide area, the losses are higher and hence a lower net efficiency. Also, the way the terminal is being used, that is it is closer to the body, would impact the radiation pattern of the antenna, hence inducing losses. Some typical antenna gains in handheld devices are:

- Frequency: 474 MHz; gain: –10 dBi.
- Frequency: 698 MHz; gain: – 07 dBi.
- Frequency: 858 MHz; gain: – 05 dBi

Sometimes, external antennae can be used to boost the reception. The expected range of values is between –3 dBi and +3 dBi with an improvement of about 7 dB with respect to an integrated antenna. This increases the quality and reduces the network complexity.

Diversity

By using the diversity receivers a significant reduction of the *C/N* is possible. The output signals received from several antenna are combined before decoding. This reduces the lower minimum field strengths and subsequent lowering of the transmitter powers but may increase the costs. An improvement of 6–9 dB is possible.

10.7.1 Site Planning

Transmitter sites are the ones where the base stations/broadcasting stations are located. In the DVB-H network; the base stations can be located at the existing broadcasting station sites, transmission link stations, mobile base station sites or even building sites. The site requirements are similar to that of radio/cellular sites – space to put on the mast for the antenna, power, space for placing transmitter/equipment, etc. There are traditional broadcast sites which are too high, that is provide a large coverage area (more than 200 metres). But these sites are few in number. There are many cellular sites available but they provide coverage in 'moderate regions' only.

10.7.2 Coverage Planning

There are three concepts/terms to be understood in coverage planning for DVB-H:

- Receiving location.
- Small coverage area.
- Coverage area.

All these three are fundamentally defined in terms of the 'amount' of the receiving area. The smallest unit is the 'receiving location' which has dimensions of about 0.5 m × 0.5 m. This comes from the assumption that a moving antenna will find optimal receiving conditions within 0.5 m in any direction. 'Small coverage' is defined typically as 100 m × 100 m. The coverage is deemed 'good' if a minimum of 99 % of the receiving locations at the edge of area are covered and are 'acceptable' if 90 % of the receiving location at the edge of area is covered. The third is 'coverage area' and is defined as the coverage area of a transmitter, or a group of transmitters, and is made up of the sum of the individual small areas in which a given class of coverage is achieved.

The power received at the receiver (P_{smin} or Minimum Receiver Signal Input Power, in dBW) is calculated as:

$$P_{smin} = P_n + C/N$$

The receiver noise figure (P_n) is calculated as:

$$P_n = F + 10\log(k \times T_0 \times B)$$

where

k is the Boltzmann's constant (1.38 × 10–23 Ws/K);
T_0 is the absolute temperature (= 290 K);
B is the receiver noise bandwidth (Hz);
F is the receiver noise figure (dB);
C/N is the RF carrier-to-noise ratio at the receiver input required by the system (dB);
P_{smin} is the minimum receiver signal input power (dBW).

The received power mentioned above is without the effect of propagation conditions. There are four conditions that need to be considered:

- Handheld portable outdoor reception – Class A.
- Handheld portable indoor reception at ground floor – Class B.
- Integrated car antenna mobile reception – Class C.
- Handheld mobile reception (i.e. terminals are used within a moving vehicle) – Class D.

$$\varphi_{min} = Ps_{min} - A_a$$
$$E_{min} = \varphi_{min} + 120 + 10\log 10\,(120\pi) = \varphi_{min} + 145.8$$
$$\varphi_{med} = \varphi_{min} + P_{mmn} + C_l + L_h \text{ (for Classes A and C)}$$
$$\varphi_{med} = \varphi_{min} + P_{mmn} + C_l + L_h + L_b \text{ (for Class B)}$$
$$\varphi_{med} = \varphi_{min} + P_{mmn} + C_l + L_h + L_v \text{ (for Class D)}$$
$$E_{med} = \varphi_{med} + 120 + 10\log 10\,(120\pi) = \varphi_{med} + 145.8$$

where

C/N is the RF signal-to-noise ratio required by the system (dB);
φ_{min} is the minimum power flux density at the receiving place (dBW/m^2);
E_{min} is the equivalent minimum field strength at the receiving place ($dB\mu V/m$);
L_h is the height loss (10 m above the ground level to 1.5 m above the ground level) (dB);
L_b is the building penetration loss (dB);
L_v is the vehicle entry loss (dB);
P_{mmn} is the allowance for man-made noise (dB);
C_l is the location correction factor (dB);
φ_{med} is the minimum median power flux density, planning value (dBW/m^2);
E_{med} is the minimum median equivalent field strength, planning value ($dB\mu V/m$);
A_a is the effective antenna aperture (dBm^2);
($A_a = G_{iso} + 10\log 10(\lambda 2/4\pi)$); G_{iso} is the antenna gain relative to an isotropic antenna.

Man-Made Noise

The impact of man-made noise is valid only for cases where the antenna gain is more than 0 dBi. The allowance for man-made noise is taken as 1 dB for bands IV and V in urban areas for adapted antennae. The man-made noise for rural areas is around 2–3 dB more than the urban areas – thus a value of 5 dB can be assumed in the case of band III.

Table 10.4 Height loss (based on ITU-R Rec P 1546)

	Receiving antenna height loss (dB)			
	Band III	Band IV	Band V	1.5 GHz Band
Urban	19	23	24	27
Suburban	12	16	18	21
Rural	12	16	17	19

Height Loss

The antenna heights are considered to be 10 m for above the ground level for network planning purposes.

Table 10.4 shows the height loss that could be used for planning purpose. The height loss may also be dependent on factors such as the environment and transmitter-to-receiver distance.

Penetration Loss

As in radio planning, losses will happen due to building infrastructure and also in-car penetration. The building penetration loss is very difficult to estimate and the values should be defined locally through measurements. The building and car penetration loss can be around 9–11 dB and 8 dB.

Appendix A

VAS Applications

Sameer Mathur
Nokia Siemens Networks

A.1 Multimedia Messaging Service

Multimedia Messaging Service accomplishes the transfer of images, audio, video, data and text from mobile to mobile, mobile to Internet and Internet to mobile. If the receiver handset is not 'MMS-capable' then it will receive an SMS stating that the MMS can be viewed at a website whose URL and login/password are mentioned in the SMS. This function is accomplished with the help of an Application Gateway. The size of an MMS message is a few kilobytes depending on the content. Figure A.1 shows the logical connectivity of an MMS Solution in a Packet Core Network.

When a MMS is sent from one PLMN to another, the MMSC of both networks are involved. Inter-MMSC connection may be defined using the SMTP protocol although this involves message conversion before delivery. Applications use the MM7 interface to send and receive the MMS, delivery and read-reply reports. Applications can connect to the MMSC using a direct leased line, VPN or the Internet using some secure transport software.

The following inputs are required for proper dimensioning of an MMS capable network:

- Assessment of existing packet core network. A connectivity diagram of the existing network, information about the h/w and s/w details, capacities, available interface types for SGSN, GGSN, DNS, firewall, routers and switches, their geographical location and backbone connectivity details should be studied. Plus what kind of services are being offered, for example, WAP, Internet/Intranet access, e-mail and their current level of usage including traffic load on the GSNs.
- Traffic mix, size (a 30 kB average MMS size is a good estimation to start with) and content, total number of existing and projected subscribers, MMS usage (MO or MT MMS) per subscriber during a busy hour and over a month, plus GPRS and MMS penetration in its own and other connecting PLMNs. Traffic distribution with respect to time, peaks during a

Cellular Technologies for Emerging Markets: 2G, 3G and Beyond Ajay R. Mishra

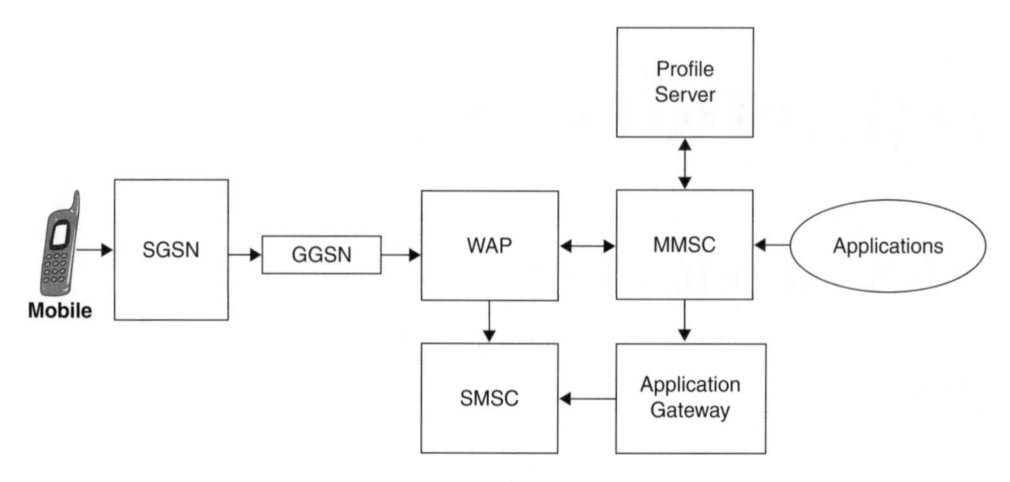

Figure A.1 MMS call flow.

busy hour, number of busy hours in a day, 'burstiness' of traffic are important factors to be considered. Subscriber growth estimates should be available for a rollout network.

- If multiple MMS servers are located at geographically different locations, the traffic flow over the backbone should be analysed. Some cities (for example, tourist locations) may provide more traffic. Some networks may throw more inter-PLMN and roaming traffic than others. All roaming scenarios should be covered for traffic dimensioning.

There can be various configurations/topologies of an MMS solution. It may consist of one or more single server nodes and one or more clusters which may be located at centralized or distributed locations. These decisions must be taken keeping in mind load sharing and distribution, traffic patterns, redundancy considerations, available capacities, commercial implications, operator policy and future evolution. An example of the configuration of a load-sharing MMS cluster is shown in Figure A.2.

IP connectivity should be provided for applications, profile server, SMSC, application gateway and terminal management servers also.

MMSC uses SMS for notifications, delivery reports and read-reply reports. Introduction of MMSC in a network will also increase SMS traffic and hence SMSC also should be re-dimensioned accordingly.

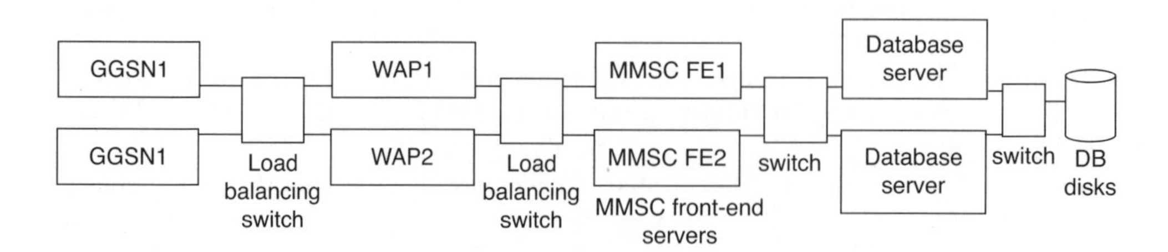

Figure A.2 MMSC connectivity.

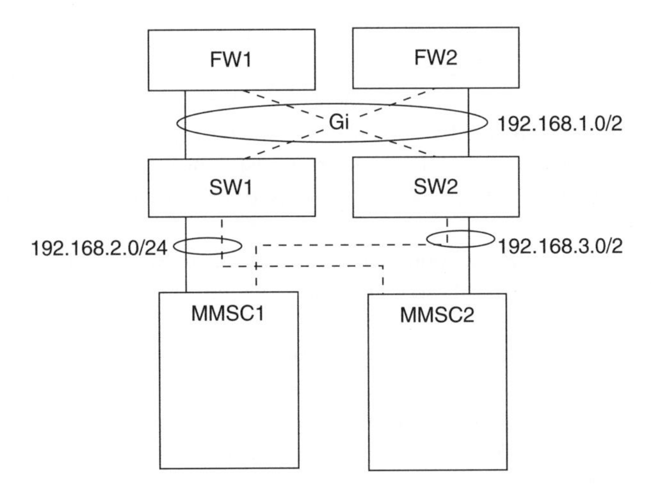

Figure A.3 Load balancing.

URI of MMSC (for example, http://mmsc.operator.com:8002) is configured in the WAP gateway as well as in mobile phone settings.

Dimensioning of the MMS solution starts with calculating the product capacities required to support the expected amount of MMS traffic. MMSC, SMSC, WAP, Profile Server and Application Gateway should support the required number of subscribers, MMS/s or transactions/s. Subsequently, the GPRS Packet Core Network (SGSN, GGSN, interfaces, backbone, etc.) should be dimensioned to support the additional MMS traffic with yearly growth projections. Once all these inputs are considered, traffic calculations are done and dimensioning results are obtained taking into account the equipment capacity constraints. Integration of MMSC in a packet core network is done on the Gi interface. Intelligent switches are used to achieve load balancing and redundancy as shown in Figure A.3.

A.2 Push-to-Talk over Cellular

Push-to-Talk over Cellular Core Networks (PoC CNs) integration consists of integrating the PoC site elements:

- PoC register;
- PoC call processor;
- cabinet switch.

and the site elements are in turn integrated to several other external systems including:

- workstations;
- GPRS network;
- charging system;
- network management system;
- domain name servers.

The PoC solution comprises of four networks to handle the different types of traffic, namely Operation and Maintenance, Bearer or Subscriber, Production and Backup Network. The O and M network consists of network elements used for operation and maintenance of the PoC. The Bearer or Subscriber network consists of the GPRS PoC terminals. The Production network consists of network elements used for provisioning, customer care and billing. The Backup network consists of the network elements used for storing backups from PoC network elements.

In the PoC System, the IP backbone connects the PoC network elements together and to the (E)GPRS Network. Among these connected elements are the PoC Call Processor, PoC Register and DNS Server.

In practice, the IP backbone is an arbitrary combination of LAN and WAN networks, which are interconnected. From a network element point of view the whole IP network is seen as a single entity. The IP backbone is an IP network with its own routing architecture, security policy, NMS and other IP services. Its basic function is to convey two types of traffic: control plane traffic (signalling) and user plane traffic (voice). Since IP is used for both signalling and voice, the main part of the traffic is user plane traffic. Its most crucial QoS requirement is low delay and low delay variation but other QoS parameters such as low packet loss are also important. The requirements for signalling delay are not as tight as those for voice traffic.

Figure A.4 illustrates the various interfaces of PoC.

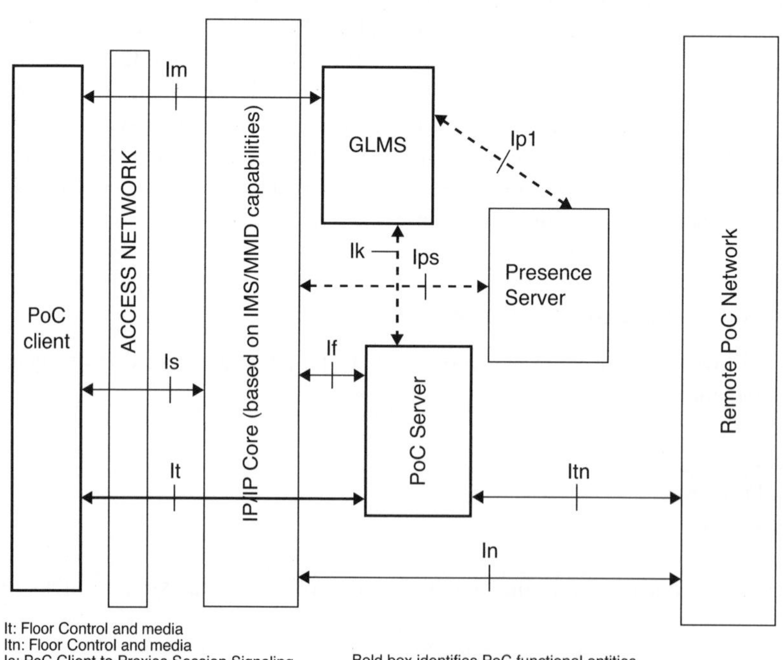

Figure A.4 POC interfaces.

In addition to actual IP network elements, the IP backbone has a set of IP infrastructure elements, such as routers, firewalls, NMS for IP elements, and possibly key management servers if, for example, IPSec is used between sites.

The most critical issue in the IP backbone routing and traffic planning is QoS provisioning, that is the careful design of the IP network's performance. Issues that should be considered are overall capacity requirements, delay, jitter and packet loss factor. Because speech traffic can vary considerably, wide design margins should be used. There are two different approaches to ensure the performance of the IP network. The first one is designing the network with margins so large that there is always enough capacity to bear even the highest traffic peaks. The other approach is to utilize different traffic classification and routing protocols in order to guarantee the required QoS, at least for the most valuable traffic.

The system traffic and capacity calculations depend on the system configuration. By using the results of the radio interface network planning, the capacity requirements of the BSCs can be calculated. The traffic can be expressed in 'erlangs' for push-to-talk traffic and in Mbit/s (in peak hour) for background data. Also, the capacity requirements for the Abis-interface should be calculated. While dimensioning the Gb- and Gn-interfaces the overhead traffic must be included due to the interface protocol layers. The capacity needed for the push-to-talk service on the radio interface is about 8 kbit/s for one connection. After adding overhead, the effective capacity needed on the Gb-interface is about 10.5 kbit/s in the downlink direction and about 9.6 kbit/s in the uplink direction. For the Gn- and Gi-interfaces, the corresponding capacity needed is about 10.6 kbit/s (IPv6) and 9.6 kbit/s (IPv4) for both directions. The types of traffic are shown in Figure A.5.

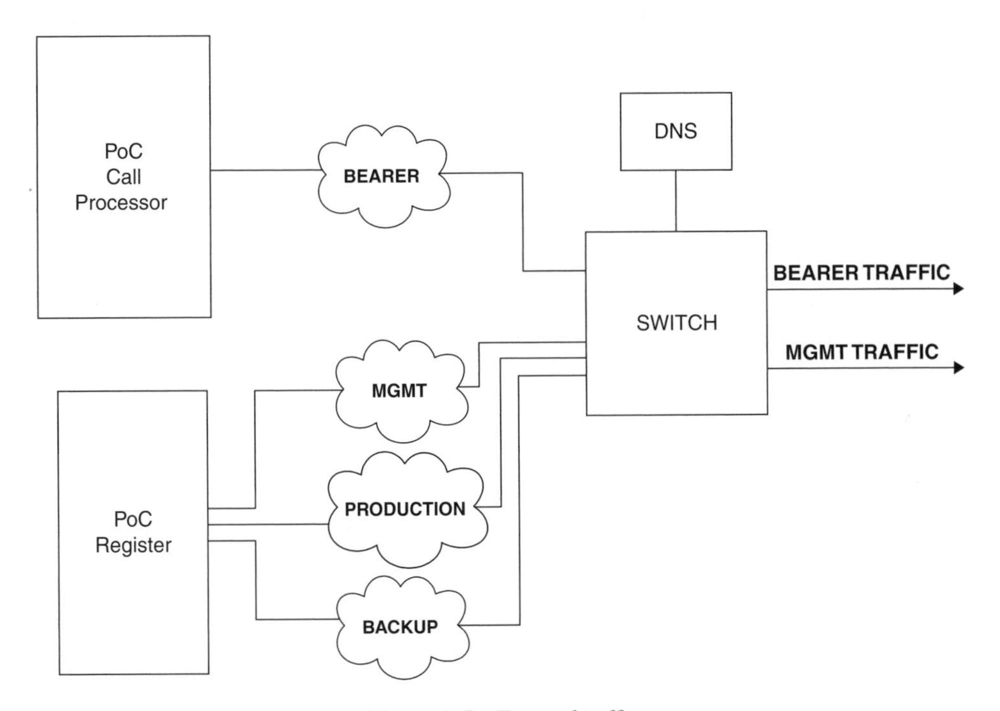

Figure A.5 Types of traffic.

In addition, the impact of GPRS on the existing SS7 network should be taken into account. The signalling-load calculations give the required number of signaling links in the Gs- and Gr/Gf-interfaces.

Detailed dimensioning should be carried out on the basis of the capacity parameters of real network elements. In BSC, these include PCU data processing capacity, PCU traffic channel/cell capacity and Abis-interface capacity while in SGSN, GGSN, subscriber (PDP) capacity and throughput capacity. As a result of the system dimensioning, the number of the PCUs, SGSNs and GGSNs needed is calculated.

PoC Call Processor provides voice services for PoC subscribers. The Call Processor is a cluster of nodes, each of which is an IP host connected with the cabinet switch via a chassis LAN switch, SWSE. Each SWSE is connected with each cabinet switch through a GBIC link for redundancy purposes on Layer 2, and RSTP is run between SWSEs and a cabinet switch for resiliency purposes.

The PoC Register (PoC R) provides subscriber database functions for the PoC and is managed using Management User Clients (MUCs). The PoC Register is responsible for data distribution to the PoC CPs. The PoC Register elements are IP hosts interconnected through the duplicated PoC Register LAN switch with each other and with the PoC cabinet switch. Each PoC Register LAN switch is connected with one cabinet switch through three 100BaseT links on L2 and RSTP is run between the PoC Register LAN switches and the cabinet switches for resiliency purposes on L2. Optionally, two of the 100BaseT links (Production and Backup VLAN links) can be replaced with two GBIC links between the PoC Register.

Each PoC Register element has multiple Ethernet interfaces. The interfaces are connected to three separate VLANs each with its own IP sub-net. Each PoC Register interface is allocated an IP address from the corresponding PoC Register VLAN IP sub-net. In each element the POC REGISTER VLAN_s HSRP group virtual IP address must be defined as either the default gateway or the next hop in a default route to the corresponding network.

A.3 Streaming Service

Streaming provides consumers with an end-to-end application that allows enjoying multimedia content. Streaming enables real-time audio and video presentation on broadcast sites, music and video jukeboxes.

There are challenges like limited bit rate and the small size of the mobile terminals, which may limit the expression of streaming media. Nevertheless, the user can still be provided with good quality of audio and video that he or she expects from the mobile terminal.

Streaming is one of the key areas to differentiate 3G from 2G, as it enables new, attractive service concepts like News and Information services, entertainment, webcam, messaging and multimedia streaming.

Multimedia files can also be completely downloaded before viewing. Streaming technologies allow users to view and/or listen to multimedia content as it is downloaded, reducing the time before playback begins. The difference between streaming and downloading is that with streaming the user can be given a controlled access to the content for a particular time whereas by downloading the user is able to replay files as many times as he or she wants.

The streaming process consists of several phases. First of all video data are captured using a camera and microphone. Media tracks are then compressed and multiplexed using different compression formats and techniques. The media files are then stored in a streaming server. In the client phase the multimedia tracks are first de-multiplexed and then decompressed. The client will use different output devices such as screen and loudspeakers for playback. The client terminal has only a small amount of buffered content at any time. After playback the data are deleted and memory is freed up for new data. The content is actually never stored in the terminal, which makes copyright and re-use matters more favorable to the content providers.

Capture → Compress → Store → Play

The key players in the streaming value chain are service provider, content provider and content producer.

The streaming solution architecture may comprise several components as follows – access methods for mobile streaming including WAP, WEB, SMS and MMS access. A GPRS, EDGE, HSCSD or 3G network is necessary for mobile radio access. A streaming server streams both live and pre-recorded media over a network. The streamed data can originate either on the Internet or within an operator's Intranet. The Gateway mobile works as a proxy for streaming data enabling content to be fetched from the Internet and streamed over the operator's networks. Client software resides in the mobile to play the streamed files.

'Best-user' experience is achieved if the operator has a streaming solution in its network. Through a Streaming Gateway the operator is able to stream the operator's own content as well as cache Internet content.

A.4 Short Message Service

Short Message Service (SMS) enables sending and receiving of short text messages between mobile phones using signalling channels. SMS can be point-to-point or point-to-multipoint. The maximum length of the message is 140 bytes. If 7-bit encoding is used this translates to 160 characters. Messages can be concatenated to form longer messages. SMS can be sent at any time even when the destined mobile station is not reachable. SMS text messaging supports all languages supported by Unicode.

SMS can also carry binary data like ringtones, pictures, operator logos, wallpapers, animations, business cards and WAP configurations to a mobile phone.

SMS is supported by all GSM mobile phones. Almost all subscription plans provided by wireless carriers include inexpensive SMS messaging service.

The Short Message Service Centre (SMSC) is the network element which is responsible for handling SMS in a GSM network. It is connected to the MSC using the SS7 protocol.

There are two different and independent point-to-point services depending on the direction of the Short Message (SM) transfer (the basic procedure is shown in Figure A.6):

- Mobile-originating (MO-SM) sent by subscriber A and transported from a Mobile Station (MS) to an SMSC.
- Mobile-terminating (MT-SM) transported from an SMSC to an MS and received by subscriber B. This message may be input to the SMSC by other mobile users (via an MO-SM) or by a variety of other sources, for example, speech, e-mail, telex or facsimile.

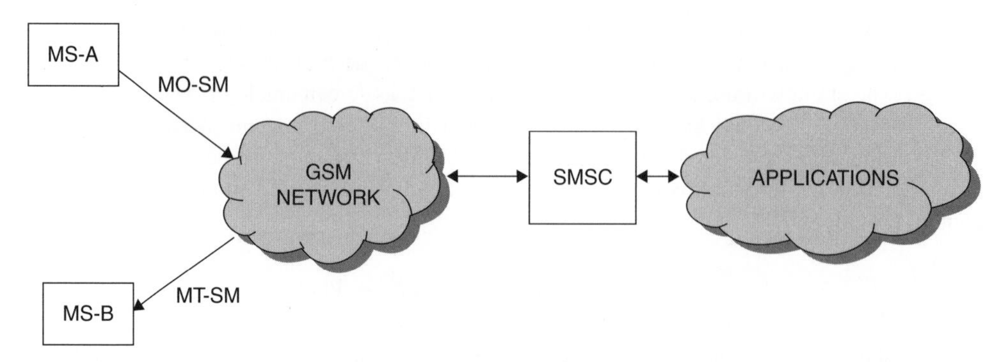

Figure A.6 Basic procedure.

The SM delivery to the SMSC is always acknowledged to the sender.

A subscriber can send an SM to subscriber B even when subscriber B's MS is switched off or not reachable. If an MT-SM cannot be delivered to the MS because it is absent, the network stores an indication of absence in the HLR and when the MS is active again, the network informs the SMSC which makes a delivery attempt.

Short Messages can be sent and received also during voice calls.

The Voice Mail System (VMS) may send voice mail alerts to a subscriber through SMSC to indicate that voice messages were left in the voice mailbox.

Various kinds of applications can be connected to the SMSC to push messages to mobile stations.

A.5 Wireless Application Protocol

The Wireless Application Protocol (WAP) is the standard for the presentation and delivery of wireless information and telephony services on mobile phones from the Internet. WAP allows carriers to strengthen their service offerings by providing subscribers with the information they want and need while on the move.

Wireless devices represent the ultimate constrained computing device with limited CPU, memory, and battery life, and a simple user interface. Wireless networks are constrained by low bandwidth, high latency and unpredictable availability and stability. The WAP solution leverages the investment in Web servers, Web development tools, Web programmers and Web applications while solving the unique problems associated with the wireless domain. WAP-enabled devices deliver timely information and accept transactions and inquiries when the user is moving around. WAP services provide pinpoint information access and delivery when the full screen environment is not available. The basic WAP connectivity is illustrated in Figure A.7.

The WAP Gateway is the network element in a GSM network which handles connectivity of the WAP-enabled mobile phones with Internet-based web servers. It converts web content using the WAP protocol and thus makes it fit to be viewed on a mobile device.

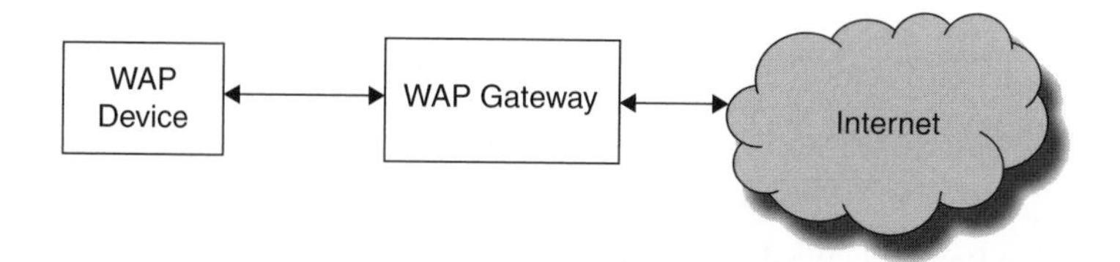

Figure A.7 Basic WAP connectivity.

Wireless Markup Language (WML) scripts are used to create WAP content. They make optimum use of small displays. The lightweight WAP protocol stack is designed to minimize the required bandwidth and maximize the number of wireless network types that can deliver WAP content.

WAP utilizes binary transmission for greater compression of data and is optimized for long latency and low bandwidth. WAP sessions cope with intermittent coverage and can operate over a wide variety of wireless transports.

WAP is based on a scalable layered architecture. Each layer can develop independent of each other. This makes it possible to introduce new bearers or to use new transport protocols without major changes in the other layers. Figure A.8 depicts the protocol stack.

WDP

The WAP datagram protocol is the transport layer that sends and receives messages via any available bearer network, including SMS, USSD, CSD and GPRS.

WTLS

The Wireless Transport Layer Security has encryption facilities that provide a secure transport service required by many applications, such as e-commerce.

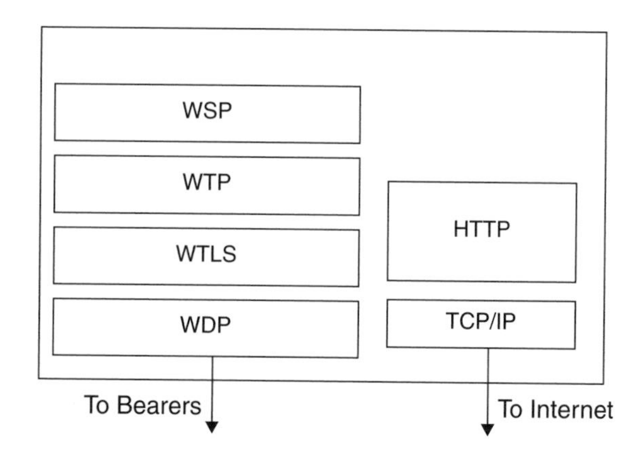

Figure A.8 WAP protocol stack.

WTP

The WAP Transaction Protocol layer provides transaction support, adding reliability to the datagram service provided by WDP.

WSP

The WAP Session Protocol (WSP) layer provides a lightweight session layer to allow efficient exchange of data between applications.

Appendix B

Energy in Telecommunications

Anne Larilahti
Nokia Siemens Networks

One of the key challenges in building mobile networks in emerging markets is energy: access to it and the price of it. It is interesting to note that people's need to communicate is so great that it exceeds their need for reliable energy. It has been estimated that about 1.6 billion people today live without access to electricity. Another one billion live in areas where electricity grids are built but the service is very unreliable. Yet, mobile networks are being built in these areas and when doing it right, operators are able to make a profit despite the difficulties.

Energy costs are playing a big role in operational expenditures of any mobile network. The portion of energy in network operating costs in developed nations can be up to 10 % while in developing countries it varies between 15 and 30 %.

The task of building an energy-efficient mobile network can be roughly divided into two parts; on the one hand ensuring the highest possible energy efficiency for the network, and on the other hand looking closely at how the energy is produced. Energy efficiency has been covered elsewhere in this book – here we will concentrate on power sources.

B.1 The Solution Exists – but It's not Very Good

Mobile users in developing markets are notoriously quick to change their service providers. Many people carry several SIM cards and will use the one that provides the cheapest service. When outside of cities, the decision will simply be made based on the coverage available. Therefore, mobile operators are 'racing' to cover even those hard to reach, unserved areas.

If there is no electricity grid available or if the grid cannot be relied on to provide power to the mobile site to satisfy the quality requirements of the operator and the users, the site needs to run independently from external power sources. The traditional way of doing this is to equip the site with twin diesel generators, running in shifts to power the site.

Cellular Technologies for Emerging Markets: 2G, 3G and Beyond Ajay R. Mishra

Diesel generators are an old and tested technology that is well trusted. Also, generators today are fairly cheap and do not have a high impact on the overall investment on a site. Despite this low capital expenditure, the total cost of ownership of the site powered by diesel will be high and the exact cost very unpredictable. The price of crude oil fluctuates strongly impacting the price of diesel.

During the past few years we've seen the price or a barrel of crude oil rise from under $25 in 2003 to an all-time high of $147.30 in 2008 and then drop back down to under $35 in 2009. Thus, at the time of writing, the price of diesel is relatively reasonable, but that is a state of affairs that is not likely to last. Although there has always been a lot of speculation involved in determining the price of oil, lately the discussion around 'peak oil' (the time when the maximum rate of oil extraction has been reached) has added a new strong element to the speculations.

No one really knows when we will reach 'peak oil', but the consensus seems to be that all the easy to access reserves have already been found. That, combined with the fact of a growing population and hence the need for oil, does imply that the price of oil will be rising in the future. Due to the uncertainty, among many other things, the price will fluctuate on its way up, making it extremely difficult to predict the operational costs of a network highly dependent on diesel.

To further complicate things, it needs to be noted that the price of diesel at a site is a lot higher than the price 'at the pump'. Typically, in those areas where off-grid (or unreliable grid) base station sites are located, also the other infrastructure is undeveloped, meaning that access to the sites is via narrow rural roads or in some cases there are no roads at all. In extreme cases, where base station sites are located on islands or comparable locations, diesel is delivered by helicopter. The distances are long, demanding clear planning and logistics processes from the operator.

As an example, Zain Nigeria covers currently over 1500 towns and 14 000 communities across all six geo-political zones of the country. Due to the physical size of the country and the fact that many communities are spread far away from bigger cities, Zain Nigeria needs to run many of its sites autonomously. They have calculated that they are burning 350 litres of diesel every minute to run their network. It is easy to understand that energy efficiency and access to power are very much on the top of the minds of their executives.

Another unfortunate issue with diesel is theft. African and Indian operators have estimated that 20–25 % of their diesel gets stolen, either during transport or from the site. Furthermore, to cover up the theft, often the culprits replace the stolen diesel with water. The mixture of diesel and water is detrimental to the generators, increasing maintenance and replacement costs.

Finally, with the increasing awareness of environmental issues, diesel generators have faced criticism due to the related CO_2 emissions. This will be discussed in some more detail later.

B.2 Renewable Energy – a Better Solution

Whereas diesel is highly polluting, cumbersome and the price is unpredictable, renewable energy has none of those challenges. Although the initial investment needed to build a renewable energy powered site can be substantially higher than a diesel powered one, the total cost of ownership 'turns the tables', making renewables an attractive choice for operators.

The most common renewable energy sources are solar, wind, biofuel, hydro and geothermal. In addition, fuel cells are often grouped into this category even though they are not renewable per se, but can be seen as an alternative energy source as well.

B.2.1 Solar

The sun is a fantastic source of energy. Every hour it provides the Earth with as much energy as it takes us all nearly a year to consume. The challenge is to harvest it efficiently enough to provide the power needed.

At base station sites solar power is harvested using photovoltaics (panels of cells containing a material that converts solar radiation into direct current electricity) installed at the site. Planning is fairly simple as good information on solar radiation is available through insolation (a measure of solar radiation energy received on a given surface area in a given time) maps to determine the solar radiation received. Intelligent site design can then increase the efficiency further by accounting for issues such as shading, tilt and tracking. In general, the new off-grid and bad-grid sites deployed are in areas where solar radiation is plentiful.

The price of photovoltaics is very high compared to diesel generators. However, after the initial investment, the costs are close to zero. As there are no moving parts, maintenance needs are minimal. The only work needed is cleaning the panels, the frequency of which depends on local conditions such as saltiness, dust and rain.

The photovoltaic market is booming. It is said to double in size every two years, making it the fastest growing energy technology. Increases in volume together with technological advances have brought the prices down while efficiencies are going up.

To sum up the pros and cons of solar energy:

+ high reliability;
+ very low operational cost;
+ no noise;
+ mature technology;
+ 20 year lifetime;
– high initial investment;
– space needed.

B.2.2 Wind

Wind power can be seen as a derivative of solar power, as the Sun's warmth causes wind. Wind power is an old invention with the first applications being sailboats and natural ventilation in buildings. Windmills that can be seen as the forefathers of modern wind turbines date back to the seventh century.

Although most of the wind power globally is produced in large scale windmill parks, wind is also suitable for more distributed power generation and works well in mobile networks. Mounted either on its own pole or on the base station tower, a wind turbine requires a smaller physical footprint than solar panels and is cheaper to acquire. The maintenance costs are nearly as low as with solar panels, making the total cost of ownership low over time.

However, wind is a bit more fickle and harder to understand than solar. There are wind maps for the globe, but the exact wind speeds vary greatly between different locations, even within a close distance. Aspects such as the terrain, the height and vegetation around the site have an effect on the power that can be obtained. Therefore, site design with exact information on the location is essential for the correct dimensioning for a wind turbine.

To sum up the pros and cons of wind energy:

+ medium initial investment;
+ low operational cost;
+ can be used to complement photovoltaics;
+ can use the antenna tower for mounting;
− erratic nature of wind;
− noise.

B.2.3 Biofuels

'Biofuels' has become a big 'buzzword' since the middle of the last decade. Quite soon it was noticed though that there were major disadvantages with the first-generation biofuels available today. Social implications such as using food crops for producing fuel, environmental implications such as deforestation and loss of biodiversity, together with uncertainty of the energy balance has cooled the enthusiasm towards biofuels. Second- (non-food crops) and third-generation (algae) biofuels are expected to address these shortcomings.

Biodiesel is a form of biofuel that can be used to power mobile networks. Most of the existing diesel generators can burn biodiesel as well as fossil-based diesel giving the operators a choice of which one to use. The challenges with both are quite similar. The logistical difficulty remains, as does theft, high operational cost and pollution. Some attempts have been made to remove the first by producing the biodiesel close to the site but the complexity of building a network of small biodiesel 'factories' close to the site has so far proven to be too much, one of the reasons being the demand for consistent quality of the fuel by the diesel generator warranty terms.

To sum up the pros and cons of biofuels:

+ easy or no installation;
+ low initial investment;
− social impact;
− acoustic noise;
− high operational expenditure.

B.2.4 Fuel Cells

Even though fuel cells are not a renewable energy source, they do provide an interesting possibility for powering remote base station sites. The main challenge in remote areas is that as opposed to wind and solar, fuel cells do need fuel to run. Basically the fuel, such as nitrogen, can be delivered to the site which brings logistical problems, or it can be produced at the site, which is complicated and can be expensive. Therefore fuel cells at the moment are best considered as an emergency back-up power source rather than the source for producing the primary power for the site.

B.2.5 Hydro and Geothermal

Hydropower as well as geothermal are old and well-understood energy sources, but due to the fact that they can be harvested only at specific locations they rarely match with the requirements of mobile networks and are therefore not at the moment considered as power sources for remote base station sites.

B.3 The Optimal Design for a Base Station Site

When designing a base station site that will use renewable energy it is important to understand that each location is different. While energy efficiency is a common requirement for all the sites, the optimum mix of energy sources depends on the amount of energy needed but also on the location and available resources.

One single power source solution is seldom the best choice, but rather a hybrid, a combination of different power sources can provide the best-end result. For example, a combination of solar and wind can be cheaper than just solar, as the wind turbine can compensate during darker periods, so that the solar panels do not need to be dimensioned according to the darkest month to ensure a 'solid performance' throughout the year.

It's not atypical that a site would get its power from a grid, solar panels, a wind turbine, a small back-up diesel generator or a battery bank. A solution this complicated is difficult to optimize, but products exist that can monitor and manage the energy input from each source as well as the status of batteries and the amount of diesel left, by using intelligent algorithms to choose the best way to obtain the energy. An example of a power solution in a 'bad-grid' mobile network site is shown in Figure B.1.

B.4 Business Case for Renewable Energy in Mobile Base Station Sites

As discussed earlier, even though the initial capital expenditure is greater in renewable power sites than the traditional diesel powered sites, the total cost of ownership is lower. A well designed site provides sustainable operational expenditure savings and can have payback time as short as one to two years. Maintenance costs are minimal and maintenance can be scheduled to take place in conjunction with other site visits or, as it is can be quite a straightforward activity, can be outsourced to a local entity. An example of a business case for renewable energy in mobile base station sites is shown in Figure B.2.

Operators still often shy away from the investments needed for renewable energy. The investment is not only the money to be paid for the equipment but also the competence that needs to be built into the operator's organization. Even though there are no high-skilled maintenance needs, the operator still needs people in the organization who can form an opinion on renewable energy sources, make choices between vendors and give the right input to designing the network. Creating and managing a group like this is an expense the operator does not necessarily want to carry.

Therefore, the new 'buzz phrase' in the industry is 'Power as a Service' (PaaS). In the PaaS model, the operator buys, for example solar power as a service, much as they would buy grid-based electricity from a utility company. The owner of the solar panels would not be the operator, but maybe a vendor, a utility company or a 'special-purpose vehicle', created

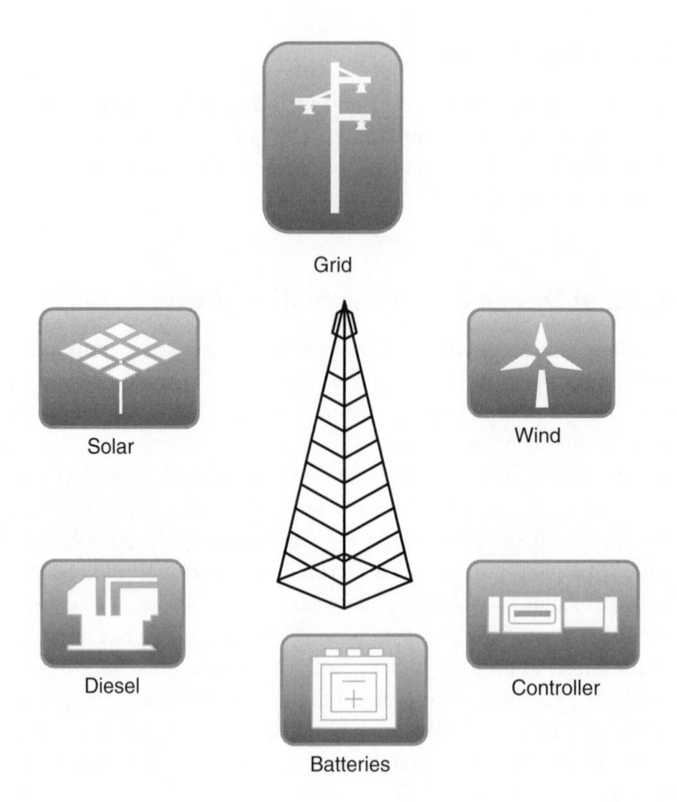

Figure B.1 Example of a power solution in a 'bad-grid' mobile network site.

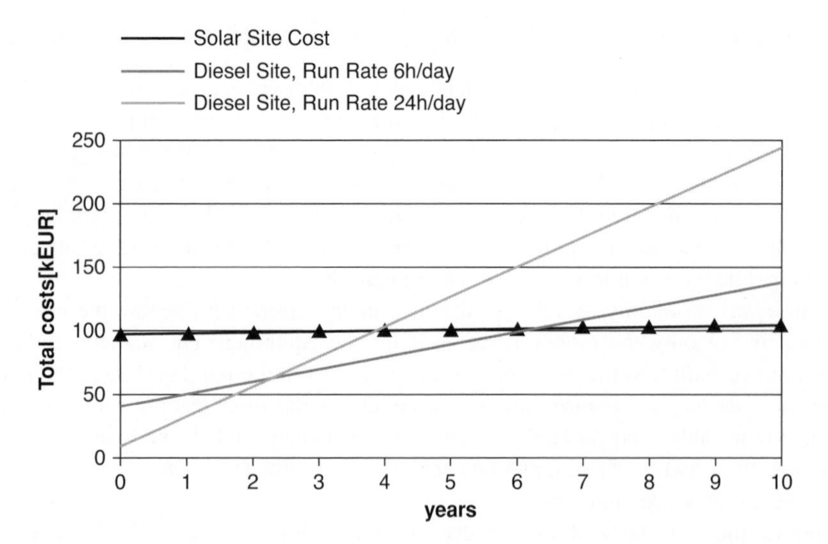

Figure B.2 Business case for renewable energy in mobile base station sites (source: Nokia Siemens Networks, 2008).

just for this purpose. Operators employing the PaaS model would see immediate gains as no investment in equipment is needed. The price can be agreed for a long time and contracts can last ten, fifteen or even twenty years. During the agreement, the price of energy will be completely predictable, making planning easier.

B.5 Effects of Climate Change on Mobile Networks

The issues of environmental degradation have been making headlines for the past years and for a good reason. The world's forests have shrunk by some 40 % since agriculture began 11 000 years ago. Three quarters of this loss has occurred in the last two centuries. As many as 90 % of all of the ocean's large fish have been 'fished-out'. 1.1 billion people lack access to drinkable water and the number is feared to grow with water shortages being experienced across the globe.

None of these issues has managed to attract as much attention as climate change. Climate change is said to be 'the biggest threat human kind has had to face'. And the reason is greenhouse gases. These gases, which are all naturally occurring, act as a blanket, trapping in the heat and preventing it from being reflected away from the Earth. They keep the Earth's average temperature at about 15 °C, warm enough to sustain life for humans, plants and animals. Without these gases, it would be too cold for most life forms.

The equilibrium between greenhouse gases being naturally produced and naturally removed from the atmosphere is almost in perfect balance. Now, however, the balance has been upset by human activity. The most important greenhouse gas is carbon dioxide, CO_2. At the moment, the burning of fossil fuels is burdening the atmosphere to a point that the balance providing the 15 °C average temperature is 'tipping'. Too much heat gets trapped in and the temperatures on Earth are climbing. This has manifold implications, such as glaciers melting, rising sea levels, spread of tropical diseases and extreme weather events.

Due to the seriousness of this issue, governments are taking an increasingly active role to fight climate change. They are pushing for stricter regulation and designing 'carbon trade systems' to curb the production of CO_2. However, global agreements are slow to make and hence many corporations are taking the lead and showing an example of reducing CO_2 emissions. Telecommunications operators in Europe have been quite active on this front.

Even though the impact of telecommunications on climate change is relatively small (telecommunications account for about 0.5 % of global CO_2 emissions), the operators are setting reduction targets for themselves and making them public. Emission reduction targets as ambitious as 50 % to decrease by 2020 (against the baseline, 2006/2007) have been communicated. For operators to reach that, they need to concentrate on their networks. For a typical mobile operator, about 86 % of the energy used is utilized by the network. Therefore, energy efficiency and renewable energy sources to power the network are key in reaching the target.

Voluntary targets, together with regulation, drive emission reductions. The Fifteenth Conference of Parties (COP15) held in Copenhagen in December 2009 was expected to produce stricter global targets for greenhouse gases. However, due to the complexity of the issue, as well as differing socio-economic concerns of developed and developing nations, no binding agreements were made.

However, an agreement dubbed the Copenhagen Accord was drawn. The accord is a non-binding document that although disappointing for many, can be seen as a first step towards stronger action. It holds out the prospect of USD 100 billion in annual aid from 2020 for developing nations for sustainable development, but does not specify where the money will come from. In any case, a considerable amount of money will be targeted to developing the emerging markets in a way that is also environmentally sustainable as well as bringing economic benefits. This aid has potential in the future to change financing opportunities, also for mobile operators considering energy efficiency and renewable energy investments.

Bibliography

3GPP, 'Combined GSM and Mobile IP Mobility Handling in UMTS', IP CN 3G TR 23.923 ver.3.0, 2000–2005.

3GPP, 'QoS Optimization for AAL Type 2 Connections over Iub and Iur Interfaces', 3G TR 25.934.

3GPP, 'Study on Rel.2000 Services and Capabilities' (3G TS 22.796), V2.0.0, 2000–0606.

3GPP, Technical Specification Group 23.907, Services and System Aspects, QoS Concepts.

3GPP R1-050587, 'OFDM Radio Parameter Set in Evolved UTRA Downlink', June 2005.

3GPP R1-050651, 'Unpaired spectrum aspects of LTE'.

3GPP TS 23.003 V8.3.0 (2008–2012), 'Technical Specification Group Core Network and Terminals: Numbering, Addressing and Identification'.

3GPP TS 23.401 V8.4.0 (2008–2012), 'General Packet Radio Service (GPRS) Enhancements for Evolved Universal Terrestrial Radio Access Network (E-UTRAN) Access'.

3GPP TS 24.301 V8.0.0 (2008–2012), 'Non-Access-Stratum (NAS) Protocol for Evolved Packet System (EPS): Stage 3'.

3GPP TS 25.221, 'Physical Channels and Mapping of Transport Channels onto Physical Channels (TDD)'.

3GPP TS 25.224, 'Physical Layer Procedures (TDD)'.

3GPP TS, 25.322, V5.3.0, '3rd Generation Partnership Project: Technical Specification Group Radio Access Network; Radio Link Control (RLC) Protocol Specification, Release 5', 2002–2012.

3GPP TS 29.228, V8.4.0 (2008–2012), 'IP Multimedia (IM) Subsystem Cx and Dx Interfaces: Signalling Flows and Message Contents'.

3GPP TS 29.229, V8.4.0 (2008–2012), 'Cx and Dx Interfaces Based on the Diameter Protocol'.

3GPP TS 29.272, V8.1.1 (2009–2010), 'Evolved Packet System (EPS): Mobility Management Entity (MME) and Serving GPRS Support Node (SGSN) Related Interfaces Based on the Diameter Protocol'.

3GPP TS 33.102, V8.1.0 (2008–2012), '3G Security; Security Architecture'.

3GPP TS 33.203, V8.5.0 (2008–2012), '3G Security: Access Security for IP-based Services'.

3GPP TS 33.210, V8.2.0 (2008–2012), '3G Security: Network Domain Security (NDS); IP Network Layer Security'.

3GPP TS 33.323.

3GPP TS 33.401, V8.2.1 (2008–2012), '3GPP System Architecture Evolution (SAE): Security Architecture'.

3GPP TS 33.402, V8.2.1 (2008–2012), '3GPP System Architecture Evolution (SAE): Security Aspects of Non-3GPP Accesses'.

3GPP TS 36.331, V8.4.0 (2008–2012), 'Evolved Universal Terrestrial Radio Access (E-UTRA): Radio Resource Control (RRC); Protocol Specification'.

3GTS, 25.213, 'Spreading and Modulation (FDD)'.

3GTS, 25.223, 'Spreading and Modulation (TDD)'.

Adachi, F., Sawahashi, M. and Suda, H., 'Wideband DS-CDMA for Next-Generation Mobile Communications Systems', *IEEE Communications Magazine*, **36**(9), 56–69, 1998.

Aguiar, R.L., *et al.*, 'Designing Networks for the Delivery of Advanced Flexible Personal Services: the Daidalos Approach', *Proceedings of the IST Mobile and Wireless Telecommunications Summit*, Lyon, France, 2004.

'Air Interface for Fixed and Mobile Broadband Wireless Access Systems', IEEE P802.16e/D12, February 2005.
Alamouti, S.M., 'A Simple Transmit Diversity Technique for Wireless Communications', *IEEE Journal on Selected Areas in Communications*, **16**, 1451–1458, October 1998.
Allen, K.C., 'Observation of Specific Attenuation of Millimetre Waves by Rain', *IEE A&P Conference*, 1987.
Anfossi, D., Bacci, P. and Longhetti, A., 'An Application of Lidar Technique to the Study of Nocturnal Radiation Inversion', *Atmospheric Environment*, **8**, 483–494, 1960.
Ansari, E., 'Microwave Propagation in Sand and Dust Storms', *IEE Proceedings F, Communication, Radar and Signal Process*, 1982.
Bean, •••, *et al.*, 'ESSA Monograph', US Government Printing Office, Washington, DC, 1966.
Beck, R. and Panzer, H., 'Strategies for Handover and Dynamic Channel Allocation in Micro-cellular Mobile Radio System', *IEEE Vehicular Conference*, 178–185, 1989.
Bellcore, S.M., 'Multi-user Detection for DS-CDMA Communications', *IEEE Communications Magazine*, **124–136**, October 1996.
Bingham, J.A.C., 'Multicarrier Modulation for Data Transmission: An Idea whose Time has Come', *IEEE Communications Magazine*, **28**(5), 5–14, 1990.
Castro, J.P., *The UMTS Network and Radio Access Technology*, John Wiley & Sons, Ltd, Chichester, 2001.
Cimini, L.J., 'Analysis and Simulation of a Digital Mobile Channel Using Orthogonal Frequency Division Multiplexing', *IEEE Transactions and Communications*, **COM-33**(7), 665–675, 1985.
Craig, K.H. and Kennedy, G.R., 'Studies of Microwave Propagation on a Microwave Line-of-Sight Link', *IEE A&P Conference*, 1987.
Crane, R.K., 'Prediction of Attenuation due to Rainfall on Satellite Systems', *IEEE*, **65**(3), 1977.
Crane, R.K., 'Fundamental Limitations Caused by RF Propagation', *IEEE*, **69**(2), 1981.
CWTS TSM 01.01, 'China Wireless Telecommunication System: 3G Digital Cellular Telecommunication System; General Description'.
CWTS TSM 04.08, 'China Wireless Telecommunication Standard: 3G Digital Cellular Telecommunication System: Mobile Radio Interface Layer 3 Specification, Part 1: Radio Resource Management'.
CWTS TSM 04.60, 'China Wireless Telecommunication Standard: 3G Digital Cellular Telecommunication System; TD-SCDMA Packet Radio Service (TD-PRS); User Equipment (UE)–Base Station System (BSS) Interface; Radio Link Control/Medium Access Control (RLC/MAC) Protocol'.
CWTS TSM 05.02, 'China Wireless Telecommunication Standard: 3G Digital Cellular Telecommunication System; Multiplexing and Multiple Access on Radio Path'.
CWTS TSM 05.03, 'China Wireless Telecommunication Standard: 3G Digital Cellular Telecommunication System; Channel Coding'.
CWTS TSM 06.41, 'China Wireless Telecommunication Standard: 3G Digital Cellular Telecommunication System; Discontinuous Transmission (DTX) for Half Rate Speech Traffic Channel'.
CWTS TSM 06.81, 'China Wireless Telecommunication Standard: 3G Digital Cellular Telecommunication System; Discontinuous Transmission (DTX) for Full Rate Speech Traffic Channel'.
Dahlman, E., *et al.*, *3G Evolution: HSPA and LTE for Mobile Broadband*, Academic Press, 2007.
Dahlman, E., *et al.*, 'UMTS/IMT-2000 based on WCDMA', *IEEE Transactions on VT*, **47**(4), November 1998.
Das, S., Viswanathan, H. and Rittenhouse, G., 'Dynamic Load Balancing through Coordinated Scheduling in Packet Data Systems', *Proceedings of IEEE INFOCOM*, San Francisco, March 2003.
Das, S., MacDonald, W.M. and H. Viswanathan, 'Sensitivity Analysis of Handoff Algorithms in Downlink CDMA', *Proceedings of IEEE Globecom*, December 2003; also submitted to *IEEE Transactions on Vehicular Technology*, 2003.
De Benedittis, R. and Buzzoni, S., 'The TD-SCDMA Radio Access Technology in the GSM Network'.
Digital Video Broadcasting (DVB): DVB mega-frame for single frequency network (SFN) synchronization, ETSI TS 101 191 V1.4.1 (2004-06), European Telecommunications Standards Institute.
Digital Video Broadcasting (DVB): DVB specification for data broadcasting, ETSI EN 301 192 V1.4.1 (2004-11), European Telecommunications Standards Institute.
Digital Video Broadcasting (DVB): framing structure, channel coding and modulation for digital terrestrial television, ETSI EN 300 744 V1.5.1 (2004-11), European Telecommunications Standards Institute.
Digital Video Broadcasting (DVB): specification for service information (SI) in DVB systems, ETSI EN 300 468 V1.6.1 (2004-11), European Telecommunications Standards Institute.
Digital Video Broadcasting (DVB): transmission system for handheld terminals (DVB-H), ETSI E N 302 304 V1.1.1 (2004-11), European Telecommunications Standards Institute.

Digital Video Broadcasting (DVB): DVB-H implementation guidelines, ETSI TR 102 377 V1.1.1 (2005-02), European Telecommunications Standards Institute.

Digital Video Broadcasting (DVB): transmission to handheld terminals (DVB-H); validation task force report VTF, ETSI TR 102 401 V1.1.1 (2005-04), European Telecommunications Standards Institute.

Dimitriou, N., Tafazolli, R. and Sfikas, G., 'Quality of Service for Multimedia CDMA', *IEEE Communications Magazine*, **38**(7), 88–94, 2000.

Dinnan, E. and Jabbari, B., 'Spreading Codes for Direct Sequence for Direct CDMA and Wideband CDMA Cellular Network', *IEEE Communications Magazine*, 124–136, October 1996.

Doble, J., *Introduction to Radio Propagation for Fixed and Mobile Communication*, Artech House, 1996.

Draft TS 101 999: 'Broadband Radio Access Networks (BRAN) HIPbRACCESS Functional Specification', European Telecommunications Standards Institute, September 2001.

DTS/BRAN030003-1, 'Broadband Radio Access Networks HIPERLAN Type-2 Functional Specification – Part 1: Physical Layer', European Telecommunications Standards Institute, Sophia Antipolis, September 1999.

Duplessis, P., 'HSOPA: Exploiting OFDM and MIMO to Take UMTS beyond HSDPA/HSUPA', *Nortel Technical Journal*, Issue **2**, July 2005.

DVB –The Family of International Standards for Digital Video Broadcasting, 2nd Edition, Springer, 2005.

Effenberger, S.J., 'The Effect of Rain on a Radome's Performance', *Microwave Journal*, May 1986.

Ekström, H., *et al.*, 'Technical Solutions for the 3G Long-Term Evolution', *IEEE Communications Magazine*, **44**(3), 38–45, 2006.

Eneroth, G., *et al.*, 'Applying ATM/AAL2 as a Switching Technology in Third-Generation Mobile Access Networks', *IEEE Communications*, **37**(6), 112–122, 1999.

ETSI Digital Cellular Telecommunication System (Phase 2+), General Packet Radio Service (GPRS): Mobile Station (MS)–Base Station System (BSS) Interface; Radio Link Control/ Medium Access Control Protocol, GSM 04.60.

ETSI, Digital Cellular Telecommunication System (Phase 2+), Radio Network Planning Aspects, GSM 03.30.

ETSI, Digital Cellular Telecommunication System (Phase 2+), Radio Transmission and Reception, GSM 05.05.

ETSI, 'The ETSI UMTS Terrestrial Radio Access (UTRA) ITU-R RTT Candidate Submission', ETSI SMG2, June 1998.

ETSI TS 102 367, 'Digital Audio Broadcasting (DAB): Conditional Access", v.1.2.1, January 2006.

Faria, G., *et al.*, 'DVB-H: Digital Broadcast Services to Handheld Devices', *Proceedings of the IEEE*, **94**(1), 194–209, 2006.

Fazel, K. and Papke, L., 'On the Performance of Convolutionally-Coded CDMA/OFDM for Mobile Communication Systems', *Proceedings of PIMRC '93*, Yokohama, 468–472, September 1999.

Foschini, G.J., 'Layered Space-Time Architecture for Wireless Communication in a Fading Environment when Using Multielement Antennas', *Bell Labs Technical Journal*, 41–59, Autumn 1996.

Foschini, G.J. and Gans, M.J., 'On the Limits of Wireless Communications in a Fading Environment when Using Multiple Antennas', *Wireless Personal Communications*, **6**, 315–335, 1998.

Foschini, G.J., *et al.*, 'Simplified Processing for Wireless Communication at High Spectral Efficiency', *IEEE. Journal on Selected Areas in Communications*, **17**, 1841–1852, 1999.

Frost, V.S. and Melamed, B., 'Traffic Modeling for Telecommunications Networks', *IEEE Communications Magazine*, 70–81, March 1994.

Grech, S. and Eronen, P., 'Implications of Unlicensed Mobile Access (UMA) for GSM Security', *Proceedings of the First International Conference on Security and Privacy for Emerging Areas in Communications Networks*, 2005.

GSM 04.60, 'Digital Cellular Telecommunication System: General Packet Radio Service (GPRS); Mobile Station (MS)–Base Station System (BSS) Interface; Radio Link Control/Medium Access Control (RLC/MAC) Protocol'.

Guangyi, L., Jianhua Z. and Ping, Z., 'Further Vision on TD-SCDMA Evolution', *2005 Asia–Pacific Conference on Communications*, Perth, Western Australia, 143–147, October 2005.

Haardt, M., Klein, A. and Schindler, J. 'UTRA TDD Physical Layer Description', *1st International Symposium on Wireless Personal Multimedia Communications*, Japan, November 1998.

Halonen, J. and Melero, R.J., *GSM, GPRS and EDGE Performance*, John Wiley & Sons Ltd, Chicester, 2002.

Harri, H. and Antti, T., *WCDMA for UMTS*, John Wiley & Sons, Ltd, Chichester, 2000.

Hata, M., 'Emperical Formula for Propagation Loss in Land Mobile Radio Services', *IEEE Transactions on Vehicular Technology*, **VT-29**(3), 317–325, August 1980.

Haykin, S., *Communication System*, John Wiley & Sons, Inc., New York, 1983.

Henkel, W. and Kessler, Th., 'Maximizing the Channel Capacity of Multicarrier Transmission by Suitable Adaptation of the Time-Domain Equalizer', *IEEE Transactions on Communications*, **48**(12), 2000–2004, 2000.

Henkel, W., *et al.*, 'The Cyclic Prefix of OFDM/DMT – An Analysis', *International Zurich Seminar on Broadband Communications*, ETH, Zurich, February 19–21.

Hillebrand, J., *et al.*, 'Quality-of-Service Signaling for Next-Generation IP-Based Mobile Networks', *IEEE Communications Magazine*, 72–79, June 2004.

Hoadley, J. and Javed, A., 'Overview: Technology Innovation for Wireless Broadband Access', *Nortel Technical Journal*, Issue **2**, July 2005.

Hunag, C.Y. and Yates R.D., 'Call Admission Control in Cellular Radio System', *IEEE Transactions on Vehicular Technology*, **I**, 1992.

IETF RFC 2401.

IETF RFC 2407.

IETF RFC 2408.

IETF RFC 2409.

IETF RFC 4303.

IKEv2 RFC-4306.

Ishimaru, 'Introduction to Wave Propagation and Scattering in Random Media', IEEE, AP-1985.

Isnard, O., *et al.*, 'Handling Traffic Classes at AAL2/ATM Layer Over the Logical Interfaces of the UMTS Terrestrial Access Network', *11th IEEE International Symposium on Personal, Indoor and Mobile Radio Communication*, London, **2**, 1464–1468, September 2000.

ITU-R P.530-9, 'Propagation Data and Prediction Methods Required for the Design of Terrestrial Line-of-Sights System'.

ITU-R P.837-3, 'Characteristics of Propagation for Propagation Modelling'.

ITU-T, 'Recommendation I.363.2, 'B-ISDN ATM Adaptation Layer Specification: Type 2 AAL', November 2000.

ITU-T G.826, 'Error Performance and Objectives for International, Constant Bit Rate Digital Paths at or above Primary Rate'.

ITU-T G.827, 'Availability Parameters and Objectives for Path Elements of International Constant Bit Rate Digital Paths at or above the Primary Rate'.

Jähnert, J., *et al.*, 'The "Pure-IP" Moby Dick 4G Architecture', *Computer Communications*, **28**(9), 2005.

Jakes, W.C., Jr (Ed.), *Microwave Mobile Communications*, Wiley-Interscience, New York, 1974.

Janaswamy, R., *Radio Propagation and Smart Antennas for Wireless Communications*, Kluwer Academic Publishers, 2000.

Jassen, J., *et al.*, 'Assessing Voice Quality in Packet Based Telephony', *IEEE Internet Computing*, May/June 2002.

Kaaresoja, T. and Ruutu, J., 'Synchronization and Cell Loss in Cellular ATM Evaluation System', *Proceedings of the 5th International Workshop on Mobile Communication, moMuc'98*, Berlin, 12–14 October 1998.

Kari, H. and Kangas, A., 'Microcell Propagation Model for Network Planning', *Proceedings of PIMRC'96*, 148–152, 1996.

Knorr, J.B., 'Guided EM Waves with Atmospheric Ducts', Microwave and RF, May 1985.

Koodli, R., *et al.*, RFC4068, 'Fast Handovers for Mobile IPv6', July 2005.

Kostas, T., *et al.*, 'Real-Time Voice Over Packet-Switched Networks', *IEEE Network Magazine*, January/February 1998.

Lach, H.-Y., *et al.*, 'Network Mobility in Beyond-3G Systems', *IEEE Communications Magazine*, **41**(7), 2003.

Ladebusch, U. and Liss, C., 'Terrestrial DVB (DVB-T): A Broadcast Technology for Stationary Portable and Mobile Use', *Proceedings of the IEEE*, **94**, 183–193, 2006.

Lawrey, E., 'Multiuser OFDM', *Fifth International Symposium on Signal Processing and its Applications, ISSPA '99*, Brisbane, 761–764, August 1999.

Lawrey, E. and Kikkert, C.J., 'Peak to Average Power Ratio Reduction of OFDM Signals Using Peak Reduction Carriers', *Fifth International Symposium on Signal Processing and its Applications, ISSPA'99*, Brisbane, 737–740, August 1999.

Lawrey, E. and Kikkert, C.J., 'Adaptive Frequency Hopping for Multiuser OFDM', *Second International Conference on Information, Communications and Signal Processing, ICICS'99*, Singapore, Proceedings: Paper Number 361, December 1999.

Lee D. and Cheun, K., 'A New Symbol Timing Recovery Algorithm for OFDM Systems', *IEEE Transactions on Consumer Electronics*, **43**(3), 766–775, 1997.

Lee, W.C.Y., *Mobile Cellular Telecommunication Systems*, McGraw-Hill Book Company, 1990.

Lee, W.C.Y., *Mobile Communication Design Fundamentals*, John Wiley & Sons, Ltd, Chichester, 1993.

Lempiainen, J., *Radio Interface Planning for GSM/GPRS/UMTS*, Kluwer Academic Publishers, 2001.

Lister, D., P., 'UMTS Capacity and Planning Issues', *First International Conference on 3G Mobile Communication Technologies*, 218–223, March 2000.

Liva, J. and Lo, T.K-Y., *Digital Beamforming in Wireless Communications*, Artech House, 1996.

McAllister, L.G., *et al.*, 'Acoustic Sounding – A New Approach to the Study of Atmospheric Structure', *Proceedings of the IEEE*, **57**, 579, 1969.

Mehrotra A., *Cellular Radio Performance Engineering*, Artech House, 1994.

Mehrotra A., *GSM System Engineering*, Artech House, 1997.

Mishra, Ajay. R., 'Observation of the Fading Phenomenon on the Western Coast of India, on a 7GHz Terrestrial Path', *IMOC Conference*, Rio de Janeiro, 1999.

Mishra, Ajay R., 'Cellular Transmission Network Optimisation', *ADVANCE (Nokia Research Centre Journal)*, March 2001.

Mishra, Ajay R., 'Transmission Network Planning and Optimisation in Third Generation Access Networks', *ADVANCE (Nokia Research Centre Journal)*, June 2003.

Mishra, Ajay R., 'EDGE Network Analysis and Optimisation', *WPMC*, Yokosuka, October 2003.

Mishra, Ajay R., *Fundamentals of Cellular Network Planning and Optimisation 2G/2.5G/3G ... Evolution to 4G*, John Wiley & Sons, 2004.

Mishra, Ajay R., *Advanced Cellular Network Planning and Optimisation 2G/2.5G/3G ... Evolution to 4G*, John Wiley & Sons, 2006.

Mitra, A.P., *et al.*, 'Tropospheric Disturbance of 17–21 December 1974 and its Effect on the Microwave Propagation', *Boundary Layer Meteor*, **11**, 103, 1977.

Mobile WiMAX – Part I: A Technical Overview and Performance Evaluation, WiMAX Forum, 2006.

Mobile WiMAX – Part II: A Comparative Analysis, WiMAX Forum, 2006.

Moose P., 'A Technique for Orthogonal Frequency Division Multiplexing Frequency Offset Correction', *IEEE Transactions on Communications*, **42**, 2908–2914, 1994.

Mouley, M. and Pautet, M-B., *The GSM System for Mobile Communications*, 1992.

Nair, G., *et al.*, 'IEEE 802.16 Medium Access Control and Service Provisioning', *Intel Technology Journal*, **8**, August 2004.

NTT DoCoMo, 'Views on Long Term Evolution', *3GPP LTE, TDD Workshop*, Tokyo, March 2005.

Oguchi, T., 'Electromagnetic Wave Propagation and Scattering in Rain and Other Hydrometers', *IEEE*, **71**(9), 1983.

OHG 'Harmonised Global 3G (G3G) Technical Framework for ITU IMT-2000 CDMA Proposal', Operator Harmonisation Group, May 1999, submitted to ITU-R in June 1999.

Ojanpera, T. and Prasad, R., *Wideband CDMA for Third Generation Mobile Communication*, Artech House, 1999.

Ojanpera, T., Prasad, R. and Harada, H., 'Qualitative Comparison of some Multi-user Detector Algorithms for Wideband CDMA', *Proceedings of the IEEE Vehicular Technical Conference*, 1, 46–50, 1998.

'Okumara–Hata Propagation Model' in *Prediction Methods for the Terrestrial Land Mobile Services in the VHF and UHF Bands*, ITU-R Recommendation P.529-2, ITU, Geneva, 5–7, 1995.

Okumara, Y., *et al.*, 'Field Strength and its Variability in VHF and UHF Land Mobile Radio Service', *Review of Electrical Communication Laboratory*, **16**(9–10), 825–873, 1968.

P802.1 laD6.0, 'LANMAN Specific Requirements – Part 2: Wireless MAC and PHY Specifications – High Speed Physical Layer in the 5 GHz Band', IEEE 802.1 1, May 1999.

P802.16dDI-2001, 'IEEE Draft Standard for Local and Metropolitan Area Networks – Part 16: Air Interface for Fixed Broadband Wireless Access Systems', November 2001.

Pajukoski, K. and Savusalo, J., 'Wideband CDMA Test System', *Proceedings of the IEEE International Conference on Personal Indoor and Mobile Radio Communication, PIMRC'97*, Helsinki, 669–672, September 1997.

Parsa, K., Ghassemzadeh, S.S. and Kazeminejad, S., 'Systems Engineering of Data Services in UMTS W-CDMA Systems', *IEEE International Conference on Communications (ICC2001)*, Helsinki, June 2001.

Parsons, D, *The Mobile Radio Propagation Channel*, Pentech Press, 1992.

Peled, A. and Ruiz, A., 'Frequency Domain Data Transmission Using Reduced Computational Complexity Algorithms', *Proceedings of IEEE ICASSP*, Denver, 964–967, 1980.

Proakis, J.G., *Digital Communications*, Third Edition, McGraw-Hill, 1995.

'Push to Talk over Cellular (PoC) – Architecture', v1.0, Open Mobile Alliance (OMA), February 2004.

Rappaport, T.S., *Wireless Communications: Principles and Practice*, IEEE Press and Prentice-Hall, 1996.

Reimers, U., 'DVB – The Family of International Standards for Digital Video Broadcasting', *Proceedings of the IEEE*, **94**(1), 173–182, 2006.

Rummler, W.D., 'A New Selective Fading Model: Application to Propagation Data', *BSTJ*, 1974.

Salvekar, A., *et al.*, 'Multiple-Antenna Technology in WiMAX Systems', *Intel Technology Journal*, **8**, August 2004.

Sari, H., 'A Multimode CDMA with Reduced Intercell Interference for Broadband Wireless Networks', *IEEE Journal on Selected Areas in Communications (JSAC)*, **19**(7), 1316–1323, 2001.

Sari, H., 'A Review of Multicarrier CDMA', in *Multi-Carrier Spread Spectrum and Related Topics*, Fazel, K. and Kaiser, S. (Editors), 3–12, Kluwer Academic Publishers, 2002.

Sari, H. and Karam, G., 'Orthogonal Frequency-Division Multiple Access and its Application to CATV Networks', *European Transactions on Telecommunications (ETT)*, **9**(6), 507–516, 1998.

Sari, H., Karam, G. and Jeanclaude, I., 'Transmission Techniques for Digital Terrestrial TV Broadcasting', *IEEE Communications Magazine*, **33**, 100–109, February 1995.

Sarkar, S.K., 'Radioclimatiological Effect on Tropospheric Radiowave Propagation over the Indian Subcontinent', *Ph.D. Thesis*, University of Delhi, India, 1978.

Shelswell, P., 'The COFDM Modulation System, The Heart of Digital Audio Broadcasting', BBC Research and Development Report, BBC RD 1996/8.

Sipila, K., *et al.*, 'Modelling The Impact of The Fast Power Control On The WCDMA Uplink', *Proceedings of VTC'99*, 1266–1270.

Siwiak, K., *Radio Propagation for Antennas for Personal Communications*, Artech House, 1998.

Stott, J., 'The Effects of Phase Noise in COFDM', *EBU Technical Review*, Summer 1998.

Sun, L., *et al.*, 'Impact of Packet Loss Location on Perceived Speech Quality', Internet Telephony Workshop, 2001.

Tarokh, V., Jafarkhani, H. and Calderbank, A.R., 'Space-time Block Codes from Orthogonal Designs', *IEEE Transactions on Information Theory*, **45**, 1456–1467, July 1999.

'TCP over 2.5G and 3G Wireless Networks', IETF Internet-draft, October 2001.

TD-SCDMA Network Planning and Optimization, TD-SCDMA Forum.

Thibault, L. and Le, M.T., 'Performance Evaluation of COFDM for Digital Audio Broadcasting, Part I: Parametric Study', *IEEE Transactions on Broadcasting*, **43**(1), 64–75, March 1997.

TIA/EIA/IS-856: CDMA2000, High Rate Packet Data Air Interface Specification.

Trends in Telecommunications Reform, ITU, 2003.

Tsurimi, H. and Suzuki, Y., 'Broadband RF Stage Architecture for Software Defined Radio in Handheld Terminal Application', *IEEE Communication Magazine*, **37**(2), 90–95, 1999.

UMA Consortium, 'Unlicensed Mobile Access (UMA), User Perspective (Stage 1)'.

UMA Consortium, 'Unlicensed Mobile Access (UMA) Architecture (Stage 2)', *Technical Specification*, September 2004.

UMA Consortium, 'Unlicensed Mobile Access (UMA), User Protocols (Stage 3)'.

UMTS 22.01, Service Aspects – Service Principles.

UMTS 22.05, Service Capabilities.

UMTS 22.25, Quality of Service and Network Performance.

UMTS 23.05, Network Principles.

Van Nee, R. and Prasad, R., *OFDM for Wireless Multimedia Communications*, Artech House, 2000.

Verscheure, O., *et al.*, 'User-Oriented QoS in Packet Video Delivery', *IEEE Network Magazine*, November/December 1998.

Vigants, A., 'Space Diversity Engineering', *BSTJ*, 1974.

Vigants, A., 'Microwave Obstructive Fading', *BSTJ*, **60**, 1981.

Vodafone Policy Paper, 2005.

Wahlqvist M., *et al.* 'A Conceptual Study of OFDM-based Multiple Access Schemes', Technical Report Tdoc 117/96, ETSI STC SMG2 Meeting no. 18, Helsinki, May 1996.

Walfish, J. and Bertoni, H.L., 'Theroretical Model of UHF Propagation in Urban Environments', *IEEE Transactions on Antenna Propagation*, **AP-36**, 1788–1796, October 1988.

Wang, F., *et al.*, 'IEEE 802.16e System Performance – Analysis and Simulation Results', *Proceedings of PIMRC*, Berlin, September, 2005.

Weinstein, S.B. and Ebert, P.M., 'Data Transmission by Frequency-Division Multiplexing Using the Discrete Fourier Transform', *IEEE Transactions on Communication Technology*, **COM-19**(5), 628–634, October 1971.

Willinger, W., *et al.*, 'Self-Similarity Through High-Variability: Statistical Analysis of Ethernet LAN Traffic at the Source Level', *IEEE/ACM Transactions on Networking*, 57–62, February 1997.

'WiMAX End-to-End Network Systems Architecture – Stage 2: Architecture Tenets, Reference Model and Reference Points', WiMAX Forum, December 2005.

Wisely, D., Eardley, P. and Burness, L., *IP for 3G*, John Wiley & Sons, Ltd, Chichester, 2002.

World Bank (Quian), 2009.

Xia, H.H., *et al.*, 'Micro-cellular Propagation Characteristics for Personal Communication in Urban and Sub-urban Environments', *IEEE Vehicular Technology*, **43**(3), 743–752, 1994.

Xiao, W. and Ratasuk, R., 'Analysis of Hybrid ARQ with Link Adaptation', *Proceedings of the Annual Allerton Conference on Communications, Control and Computing*, 1618–1619, October 2002.

Yang, S.C., *CDMA RF System Engineering*, Artech House, 1998.

Yagoobi, H., 'Scalable OFDMA Physical Layer in IEEE 802.16 WirelessMAN', *Intel Technology Journal*, **8**, August 2004.

Yee, N., Linnartz, J.-P. and G. Fettweis, G., 'Multicarrier CDMA for Indoor Wireless Radio Networks', *Proceedings of PIMRC '93*, Yokohama, 109–113, September 1999.

Yin, H. and Liu, H., 'An Efficient Multiuser Loading Algorithm for OFDM-based Broadband Wireless systems', *IEEE Global Telecommunications Conference*, 2000.

Yin, H. and Alamouti, A., 'OFDMA – A Broadband Wireless Access Technology', *IEEE Proceedings of Sarnoff Symposium*, March 2006.

Zander, J., 'Radio Resource Management – An Overview', *Proceedings of the IEEE Vehicular Technology Conference, VTC-96*, Atlanta, 661–665, 1996.

Index